5-23-83

743804/

AMATEUR
ASTRONOMER'S
HANDBOOK

J. B. SIDGWICK
F.R.A.S.

SECOND EDITION
prepared by
DR. GILBERT FIELDER

THIRD EDITION
prepared by
R. C. GAMBLE
B.Sc., F.R.A.S.

DOVER PUBLICATIONS, INC.
NEW YORK

The publisher of this edition wishes to express his appreciation to Herbert A. Luft, who suggested republication and oversaw preparation of this reprint.

This Dover edition, first published in 1980, is an unabridged and unaltered republication of the third (1971) edition as published by Faber and Faber, London, except that former section 40 (Manufacturers and Suppliers), which dealt exclusively with British sources of materials and is now out-of-date, and the Appendix, which consisted of the complete table of contents to the companion volume by the same author, *Observational Astronomy for Amateurs*, have been omitted, the latter item because of space limitations. A new Supplemental Bibliography listing books published since 1955 (the date of the first edition) has been especially prepared for the present edition by Herbert A. Luft. An up-to-date list of radio time signals has been added on page 559, courtesy of the British Astronomical Association.

International Standard Book Number: 0-486-24034-7
Library of Congress Catalog Card Number: 80-66643

Manufactured in the United States of America

Dover Publications, Inc.
180 Varick Street
New York, N.Y. 10014

CONTENTS

6

CONTENTS

CONTENTS

20. PHOTOGRAPHY

21. THE MEASUREMENT OF RADIATION INTENSITIES

22. ACCESSORY INSTRUMENTS AND EQUIPMENT

CONTENTS

TABLE OF ABBREVIATIONS

The following commonly occurring abbreviations and symbols are used in the sense quoted unless the contrary is specified or is obvious from the context.

Å	Ångström unit$=10^{-8}$ cms
A.E.	*Astronomical* (or *American*) *Ephemeris*
AT	Apparent Time
A.U.	Astronomical Unit
α	Right Ascension
B	Latitude
B.	(followed by number): reference to Bibliography, section 41
B.A.A.	Bibliographical abbreviations: see section 41
	British Astronomical Association
c/i	colour index
CM	central meridian
D	aperture of objective in inches
Dec	Declination
d/v	direct vision
δ	Declination, or diameter of the pupillary aperture of the eye
ET	Ephemeris Time
E/T	Equation of Time
F	focal length of objective
f	focal length of ocular, or "following"
f/	focal ratio (F/D). $f/10$ is said to be a larger focal ratio and a smaller relative aperture (D/F) than, e.g. $f/5$
𝖩	equivalent focal length
GMAT	Greenwich Mean Astronomical Time
H	hour angle
HP	high power (of oculars)
HR	hourly rate
I	intensity (of radiation)
I.A.U.	International Astronomical Union
JD	Julian Date
L	longitude

LP	low power (of oculars)
LPV	long-period variable
λ	wavelength
M	magnification, or integrated magnitude, or absolute magnitude
M'	minimum useful magnification
M''	maximum useful magnification
M_r	minimum magnification for full resolution
m	apparent stellar magnitude
m_b	bolometric magnitude
m_p	photographic magnitude
m_v	visual or photovisual magnitude
MSL	mean sea level
MT	Mean Time (prefixed by L=Local, or G=Greenwich)
μ	refractive index, or micron=10^{-4} cms
NCP	North Celestial Pole
NPS	North Polar Sequence
ν	constringence
O.A.A.	*Observational Astronomy for Amateurs*
OG	object glass
ω	deviation
p	preceding
p.a.	position angle
π	stellar parallax, or transmission factor of an optical train
ϕ	latitude
R	theoretical resolution threshold
R'	empirical resolution threshold
r	angular radius of a given diffraction ring or interspace, or angular separation of components of a double star
r'	linear radius of a given diffraction ring or interspace
RA	Right Ascension
SPV	short-period variable
ST	Sidereal Time (prefixed by L=Local, or G=Greenwich)
t	time, or turbulence factor
UT	Universal Time
ZD	Zenith Distance
ZHR	zenithal hourly rate

FOREWORD

In 1950 the membership of the British Astronomical Association (the only national association of amateur astronomers in the country) passed the 2000 mark, and at the time of writing it is still growing. A large proportion of these amateurs are active, in the sense that they possess telescopes of some sort and engage in more than desultory observation; of these, a hard core of 'specialists' take a regular part in the organised programmes of the various Observing Sections.

Why the Association's membership, which maintained a fairly constant level at about the 1000 mark during the first forty years of this century, should suddenly have doubled itself within the last ten years I do not know. Presumably the war had something to do with it. But considering the obvious disadvantages of astronomy as a hobby—chief of which are the expense and size of its equipment, and the discomforts and inconveniences of its pursuit—it is a remarkable fact that over 2000 men and women are sufficiently engrossed by it to take up membership of an association of their fellows.

In so far as there is such a thing as a typical amateur astronomer, his evolution might run somewhat as follows. The initial spark of interest may be struck in one of a thousand ways—a book or a magazine article, a radio talk, a snippet of news from one of the great observatories, a personal contact, an obvious astronomical event such as eclipse or a naked-eye comet. Interest once aroused, it will be both fed and increased by a wide range of reading. The first milestone is reached when the amateur realises that the armchair is no place to study astronomy, and he determines to buy himself some sort of telescope, no matter how cheap and unprofessional, in order to see things for himself. From this moment he will be lost to the domestic comforts of the fireside, and is likely to become an increasing trial to his wife. Nine times out of ten his first instrument will be a small refractor, supplied with perhaps three or four oculars and mounted on a tripod altazimuth. Then will follow a period—which can afford to be at least as long as a year, and will probably be longer—during which he 'sight-sees' over the expanse of the night sky. This may at the time appear to him to be primarily a pleasurable

15

and exciting exploration, more or less without direction, but he is in fact laying the foundation of that first-hand familiarity with the night sky which is one of the characteristics of the amateur astronomer, and also one of the most valuable components of his mental equipment. Sooner or later two realisations will be borne in upon him: that he has exhausted the potentialities of his small telescope, and that his now fairly wide though shallow experience of observing the Sun, Moon, planets, double stars and variables, comets, meteors, nebulae, and the rest, is crystallising into an interest directed upon one specific branch of the subject.

This marks the second milestone in the career of this supposedly typical amateur astronomer. In a field as vast as that opened up by the telescope, specialisation is unavoidable if the observer's energies are not to be hopelessly and fruitlessly dissipated. Not only is the amount of time that anyone can devote to telescopic work necessarily limited, but the further pursuit of each branch of observation beyond the 'sight-seeing' level requires specialised techniques, the training of the eye to excel in some particular function, and to a large extent specialised instruments. Great care needs to be devoted to the choice of the best equipment—for, financially, it often means putting all one's eggs in one basket—as well as of the most suitable field of specialisation: one who lives in a great city, for example, would be unwise to devote his time to comet-seeking or the study of the Zodiacal Light, or one who for reasons of finance or available space cannot install a telescope of more than 6 ins aperture, to specialise in planetary observation.

And so the amateur, now in possession of a thorough working knowledge of the night sky and of his instrument, settles down to the regular and intensive observation of a single object or class of related objects. Already he is on the road to becoming an expert in his own chosen field. The leading amateur astronomers are, indeed, men of international repute: one has only to think of such names as Denning, Ellison, Goodacre, Hargreaves, Maunder, Phillips, Steavenson, Stanley Williams—so many might be mentioned that any selection, however random, must be invidious.

Some examples of amateur achievement may not be out of place here: they will put flesh on the bones of the biographical skeleton described above, and at the same time provide both a spur and an encouragement. W. F. Denning, an accountant by profession, became interested in astronomy at the age of twenty-one, when he happened by chance to witness a spectacular Taurid fireball. During the subsequent sixty-two years till his death in 1931 he built for himself a reputation as one of the world's leading authorities on meteor radiants; he was also one of the foremost planetary observers of his time, his work on the rotations of the surface features of Jupiter being of fundamental importance. He discovered five comets and two novae. His classic work on radiants culminated in his *General Catalogue of the Radiant Points of Meteoric Showers*, published in the *Monthly Notices* of the Royal Astronomical Society in 1899, which is still a standard reference work on radiants. He was the first observer of a radiant that moved from night to night, in accordance with theory; and he was the first worker to publish data relating to stationary radiants in a form that permitted profitable discussion of these strange contradictions between theory and observation. The accuracy and reliability of his observations were outstanding—yet he never used more elaborate equipment than a 12·6-in altazimuth reflector.

Burnham, one of the greatest observers of double stars, was an official reporter in the U.S. District Court, Chicago, until eleven years before his death in 1921. His first catalogue—of 81 doubles discovered by himself with a 6-in refractor during the previous three years—was published in 1870, only nine years after he had taken up observing (with a modest 3-in). Two supplementary lists rapidly raised the number of his published discoveries to 182, and his great *General Catalogue of Double Stars* appeared thirty-five years later, in 1906. In all, he discovered 1340 new binaries, besides making many thousands of individual measures.

J. Franklin-Adams (1843–1912) became so engrossed in his astronomical hobby that at the age of forty-seven he forsook his business career—he was an insurance broker at Lloyd's—in order to have more time to devote to it. His name, associated with his photographic chart of the whole sky down to magnitude 14·5, is assured of immortality.

Going back somewhat further—to the early years of last century—Pons and Schröter demand mention. Schröter was by profession a lawyer, but this twenty-eight years' study of the lunar surface earned him the title of Father of Selenography. Pons, who at one time was concierge at the Marseilles Observatory, independently discovered no fewer than 37 comets during the twenty-six years from 1801 to 1827—a record of achievement that has never been broken.

Today the greater part of the observational data on which rests our knowledge of, among others, long-period variables, meteors, and the motions of Jupiter's surface markings is provided by amateur observers. The reverse side of the picture is, of course, that certain branches of work are peculiarly unsuited for amateurs, being the natural stamping ground of the professional: generally speaking, they are those that require large

17

apertures, long focal lengths, great precision, or all three, and as examples may be cited the measurement of radial velocities, the photometric study of the extragalactic nebulae, and fundamental positional work.

At some stage during his progress from armchair enthusiast to observational specialist, our typical amateur may be presumed to have joined the B.A.A. and, after his interest has focused on a particular branch of work, to have joined the appropriate Observing Section of the Association. Here he will be brought into contact with other observers in the same field, and a channel will be provided for the publication of his observations. Observations whose records are kept locked away in a drawer might just as well never have been made, so far as their value to science is concerned. They must either be published or periodically sent in to the Director of the Section for correlation and combination with the observations of other members and eventual inclusion in the interim *Reports* and *Memoirs* which are published by each of the Sections from time to time. The B.A.A. not only facilitates contact with other observers engaged on the same work, but by means of its geographical list allows any observer to get in touch with other members in his own locality, irrespective of their work and specialised interest. Furthermore it provides an organisation for arranging cooperative observation. This is not only the *sine qua non* of such work as the determination of the real paths of meteors—where parallactic observations from separated stations are required—but in general provides a safeguard against the needless duplication of observations and against gaps in the observational record due to bad weather or bad seeing, while to some extent smoothing out the accidental errors which are ineradicable from the observations of a single observer. The value of cooperation in the pooling of observations has perhaps never been more clearly and convincingly demonstrated than in the work of the Variable Star Section of the B.A.A.

There is, however, more to astronomical research than the mere piling up of observational data, however essential this basic supply of material for discussion may be. An accumulation of facts, as Poincaré remarked, is no more a science than a pile of bricks is a house. It is possible that some observers in this country have got into something of a groove, grinding on and on with the production of observational material without bothering too much about its integration or significance. There is always room for individual purposive research designed to reach a definite objective and solve, or at least throw some light on, a specific problem, no matter how limited it may be. Even a casual flip through such periodicals as the *Journal* of the B.A.A. or *Popular Astronomy* for the past twenty years will bring to light examples of the phenomenal success and of the importance of the results that can be achieved when amateurs do break away from established routine.

The willingness to develop new techniques and to explore and exploit new instrumental developments is perhaps more characteristic of the Americans than of ourselves. It would be interesting, for example, to know how many Schmidt cameras are in amateur hands on the two sides of the Atlantic. The 4-in refractor and the 12-in Newtonian by no means exhaust the instrumental possibilities for the amateur, and a greater degree of daring in the use of available instrumental resources would certainly be an invigorating influence in amateur astronomy. There is no need to go to the other extreme of instrument-fixation exemplified by so many American telescope makers, who become hypnotised by their products as ends in themselves, and while turning out first-class equipment cannot find time away from their laps and their lathes to direct their own telescopes towards the sky. There have been plenty of instrumental developments during the last twenty years which could be successfully exploited by amateurs, but which have, in fact, been to a surprising extent neglected by them. Examples spring to mind: the greater use of photoelectric methods in photometry, other types of reflector than the usual Newtonian, the new and varied offerings of electronics—automatic guiding, the automatic registration of occultations or transits by photocell and chronograph, etc—long-wave solar research (in which amateurs are cooperating usefully in Australia), new optical systems such as the Schmidt and Maksutov, the monochromator, or comparatively rare instruments of enormous scope such as the spectrohelioscope. The astronomical sightseer turning specialist would be very well advised to consider the adventurous outlay of his capital rather than the mere duplication of instruments already in prolific existence.

It is many years since the last edition of Webb's *Celestial Objects for Common Telescopes* appeared, and it is now badly out of date, though its list of objects is still useful to the amateur who is exploring the sky with his first telescope. The inter-war period saw the publication of Hutchinson's *Splendour of the Heavens*, edited by T. E. R. Phillips and W. H. Steavenson, which contained a great deal of information valuable to the amateur astronomer, though it was primarily a descriptive work. The most useful companion to practical observation that has appeared in fairly recent years is undoubtedly the *Astronomisches Handbuch*, edited by Henseling (1921); this, however, has never been translated, and in any case is nowadays virtually unobtainable. There is, therefore, a gap on the astronomical shelf at just that point where the amateur looks for information and guidance, and it is hoped that the present *Handbook* will go at least some of the way towards filling it.

It may be useful to emphasise that this is neither a formal textbook designed for consecutive reading, nor, on the other hand, a narrowly practical manual of instructions. It is a reference handbook, and a number of its topics constitute the background to telescopic observation rather than the immediate concern of the observer while actually at work: they provide the 'why' behind the 'how'. It is not difficult, I think, to justify the rather detailed treatment given, for example, to telescopic function or the aberrations. The man who understands the operation of his instrument or machine—whether it be a telescope or a motor-car—will use it more efficiently, because more intelligently, than the man who trusts to rule-of-thumb methods. The solution of many of the purely practical problems which confront the amateur observer leads direct to a more thoroughgoing account of the optics and mechanics of his instrument than can be found in most of the books ostensibly written with him especially in mind. That an astronomer is an amateur by no means implies that he wishes, or needs, only to read lists of severely practical instructions on the use of his instrument, unrelated to the so-called 'theoretical' background from which they are derived.

It will be found, therefore, that the various topics figuring in the Table of Contents are treated with different degrees of thoroughness according to their theoretical importance to the telescopist, their immediate value to the practising observer, their intrinsic complexity, and whether or not they are dealt with elsewhere in readily accessible books or journals. Thus no attempt has been made to present more than an introductory bird's-eye view of lens-and-mirror systems, supplemented by a fairly extensive bibliography; the relation between atmospheric condition and telescopic 'seeing', on the other hand, is discussed in more detail, with a correspondingly restricted bibliography. The aim, generally speaking, has been to take up the treatment of a subject at a more elementary level than that of the average technical textbook, and to carry it further than the average 'general reader' book. The Bibliography forms an integral part of the whole, being complementary and supplementary to the text. The extensiveness of the Bibliography may derive some justification from Joad's definition of the educated man as one whose head is not crammed solid with a mass of isolated facts, but who knows exactly where to put his hand on any particular piece of information that he may require.

Inevitably, many topics have been skimpily treated, and many more have been omitted altogether. No mention will be found of telescope making, for example: an adequate treatment of this technical subject would have swollen an already over-long book to prohibitive proportions and, furthermore, a number of excellent books dealing with the subject are easily obtainable; references to these will be found in the Bibliography. The theodolite and sextant, again, are not described, since their uses are primarily non-astronomical and they are in any case of comparatively little interest to the amateur. The meridian transit telescope is excluded, because it is an instrument that few amateurs possess—or need to possess, since its functions can be performed in other ways. Some instruments which, like the coronagraph, are of little practical significance to the amateur, but are of outstanding theoretical interest, are glanced at briefly in passing.

A device, gadget, instrument, or accessory is often described as being 'easily homemade'. The amateur who can handle ordinary carpenter's tools and is generally dexterous has a very definite advantage over one who is unpractical and keck-handed. There is nevertheless a great deal in the way of ordinary maintenance, modifications, and the construction of accessories of various kinds that anyone can do with simple tools and a little ingenuity and determination. If, of course, the amateur has access to a well-equipped machine-shop, or even to a small lathe, he is to be envied. Some quite astounding examples of ingenuity and improvisation in the building of, for instance, equatorial mountings from such materials as old car parts or water-pipe fittings are to be found described in the pages of the amateur journals.

Amateur Astronomer's Handbook deals solely with the theoretical and instrumental background to observation, together with various matters with practical implications. The actual techniques employed in the observation of the various types of celestial object are described in *Observational Astronomy for Amateurs*, which may be regarded as a companion volume to the present work. The two books deal with complementary aspects of the amateur astronomer's work, while being individually self-sufficient and mutually independent; cross references between the two volumes* have therefore been reduced to a minimum.

It would be pleasant to think that amateurs of all classes will find something of interest or value in this *Handbook*, but the reader for whom it was specifically written is that amateur astronomer who has been described as approaching his second milestone—the observer who, while

* *Observational Astronomy for Amateurs*, also reprinted by Dover (24033-9), is hereafter referred to as *O.A.A.*

gaining practical experience with a small instrument, is beginning to consider in what branch of work he will ultimately specialise.

By way of envoi I cannot do better than quote the words with which a professional astronomer (R. O. Redman), writing in 1945, concluded an article on the post-war prospects of astronomy: 'To attempt at the present juncture a discussion of post-war professional astronomy is to risk a charge of futility; to try to deal with post-war amateur astronomy would probably ensure not only the accusation but also the verdict of guilty. It is extraordinarily difficult to say anything useful on the subject at present, and hence a far from unimportant branch of astronomy has been omitted in this article. The enthusiasm and ability of amateurs, to which we have owed so much in the past, may yet prove the principal means of salvation for astronomy in a difficult future.'

It is with pleasure and gratitude that I acknowledge the assistance I have received from various sources: Mr R. H. Tucker, of the Abinger Magnetic Station, for information regarding the Greenwich Time Service; Mr M. J. Moroney for invaluable suggestions regarding section 29; Dr R. L. Waterfield for allowing me to examine Will Hay's plate-measuring machine; the staff of the Royal Astronomical Society's Library. And I must express my particular gratitude to Dr W. H. Steavenson, who undertook to read the complete MS, and to whom I am indebted for innumerable suggestions, corrections, and criticisms. It is hardly necessary to add that the book's errors and shortcomings are my own responsibility.

London,   J. B. SIDGWICK
June 1954.

EDITOR'S PREFACE TO THE THIRD EDITION

This book is basically not a book about astronomy, but a compendium of things an astronomer needs to know. As such, its contents are, for the most part, essentially physical and mathematical in nature, and their truth is sufficiently fundamental that they will not change with time. In consequence, the changes in this third edition are confined to the chapter on photography, which, dependent as it is upon the properties of the available materials, which are themselves subject to continuous evolution and improvement, will remain subject to review for many years to come.

I am indebted to Mr R. S. Scagell for much valuable advice and assistance.

R. C. GAMBLE, *Editor*

EDITOR'S PREFACE TO THE SECOND EDITION

Only a few changes have been made, in this edition, and these have been necessary mainly because of the introduction of Ephemeris Time in 1960.

G. FIELDER, *Editor*

SECTION 1

TELESCOPIC FUNCTION: LIGHT GRASP

1.1 Light grasp and image brightness

Light grasp is a term, often used rather vaguely, that indicates the quantity of light from a given source that is concentrated by an objective into the image at its primary focus. It is a function of the objective's aperture, D, and is independent of focal length, magnification, or any other factor. More precisely it may be defined as the relative intensity of the light from a point source passed to the retina by the telescope's optical train and by the naked eye. Provided that the diameter of the eye's pupillary aperture, δ, is not smaller than the exit pupil of the telescope, then the instrument's light grasp is equal to

$$\left(\frac{D}{\delta}\right)^2 . \pi$$

where π is the transmission factor of the system. A fair value, in the case of an astronomical telescope, is 60% to 65%, the remaining 35% to 40% of the incident light being lost to the final image through absorption in lenses, reflections at lens surfaces, and imperfect reflection at mirror surfaces (section 6.8).

A convenient criterion of the light grasp of an instrument is the magnitude of the faintest star that it will show to averted vision. Most textbooks contain a table of the minimum magnitudes revealed by various apertures. There is a variety of reasons why such tables tend to be misleading, among them:

(*a*) The observation, or theoretical assumption, which forms the basis of the extrapolation may be inaccurate or inapplicable to observing conditions at another time.

(*b*) Atmospheric factors, such as turbulence (which discriminates in favour of small instruments, with, in extreme cases, such anomalous results as the clear visibility of a star with half the aperture of a telescope

25

in which it is practically invisible),* and thin mist or high cloud, which may affect the instrument's performance while still being too diffuse to be detected with the naked eye.

(c) Most objectives, and probably all observers, depart in some degree from the 'normal'.

(d) The magnification employed will (for a reason discussed below, which does not invalidate what was said in the first paragraph) affect the visibility of the faintest stars.

(e) The use of direct or of averted vision.

(f) The use of a reflector or a refractor, the former being superior, aperture for aperture, providing the silver or aluminium film is in good condition.

(g) The presence of another and brighter object in the field. (It is assumed that such precautions as the thorough dark adaptation of the eye and the choice of a moonless night are taken.)

Under the most favourable conditions the normal human eye is just capable of glimpsing stars of mag 6·5,† whence it can be deduced that the magnitude of the faintest star visible with a telescope of aperture D is‡

$$m = 6·5 - 5 \log \delta + 5 \log D$$

(If a star in the telescopic field has an apparent brightness other than that of bare visibility, its real magnitude can be derived from this expression by substituting its equivalent naked-eye magnitude—i.e. that of a star which, observed with the naked eye, has the same apparent brightness—for the first term on the right-hand side.)

Now substituting the accepted value of $\delta = 0·3$ ins, we have that

$$m = 9·1 + 5 \log D$$

which agrees closely with the result arrived at by assuming that a 1-in objective will show stars to the 9th magnitude.

In the following Table (whose data are represented graphically in Figure 1) column (a) tabulates the magnitude thresholds derived from the last expression. Their close correspondence with those in column

---

* The 100-in Mt Wilson instrument, with a theoretical threshold of about mag 18·8, loses nearly 3 mags under bad seeing conditions.

† The value that may be expected with a normal eye whose pupillary aperture is expanded to 0·3 ins. Artificial stars have been seen in the laboratory, against a completely dark background, at mag 8·5. The 2-mag difference is accounted for by the reduced contrast between the natural star and its background of the night sky.

‡ If mag 6·5 is taken as the naked-eye threshold, then $2·512^{6·5} = k.\delta^2$, whence $\log k = 2·6 - 2 \log \delta$. But the telescopic brightness of a point source $= k.D^2$. The expression for m is derived by combining these last two expressions.

(*b*), based on the assumption that mag 9 is the threshold for a 1-in object glass, will be seen.

D (ins)	Minimum mag (uncorrected)				Correction for absorption	Corrected mag	
	(a)	(b)	(c)	(d)		(e)	(f)
2	10·6	10·5	12·1	12·2	−0·1	12·0	12·1
3	11·5	11·4	13·0	13·1		12·9	13·0
4	12·1	12·0	13·6	13·7		13·4	13·5
5	12·6	12·5	14·1	14·2	−0·2	13·9	14·0
6	13·0	12·9	14·5	14·6		14·3	14·4
7	13·3	13·2	14·8	14·9		14·5	14·6
8	13·6	13·5	15·1	15·2	−0·3	14·8	14·9
9	13·9	13·8	15·4	15·5		15·1	15·2
10	14·1	14·0	15·6	15·7		15·2	15·3
11	14·3	14·2	15·8	15·9	−0·4	15·4	15·5
12	14·5	14·4	16·0	16·1		15·6	15·7
15	15·0	14·9	16·5	16·6	−0·5	16·0	16·1
20	15·6	15·5	17·1	17·2	−0·6	16·5	16·6
Symbol in Figure	△	⊙	⊡	×		———	------

However, these curves do not agree well with the results of observation, yielding results which are consistently high by about 1·5 mag throughout the aperture range from 2 to 20 ins. In fact, a 1-in OG has a threshold nearer to 10·5 than 9·0. If each of the entries in column (*a*) is increased by 1·5, we have the series of thresholds given in column (*c*). This curve agrees very much more closely with that derived observationally, while approximating to the first two curves at their naked-eye origin ($D=0·3$ ins).

Column (*d*) gives figures based upon the observation by W. H. Steavenson of a mag 11·9 star with a 1·7-in OG. The two curves (*c*) and (*d*), whilst now more nearly representing the actual performance of small refractors, become increasingly inaccurate with apertures larger than about 5 ins. This is due to neglect of the factor of absorption in the increasingly thick object glasses, which is equivalent to a light-loss of about 0·2 mag with even a 4-in or 5-in, and approaches 2 mags in the case of the 40-in refractors.* In columns (*e*) and (*f*) are tabulated the figures in (*c*) and (*d*), respectively, corrected for absorption.

---

* The increase in light grasp occasioned by any increase of *D* beyond about 75 ins (supposing such an objective could be satisfactorily manufactured and mounted) would be swallowed up by the loss due to absorption in the great thickness of glass.

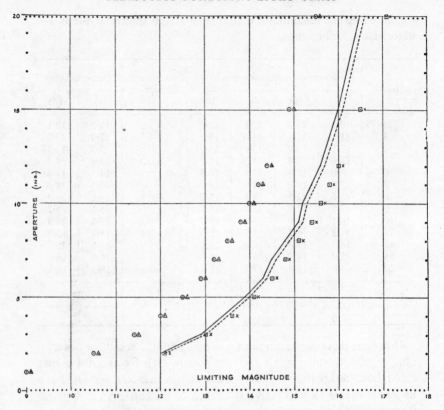

FIGURE 1
Visual magnitude thresholds

These last two curves very nearly represent the actual thresholds for refractors, operating under optimum conditions; a trace of haze or diffused artificial light can easily raise the threshold by half a magnitude or more. The thresholds for reflectors are more nearly represented by the uncorrected curves (*c*) or (*d*). (See also sections 6, 23.)

The superiority of reflectors in this respect is principally due to two causes: the light-loss in refractors by absorption; and the leakage of light into the secondary spectrum where it contributes nothing to the brightness of the image—where, in fact, it does the reverse by reducing contrast. Dennis Taylor and Linfoot have shown that such leakage may, in some large refractors, amount to about 50% of the light leaving the objective.

Objects most suitable for testing the light grasp of a telescope are the stars of the North Polar Sequence (section 37) and the comparison stars of variables, for which charts exist and whose magnitudes have been established with accuracy. The latter should be observed on the meridian, and the NPS stars when at the same altitude as the pole.

1.2 Brightness of point source seen with the naked eye

The apparent brightness of most 'point' objects is in fact determined by the law governing the brightness of extended sources, since it is only if the retinal image is small enough to fall upon a single cone that it can be treated as a true point image.

If this condition is fulfilled, however, its apparent brightness is directly proportional to the area of the eye's pupillary aperture (i.e. to δ^2).

1.3 Apparent brightness of extended images

The apparent intensity per unit area of an extended source viewed with the naked eye being proportional to δ^2, the amount of light passed by the objective to the image in its focal plane will be greater than that received on the retina of the naked eye by a factor of $\dfrac{D^2}{\delta^2} . \pi$ (where π is the telescope's transmission factor). When the telescope is used with a magnification M, the illuminated retinal area is increased by a factor M^2. Hence the apparent brightness of the image is increased by a factor

$$\frac{D^2}{M^2\delta^2} . \pi$$

The brightness of the telescopic image is therefore proportional to $\dfrac{D^2}{M^2} . \pi$, and the increase of the telescopic over the naked-eye brightness is:

$$\frac{\text{brightness of telescopic image}}{\text{brightness of naked-eye image}} = \frac{\pi D^2}{\delta^2 M^2} \quad \cdot \quad \cdot \quad \cdot \quad (a)$$

We shall see later (section 3.3) that the minimum magnification (M') that can fully employ the available light grasp of an objective is equal to the aperture divided by that of the eye, or $M' = \dfrac{D}{\delta}$. From this, and from what has gone before, it follows that the brightness of the telescopic image is π times that of the naked-eye image. This means that even under optimum magnification, and neglecting the transmission factor, the telescopic image can only equal, and never exceed, the naked-eye image in brightness. In fact the transmission factor is always less than unity, and hence the apparent brightness of an extended object is always reduced by viewing it telescopically.* Note, in passing, that the relative brightness of the telescopic and naked-eye images of an extended object is independent of the focal length of the objective.

But while the telescopic image of an extended object is always fainter,† area for area, than the naked-eye image, what the telescope can accomplish is the enlargement of the image from an inappreciable to an appreciable size. Also, the sky appears less bright through a telescope than to the naked eye. This fact has practical applications in the observation both of objects with a superabundance of light (e.g. the Moon), and of those whose images lose by lack of illumination if magnification is pressed too far (e.g. the planets).

The extent to which the brightness of extended images is reduced by magnification is brought out in the following Table, which sets out the results of substituting various values for M in equation (b) below.

Magnification (D expressed in ins)	Relative telescopic/ naked-eye brightness (theoretical)	Relative telescopic/ naked-eye brightness (corrected for 60% transmission factor)
$3D=M'$‡	1·0	0·6
$9D$	0·11	0·07
$12D=M_r$‡	0·06	0·036
$18D$	0·028	0·017
$27D$	0·012	0·007
$36D$	0·007	0·004
$45D$	0·0045	0·0027
$54D$	0·0031	0·0019
$60D=M''$‡	0·0025	0·0015

* Alternatively, from expression (a), since $\dfrac{D}{\delta}=M$

$$\frac{\text{telescopic brightness}}{\text{naked-eye brightness}}=\frac{\pi(M')^2}{M^2} \quad \cdots \cdots \cdots \quad (b)$$

which reveals the same fact—that $\dfrac{\text{telescopic brightness}}{\text{naked-eye brightness}}$ cannot be larger than unity (even neglecting to take π into account), since if M, the magnification used, is less than M', the minimum useful magnification, then the exit pupil will be too large to be wholly accepted by the eye. This topic is taken up again in section 3.3.

† This is not true of the Galilean telescope, however; for under all circumstances the diameter of the exit pencil is larger than that of the pupillary aperture, and except for the transmission factor, unit areas of the telescopic and naked-eye images are of equal brightness. (See section 9.2.)

‡ These particular values of M are explained in sections 3.3, 3.4.

The reverse side of the picture is exhibited in the following Table, which tabulates the magnifications corresponding to the selected values of $\dfrac{\text{telescopic image brightness}}{\text{naked-eye image brightness}}$ given in the left-hand column, for a 1-in, 3-in, 6-in, and 12-in telescope.

	$D=1$ in	3 ins	6 ins	12 ins
1·0	3	9	18	36
0·1	9	28	57	114
0·01	30	90	180	360
0·001	95	284	569	1139
0·0001	300	900	1800	3599

It can readily be seen that a large aperture gives a brighter extended image than a smaller one using the same magnification, and that the large instrument can use a higher magnification than the smaller in the production of an equally bright image.

1.4 Brightness of a point image formed by a lens

The apparent brightness of a point source, unlike that of an area, is increased by passing its radiation through a lens, or mirror and lens, system. (Practical application of this difference: telescopic vision of stars by daylight, when sky brightness is diluted and their own increased to a level at which the contrast is sufficient for them to be seen —for whereas the brightness of a point image varies solely with D, that of the extended image, as we have just seen, varies inversely as M, and can therefore be reduced by increasing M.)

Compared with the naked-eye brightness of a point source, that of its telescopic image is increased by $\dfrac{D^2}{\delta^2}$, since in each case the amount of light received by the retina is a function of the area of the collecting lens. But $\dfrac{D}{\delta}$, as will be seen in section 3, is a measure of the minimum magnification that can be usefully employed. Hence, for point sources,

$$\frac{\text{brightness of telescopic image}}{\text{brightness of naked-eye image}} = (M')^2 \quad \cdots \quad (c)$$

Thus the brightness of the telescopic image is greater than that of the naked-eye image by a factor equal to the square of the minimum useful magnification—which, in turn, depends solely upon aperture.

31

Magnification may nevertheless make a stellar image appear to be brighter by darkening the background, when the increased contrast between the star and the sky is wrongly interpreted as a brightening of the stellar image.

1.5 Point image brightness and magnification

The relative brightness of the naked-eye and telescopic images of a point object being determined by the relative areas of the objective and the pupillary aperture of the eye, magnification theoretically has no effect upon the observed brightness of a star, nor on the value of the magnitude threshold—provided always that it is not less than the minimum useful magnification, when the aperture is in effect reduced. In practice, however, if M is considerably greater than the resolving magnification M_r,* or if the atmospheric turbulence is large, a new factor enters the picture: the expansion of the spurious disc past the size at which it can be considered as a point, to a sensible disc whose brightness is then determined by expression (b), not (c), and is illustrated in the Table on p. 30. The effect of this is to reduce its brightness in proportion to M, and thus to modify the theoretical figures relating aperture and minimum visible magnitude, when the atmosphere is poor or excessive magnifications are used.

The value of M at which this occurs cannot be precisely defined, since the expansion of the spurious disc with increasing magnification is gradual, and at no stage does it suddenly cease behaving like a point and start behaving as an area. Within this upper magnification range, however, the brightness of the image is in inverse proportion to M^2, and it is for this reason that stars near the telescope's light-grasp threshold may be rendered altogether invisible by gross over-magnification.

1.6 Photographic intensity of the image

The photographic and visual intensities of an extended image are not to be confused. We have seen that the visual, or subjective, brightness of an extended image is controlled by the magnification and the aperture (section 1.3). The photographic intensity of such an image in an instrument of given aperture, on the other hand, is a function of its focal ratio: for the area over which the light is spread (proportional to F^2) is as relevant a factor as the amount of light (proportional to D^2). Hence the photographic intensity of the image of an extended object is propor-

* The lowest magnification capable of showing all the resolved detail in the primary image—see section 3.3.

tional to $\left(\dfrac{D}{F}\right)^2$, and may be increased by reducing the focal length, or increasing the aperture, or making both changes simultaneously.

It is for this reason that objectives of small focal ratio (large relative aperture) are faster than those of large: since the proportionality is with the square of the relative aperture, an instrument working at $f/8$ is four times as fast as one working at $f/16$. And since the image intensity is independent of the aperture as such, a 1-in lens working at $f/2$ is faster than the 200-in objective of the Mt Palomar reflector working at $f/3\cdot3$.

The photographic speed of an objective where point images are concerned is, like the visual brightness (section 1.4), a function of aperture only. Thus a 6-in lens at $f/5$ is no faster than a 6-in lens at $f/15$.

1.7 Relative brightness and the scale of stellar magnitudes

The basis on which the scale of stellar magnitudes is constructed is the definition of magnitude such that a star of magnitude m is $2\cdot512$ (or $\sqrt[5]{100}$) times as bright as a star of magnitude $m+1$.

Since $\log_{10} 2\cdot512 = 0\cdot4$, we may say of two stars—the brightness and magnitude of the brighter of which are b and m, and of the fainter b' and m'—that

$$\frac{b}{b'} = 10^{0\cdot4(m'-m)}$$

or, taking logs,

$$\log\frac{b}{b'} = 0\cdot4(m'-m) \ . \quad . \quad . \quad . \quad . \quad . \quad . \quad (d)$$

This relationship may be expressed in a variety of forms, some of which it may be useful to have:

$$\frac{b}{b'} = 2\cdot512^{\Delta m} \quad . \quad . \quad . \quad . \quad . \quad . \quad . \quad (e)$$

$$\log\frac{b}{b'} = \log 2\cdot512 . \Delta m$$

$$= 0\cdot4 . \Delta m$$

$$\Delta m = \frac{\log b - \log b'}{0\cdot4}$$

$$\Delta m = 2\cdot5 \log\frac{b}{b'} \quad . \quad . \quad . \quad . \quad . \quad . \quad (f)$$

If M is the integrated magnitude of two stars, of magnitudes m and m', then

$$2\cdot512^{-M} = 2\cdot512^{-m} + 2\cdot512^{-m'}$$

Also $$M=m-x \qquad \qquad \qquad (g)$$

where $$x=\frac{\log\left(1+\dfrac{1}{2 \cdot 512^{(m'-m)}}\right)}{0 \cdot 4}$$

The following Table relates $m'-m$ to b/b' (equation (d)) and to x (equation (g)). Hence it can be used to derive (i) the relative brightness of two stars of known magnitudes, (ii) the magnitude difference of two stars of known brightness difference, (iii) the individual magnitudes of two stars whose integrated magnitude and relative brightness are known (iv) the integrated magnitude of two stars whose individual magnitudes are known.

$m'-m$	b/b'	x	$m'-m$	b/b'	x
0·00	1·00	0·75	1·75	5·01	0·20
0·10	1·10	0·70	1·80	5·25	0·19
0·20	1·20	0·66	1·83	5·40	0·18
0·30	1·32	0·61			
0·35	1·38	0·59	1·85	5·50	0·18
			1·87	5·60	0·18
0·40	1·45	0·57	1·90	5·75	0·18
0·45	1·51	0·55	1·93	5·92	0·17
0·50	1·58	0·53	1·95	6·03	0·17
0·55	1·66	0·51			
0·60	1·74	0·49	2·00	6·31	0·16
			2·10	6·69	0·15
0·65	1·82	0·48	2·20	7·59	0·13
0·70	1·91	0·46	2·30	8·32	0·12
0·75	2·00	0·44	2·40	9·12	0·11
0·80	2·09	0·42			
0·85	2·19	0·41	2·50	10·00	0·10
			2·60	11·00	0·09
0·90	2·29	0·39	2·70	12·00	0·09
0·95	2·40	0·38	2·80	13·00	0·08
1·00	2·51	0·36	2·87	14·00	0·08
1·05	2·63	0·35			
1·10	2·75	0·33	2·90	14·50	0·08
			2·95	15·00	0·08
1·15	2·88	0·32	3·00	15·85	0·07
1·20	3·02	0·31	3·10	17·50	0·06
1·25	3·16	0·30	3·25	20·00	0·05
1·30	3·31	0·29			
1·35	3·47	0·28	3·50	25·00	0·04
			3·70	30·00	0·04
1·40	3·63	0·27	4·00	39·80	0·03
1·45	3·80	0·25	4·25	50·00	0·02
1·50	3·98	0·24	4·45	60·00	0·02
1·55	4·17	0·23			
1·60	4·37	0·22	4·60	69·00	0·02
			4·75	79·00	0·02
1·65	4·57	0·21	4·80	91·00	0·01
1·70	4·79	0·21	5·00	100·00	0·01

Example (i)

How much brighter is a star of the 1st magnitude than one of the 10th?

$$m'-m=9$$

From the Table, if $m'-m=5$, then $b/b'=100$,
if $m'-m=4$, then $b/b'=39\cdot8$.

Hence if $m'-m=9$, $b/b'=100\times39\cdot8$
$$=4000 \text{ nearly.}$$

Example (ii)

If star A is fifteen times as bright as star B, what is their magnitude difference?

From the Table, if $b/b'=15$, then $m'-m=2\cdot95$.

Hence A is nearly 3 magnitudes brighter than B.

Example (iii)

If the combined magnitude of a close stellar pair is 4·8, and the estimated magnitude difference of the components is 0·5, what are the magnitudes of the two stars?

$$m'-m=0\cdot5$$
$$\therefore \quad x=0\cdot53$$
$$\therefore \quad m=4\cdot8+0\cdot53$$
$$=5\cdot33$$
$$\therefore \quad m'=5\cdot33+0\cdot5$$
$$=5\cdot83$$

Hence the magnitudes of the two stars are 5·3 and 5·8.

Example (iv)

What is the integrated magnitude of a close double, whose individual magnitudes are 2·7 and 3·0?

$$m'-m=0\cdot3$$
$$\therefore \quad x=0\cdot61$$

and
$$M=2\cdot7-0\cdot61$$
$$\simeq2\cdot1$$

Hence the integrated magnitude of the pair is 2·1.

1.8 Other magnitude relationships

If $m_p=$photographic magnitude,

$m_v=$visual or photovisual magnitude,

$m_b=$bolometric magnitude,

$M=$absolute magnitude $\Big\}$ on same magnitude scale,
$m=$apparent magnitude

$\pi=$stellar parallax (in $''$ arc),

then,

$$\text{colour index} = m_p - m_v$$
$$\text{heat index} = m_v - m_b$$
$$M = m + 5 + 5 \log \pi$$

In the past the value of the c/i has, by definition, been taken as zero for an average star of spectroscopic type A0. Thence it increased with a negative sign for stars of increasing temperature and blueness, and with a positive sign for stars of decreasing temperature and increasing redness:

Spectroscopic type	B	A	F	G	K	M
Colour index	−0·31	0·0	+0·32	+0·71	+1·17	+1·68

Thus a negative c/i indicates that the star is brighter photographically than visually; a positive c/i that it is brighter visually than photographically.

In recent years, however, the arbitrary coincidence of the photovisual and photographic scales at A0 has proved unsatisfactory, and instead both scales are now established empirically by reference to the NPS. (See section 37.)

The colour indices of the Sun, Moon, and planets are all positive ranging from +0·50 for Jupiter to +1·38 for Mars. Two stars note, worthy for their abnormally large c/i are Y CVn, a type N variable with an index of +4 ($m_p = 9·5$, $m_v = 5·5$), and a 9th-magnitude star in the field of T Mon, c/i = +5 ($m_p = 14$, $m_v = 9$).

SECTION 2

TELESCOPIC FUNCTION: RESOLUTION

2.1 Introduction

Diffraction is a phenomenon which, though similar to interference in its general effects, occurs within a single wave system. In a wavefront every point may be regarded as a secondary radiator, itself generating a wave system. Within a pencil, however, these secondary vibrations are eliminated by destructive interference. Only at the edge of the pencil are they not eliminated. For this reason the intensity at the edge of a pencil does not fall sharply to nil, but rhythmically.

It will be seen from equation (*d*), p. 38, that the angular size of the spurious image of a point source, the basis of all diffraction effects observed with a given telescope, is a function of that telescope's aperture. It should be emphasised that throughout the following discussion the *used* aperture is referred to. If the full aperture is not used, the diffraction pattern will be determined by that fraction of it which is responsible for forming the image. For example, a slit in the focal plane produces the same diffraction effects as a conjugate slit in front of the objective; again, a camera shutter for solar photography, consisting of a slit in an opaque screen that is allowed to fall across the lens, will produce diffraction effects appropriate to its width, irrespective of the aperture of the lens.

Since it is only at the boundary of the wavefront that interference is asymmetrical and diffraction is not eliminated, its effects are most noticeable in those pencils which, so to speak, consist mostly of boundary: in the images of point sources, of boundaries (as in the edges of shadows), and of minute detail generally. The observer should be aware of the fact that diffraction is capable of profoundly modifying the details of the image in shape, size, intensity, and visibility.

2.2 Structure of the star image

The laws of geometrical optics would lead one to expect that a perfectly corrected objective would produce a point image of a point

37

object. This is not so: diffraction within the converging pencil produces an image of finite dimensions and characteristic structure. This diffraction pattern consists (assuming a circular objective) of a central bright disc, termed the spurious disc or Airy disc, surrounded by alternate concentric rings of light and darkness—the diffraction rings and interspaces, respectively—the latter resulting from destructive interference.

The linear radii of the maxima and minima within the image at the focus are give by

$$r' = k \cdot \frac{\lambda F}{D} \quad \ldots \ldots \ldots \quad (a)$$

where k is a coefficient whose value is different for each ring and interspace; for the first interspace, $k = 1 \cdot 220$; for the first ring, $k = 1 \cdot 620$. Hence the theoretical linear radius of the first interspace is given by

$$r'_1 = \frac{1 \cdot 22 \lambda F *}{D} \quad \ldots \ldots \ldots \quad (b)$$

In angular measure this (the angle subtended by the image at the centre of the objective) becomes

$$\left. \begin{array}{l} r_1 = \dfrac{1 \cdot 22 \lambda}{D} \text{ radians} \\[2mm] \text{or} \qquad r_1 = \dfrac{1 \cdot 22 \lambda}{D} \cdot 206,265 \text{ '' arc} \end{array} \right\} \quad \ldots \ldots \quad (c)$$

And finally, taking a value for λ corresponding to the yellow region of the spectrum to which the eye is most sensitive (5500Å), the above expression becomes

$$r_1 = \frac{5 \cdot 45*}{D} \quad \ldots \ldots \ldots \quad (d)$$

D being expressed in inches.

Using the same value of λ, we have from equation (a) above,

$$r'_1 = 2 \cdot 642 \cdot 10^{-5} \cdot \frac{F}{D} \quad \ldots \ldots \ldots \quad (e)$$

for the linear radius (ins) of the first minimum.

Hence the angular radius, r, is determined solely by D, to which it is inversely proportional; r', the linear radius, on the other hand is a function of F/D, the focal ratio.

* See Table, p. 46.

Theoretically, about 85% of the light forming the whole image is concentrated in the spurious disc. The rings are therefore comparatively inconspicuous—even the first is only 1·7% as bright as the disc itself—and may be disregarded in a preliminary discussion of resolving power; they are altogether absent from faint images (see Figure 2). Since λ appears in expression (c), above, it follows that the rings are coloured, their inner edges being blue and their outer red, but this is entirely imperceptible in normal astronomical observation.

The intensity of the spurious disc is theoretically greatest at its centre, falling off rapidly to zero in the first interspace, there being no clearly

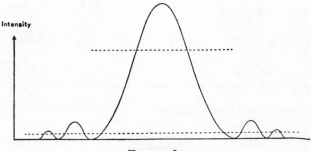

FIGURE 2

defined 'edge' to the disc. Although f/ratio has no effect upon the angular size of the diffraction pattern it does affect its appearance, and the theoretical appearance just described is in fact only to be seen with f/ratios considerably greater than are usually to be found in astronomical telescopes. At normally encountered f/ratios (say, $f/5$ to $f/20$) no intensity gradient across the disc is perceptible, the border between the disc and the first minimum appears nearly sharp, and the rings are brighter than theory would indicate, the first ring being not much fainter than the disc itself. With good seeing and a 1st-mag star, as many as 5 or 6 rings may be seen with an aperture of about 6 ins.

Theory can define the position of the zone of zero intensity in each interspace—the locus of a point maintaining a constant distance r from the centre of the spurious disc—but cannot define the point at which the intensity of the disc falls below the threshold of visibility: the visible extent of the disc, like the number of rings visible, varies for a given instrument with the brightness of the source, although the discs are in fact the same size, irrespective of brightness. For this reason r_1 cannot

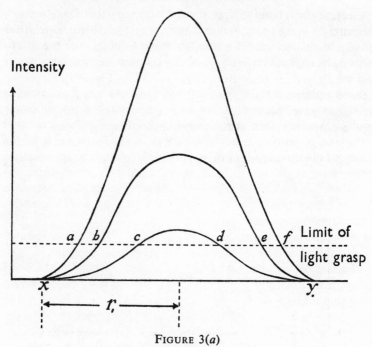

FIGURE 3(*a*)

Visible diameters of the three discs are
$af>be>cd$
Diameter of the first interspace is xy in each case

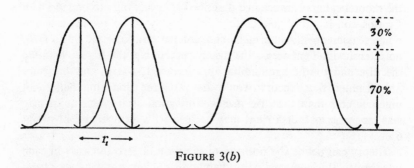

FIGURE 3(*b*)

be taken as the radius of the spurious disc; a closer approximation would be $\frac{1}{2}r_1$ (Figure 3 (*a*)).

The complete profile of the diffraction pattern of a point source is shown in Figure 2. It can be seen that, with a given instrument, the visibility and number of the rings depends upon the brightness of the star; alternatively, that for a given star it depends upon the light grasp, and therefore aperture, of the instrument employed. Also, more of the disc of a bright star than of a faint star will be visible; this, it should be realised, does not affect the value of r_1, which can only be increased by reducing D.

The appearance of the diffraction pattern as described above assumes that the image is accurately in focus—i.e. that the focus of the ocular coincides precisely with the focal plane of the objective—and that aberrations, if present, are imperceptible. If the focal planes of the ocular and objective are separated, the appearance of the image suffers a progressive modification. The first ring (and to a lesser extent the others) brightens at the expense of the disc, while the first interspace becomes faintly luminous. As the separation of the focal planes increases, a stage is reached when all the light has been drained from the disc into the rings; the image then consists of a set of exaggerated rings enclosing a dark central disc. With further separation, a new development begins. The rings start to expand, and a bright point appears at the centre of the image; this point expands into a bright ring. Again a bright point appears at the centre of the pattern, and in turn expands into a ring. Thus successive new rings are thrown off from the centre of the image. In the case of a reflector a stage is reached at which the throwing out of new rings from the centre is succeeded by the appearance of a ring of characteristic appearance (broad, with an irregular edge) which then remains, further rings being projected from its outer edge. These extra-focal rings, being extremely sensitive to aberrations in the objective, are usefully employed in the testing of OGs and mirrors.

The structure of the in-focus diffraction pattern produced by a Newtonian is modified by the central obstruction in the way of the incident light caused by the diagonal. The diameters of the disc and rings are slightly reduced, while the rings are relatively thicker and noticeably brightened at the expense of the disc. The interspaces being unaffected, the apparent brightness of the rings is increased still further by contrast. With a given instrument, an obstruction of $D/2$ reduces the diameter of the first interspace to 0·818 of its value with full aperture; at the same time the brightness of the first ring is increased by a factor of 4 and that of the disc reduced by a factor of 2. The brightening of the rings, and the correlated reduction of the intensity of the disc, varies with the focal ratio and with the size of the obstruction; it is also increased by residual spherical aberration.

The modifications of the image caused by diffraction at the supports of the flat may conveniently be mentioned here. In the case of a bright image, a four-armed spider produces a four-pointed cross centred upon the diffraction disc. The length of the arms of this cross is inversely proportional, and their brightness directly proportional, to the thickness of the supports. The diffraction cross causes slight loss of contrast in an extended image.

The effect of this relative thickening and brightening of the rings upon the definition of the reflector's image is to some extent controversial: Pickering, for example, considered it to be one of the causes of the reflector's slightly inferior definition, while Hargreaves denies that it has any appreciable effect upon planetary definition whatever. It is probable that this difference of opinion springs from a consideration of different cases, for Dall, in a careful investigation (B. 89),* has clearly shown that if the obstruction is relatively small (for an instrument of given f/ratio) little or no deterioration of planetary definition is encountered; if it exceeds about $D/5$ an impairment of definition becomes increasingly evident, especially through the loss of contrast near the limit of vision.

The visibility of double stars of equal, or nearly equal, magnitude is certainly facilitated by the type of diffraction pattern produced by the reflector,† though this is not accompanied by improved accuracy in their measurement—an observational field in which the refractor is certainly superior. Close unequal doubles, on the other hand, tend to be more easily resolved with a refractor than with a reflector of equal aperture, owing to the exaggeration of the rings (especially the first) in the latter's diffraction pattern. A *comes* much more than 1 mag fainter than its primary will be imperceptible if its image falls on the first ring; with a smaller magnitude difference it may still appear as nothing more than a slight local brightening of the ring, and as such be very easily overlooked. Again, a very faint *comes* close to its primary may on occasions be rendered invisible by *increasing* the aperture, it falling in the first interspace of the smaller telescope's image of the primary, but on the first ring of the larger telescope's.

It follows from expression (*c*), p. 38, that the angular size of the spurious disc varies inversely as aperture: a 2-in objective produces star images four times the angular size of those produced by an 8-in. The

* Reference to Bibliography (section 41).

† For work requiring maximum resolving power, increasing the diameter of the central obstruction might profitably be tried. In the case of a refractor, however, it will exaggerate the secondary spectrum, as also the abaxial aberrations. Sir John Herschel long ago reported an improvement in a telescope's resolving power when used with a $D/5$ to $D/6$ central stop, despite the brightening of the rings.

linear size of the image, on the other hand, is a function of the f/ratio of the objective (equation (b), p. 38). Since it is the angular size of the images that is of importance in the resolution of closely adjacent images, large apertures are the desideratum and focal ratios are irrelevant.

Summarily,

$$\text{linear size} \propto \frac{F}{D}$$

$$\text{angular size} \propto \frac{1}{D}$$

and in addition,

$$\text{apparent size} \propto \text{brightness of image}$$
$$\text{brightness of image} \propto D.$$

2.3 Structure of extended images

All linear and area detail of the image is falsified by diffraction to an extent inversely proportional to the difference between its size and the size of the diffraction pattern of a point image. It varies from a slight distortion of shape, size, or intensity, to complete suppression. Conversely, diffraction can 'create' detail in an image which has no objective counterpart.

Diffraction at the edge of a bright area of the image results in blurring of that edge. Encroachment of light from the bright into the dark causes the dimming of a narrow strip of the former and the 'greying' of an equal strip on the dark side. This blurring of an edge can be observed in the shadow of a ruler (even with parallel light), at both jaws of a spectroscope slit, and among such astronomical objects as planetary limbs and lunar shadows.

A bright line of negligible width, crossing a dark area, may be regarded as consisting of a very large number of contiguous points. Each of these will produce its own diffraction pattern, with the result that the image of the line will be thickened by a fringe on either side.

If the line is turned round upon itself to form a circle, the diffraction band on the inner and outer sides of the image will only be of equal density if the radius of the circle is large compared with r_1. As it approaches r_1 there will be progressive overlapping of the bright rings of the internal fringe, building up a brighter central area, intermediate in intensity between the band itself and the outer fringe. When the radius of the ring is equal to r_1 the first internal rings will all intersect at the centre of the image, with the formation of a bright, star-like point.

The images of small dark spots on bright grounds, and of small bright spots on dark grounds, are modified in respect of both brightness and size: the former being 'greyed' (in extreme cases being rendered invisible through loss of contrast) and reduced, the latter enlarged and brightened centrally. Under certain circumstances a bright patch or point will form at the centre of a small dark spot, and a dark one at the centre of a bright spot on a dark ground.

Astronomical objects which are subject to such diffraction effects include planetary limbs, Martian and lunar 'canali' and fine lunar and planetary detail generally, satellites of sensible disc, shadows of satellites on primaries, inner planets in solar transit (the 'volcano' on Mercury is visible with apertures not larger than about 1 in), and the 'black drop' (visible with magnifications above about $15D$).

2.4 Resolving power of a telescope

Whereas light grasp and magnification are explicable in terms of geometrical optics, an understanding of resolution thresholds and resolving power involves recognition of the undulatory nature of light.

The separation of the images of a close pair of stars is more than a mere matter of magnification, and is not assisted by increasing F. Since, as has just been seen, these images are not points, but discs, it can happen that they are so large, with a given aperture, as to overlap to an extent that prohibits their visual separation, even though high magnification may have separated their centres to a sufficient distance for them to have become individually visible were they point images. In such a case resolution can only be achieved by reducing the size of the spurious discs, i.e. by increasing D. To increase the magnification alone will achieve nothing, since the size of the discs will increase with their separation, their resolution being therefore unaffected. Increasing M can only result in resolution if the primary images in the focal plane are resolved.

It is generally agreed that the intensity drop midway between the central maxima of two equally bright, overlapping diffraction patterns is relatively deep enough (about 30%) to appear clearly marked to the eye when the distance separating their centres is equal to the radius of the first interspace. The centre of each pattern then lies in the first interspace of the other, and the areas defined by the two first interspaces are overlapping one another by one half (Figure 3 (*b*)). Owing to the character of the light-distribution across the diffraction pattern it is less easy to define precisely the diameter of the 'visible disc' than that of the first minimum, and this has led to some confusion in the literature: some writers speaking of the spurious disc as though bounded by the first interspace, others as though the diameter of the visible disc is only half that of the first interspace.

From equation (d), p. 38, we therefore have for the theoretical resolving power, or threshold of resolution, of a telescope:

$$R = \frac{5''45}{D} \qquad \qquad (f)$$

In practice it is found that a separation of R is clearly perceptible, and that under conditions of good seeing a normally acute eye can just perceive a minimum only 5% less than the intensity of the maxima of the two patterns. This yields a value for R (designated by R' to distinguish it from the theoretical value) which agrees closely with Dawes' empirical criterion,* that a pair of mag 2 stars, separation 4''56, are just resolvable with a 1-in objective:

$$R' = \frac{4''56}{D} \qquad \qquad (g)$$

This is also in accordance with the results of direct micrometrical measures of spurious discs, e.g. Pickering's value of 0''46 for a 10-in instrument.

In the following Table of linear and angular resolutions, column (1) is derived from equation (e) above, column (3) from (f), and column (4) from (g). The figures in column (2) represent focal ratio when taken in conjunction with column (1), and aperture when applying to columns (3) and (4).

In practice, the performance of a given instrument and observer may differ from both the theoretical figure and the Dawes criterion, for a number of reasons:

(1) Abnormal acuity or experience of the observer.

(2) The Dawes criterion implies a clearly perceptible intensity minimum between the two maxima. But the existence of two overlapping images may be inferred from the elongation of the residual image although there may be no perceptible intensity drop between two maxima. In this case the pair is not, in the theoretical sense, resolved. Elongation of the image is visible to normal eyes when the separation is not less than about $R'/2$; Burnham was able to detect it with his 6-in for separations of 0''2 ($R'/3\cdot8$). The following separations, achieved by

---

* Reached as a result of tests with various apertures, and not, as sometimes stated, based on actual observations with a 1-in OG (B. 40).

(1)	(2)	(3)	(4)
Theoretical radius of 1st interspace (ins) r_1'	Aperture (ins) D———→ ←———F/D Focal ratio	Theoretical resolution threshold $R=r_1$	Empirical resolution threshold (Dawes) R'
$2\cdot6\times10^{-5}$	1	$5''45$	$4''56$
$5\cdot3$	2	$2\cdot73$	$2\cdot28$
$7\cdot9$	3	$1\cdot82$	$1\cdot52$
$1\cdot06\times10^{-4}$	4	$1\cdot36$	$1\cdot14$
$1\cdot32$	5	$1\cdot09$	$0\cdot91$
$1\cdot59$	6	$0\cdot91$	$0\cdot76$
$1\cdot85$	7	$0\cdot78$	$0\cdot65$
$2\cdot11$	8	$0\cdot68$	$0\cdot57$
$2\cdot38$	9	$0\cdot61$	$0\cdot51$
$2\cdot64$	10	$0\cdot55$	$0\cdot46$
$2\cdot91$	11	$0\cdot49$	$0\cdot41$
$3\cdot17$	12	$0\cdot45$	$0\cdot38$
$3\cdot96$	15	$0\cdot36$	$0\cdot30$
$5\cdot28$	20	$0\cdot27$	$0\cdot23$
—	36	$0\cdot15$	$0\cdot13$
—	100	$0\cdot05$	$0\cdot05$
—	200	$0\cdot03$	$0\cdot02$

experienced observers, are among the many scattered through the literature:

$3\frac{1}{2}$-in	$R'=1''30$	Separation down to $1''16$.
12-in	$R'=0''38$	Separation down to $0''25$.
36-in (Lick)	$R'=0''13$	Separation down to $0''09$.
40-in (Yerkes)	$R'=0''11$	Separation clear enough for measurement at $0''2$; estimation of position angle by direction of disc elongation possible below this.
82-in (McDonald)	$R'=0''06$	Clear separation down to about $0''1$; measurable elongation below this limit.

(3) It might be expected from theoretical considerations—less of the total spurious disc of a faint star being visible than of a bright one*— that faint pairs should be more easily resolved than bright. This is not

---

* The diameters of a bright and a faint stellar spurious disc, as measured by Pickering with a 10-in objective under good seeing conditions, were $0''60$ and $0''35$ or $\dfrac{6''0}{D}$ and $\dfrac{3''5}{D}$ respectively.

borne out by experience. From a discussion of a large number of observations, Lewis* found, for example, that whereas bright pairs of roughly equal magnitudes, lying within the mag range 5·7 to 6·4, could be separated at $\dfrac{4\rlap{.}''8}{D}$, equal faint pairs (mags between 8·5 and 9·1) could not be resolved unless their separation was $\dfrac{8\rlap{.}''5}{D}$. For a very similar magnitude range (8·8 to 9·0) Aitken found that he required a separation of $\dfrac{6\rlap{.}''1}{D}$. Similarly, very bright pairs are more difficult than those of moderate brightness, as exemplified by Lewis's first group. If one or both components are bright enough to 'dazzle', the effect of irradiation will be added to the already reduced acuity of the eye, and the Dawes criterion will be (in extreme cases, considerably) too low.

(4) The theoretical figure and the Dawes criterion are only strictly applicable to equally bright images;† the greater the magnitude difference between the two stars, the greater will be the discrepancy between either of these figures and the actual telescopic performance. Lewis found that for pairs differing by about 3 mags (the mean mags of the pairs included in his study were 6·2 and 9·5), the resolution threshold was raised to $\dfrac{16\rlap{.}''5}{D}$; while for a group of doubles, the mean mags of whose components were 4·7 and 10·4, it had risen to $\dfrac{36\rlap{.}''0}{D}$.‡

(5) Expression (c), p. 38, shows that the size of the disc varies with the wavelength of the radiation, the red disc being considerably larger than the blue:

for violet light ($\lambda = 3600\text{Å}$), $r_1 = \dfrac{3\rlap{.}''57}{D}$

for red light ($\lambda = 7700\text{Å}$), $r_1 = \dfrac{7\rlap{.}''63}{D}$

Hence the resolution of a pair of blue stars may be expected at smaller separations than a pair of red stars.

(6) The resolving power of a reflector is some 5% greater than that

---

* B. 45.

† The figure adopted by Dawes was stated by him to apply to two 6-mag stars seen in a 6-in telescope. Any departure from this magnitude-aperture relation (even when equally bright images are involved) will, strictly speaking, introduce some modification of the criterion.

‡ See also B. 61.

of a refractor of the same aperture, owing to the smaller spurious disc of the former (see p. 41), though this depends upon the size of the flat.

(7) It has been suggested that slight residual spherical aberration may account for some instances of the theoretical threshold being passed, in the case of stellar resolution. Atmospheric turbulence would probably, however, more than counteract whatever advantage might be gained in this way.

(8) Even slight atmospheric turbulence will seriously impair the instrument's performance. This, more than any other single factor, is the commonest cause of actual performance falling below that suggested by the Table. Owing to their reduced susceptibility to atmospheric effects, small instruments generally are more efficient than large —both, of course, in relation to the appropriate value of R'. Thus with a 6-in Burnham achieved, on the average, $R'/2$, but with a 9·4-in less than R', and with apertures between 12 and 36 ins, from $1·15R'$ to $1·6R'$.

(9) The combination of atmospheric turbulence and the low resolution of the emulsion may easily reduce the photographic resolving power by a factor of 20.

The resolving power of prisms and gratings is discussed in section 19.3.2.

2.5 Determination of telescopic resolving power

(1) By the observation at a high altitude of double stars whose separation lies in the neighbourhood of the figures quoted in the Table on p. 46.* The factors listed above must be taken into account, the most important being the state of the atmosphere: no instrument can be fairly tested if the turbulence is marked. Unequal magnitudes and visual inexperience can also combine to make an instrument appear misleadingly inferior.

(2) By the use of a test object. The factor of turbulence can be largely eliminated by reducing the length of the light path, but results so obtained are likely to be misleading for the very reason that conditions do not resemble those under which the instrument is actually used.

(a) For stellar sources: a tinfoil sheet, with two minute holes punctured in it, mounted in front of a light source from which it is separated by a sheet of ground glass. Measurement of the distance from the telescope at which the first trace of elongation of the image, the first appearance of a perceptible intensity minimum, and clear separation occur, yields results tolerably near those obtained under normal

* See *B.A.A.H.*, 1922.

observational conditions. They tend nevertheless to be on the optimistic side, owing to reduced turbulence.

(b) For planetary objects: a fine wire gauze placed in front of a light source and separated from it by a sheet of ground glass. From the distance, a, screen to objective, at which the wires are first separated, and the linear separation of the wires, b, which for a uniform grid can be determined by dividing its overall width by the number of wires it contains, their angular separation can be obtained from

$$R = \frac{b}{a} . 206,265$$

in " arc. Alternatively a grid of equal black bars and interspaces can be ruled on a sheet of paper; this is observed with a high-power ocular and brought towards the telescope until the individual black lines begin to emerge from the uniform grey tone. If each rule and interspace together measure 1 mm, and the grid is placed 10 m from the objective, then from the expression above, the angular separation is approximately 20"; it is in direct linear proportion for other grid dimensions, and inverse linear proportion for other distances.

As shown by the figures quoted in section 2.6, the resolution threshold obtained in this and similar ways bears little relation to that applicable to stellar images.

2.6 Definition: the resolution of extended detail

The definition—i.e. sharpness, and smallest detail visible—of an extended surface also varies with D, and is heavily dependent on atmospheric state. The image is composed of the overlapping diffraction patterns formed by light emanating from every point in the object; if the radii of these diffraction discs were effectively of the full value r_1 there could be no resolution, however high the power used: but the low intensity of the light, compared with that of stars, reduces their effective size, and the Dawes criterion therefore cannot be applied to the resolution of detail in extended images.

This is clearly demonstrated by the following actual performances of instruments of varied apertures on three types of extended detail:

A. BLACK SPOT ON A WHITE GROUND

(i) W. H. Pickering, while testing his 10-in reflector under good seeing conditions in Jamaica, found that a circular spot became visible when its angular diameter exceeded 0".20 $(R'/2\cdot3)$.*

---

* The position regarding the visibility of a light disc, as such, on a dark ground is less easy to define, but records of the visibility of Neptune's disc suggest that only about half this aperture is required. See O.A.A., section 10.2.

(*ii*) Experiments by W. H. Steavenson showed that $R'/3$ is a fair average figure.

(*iii*) The naked-eye visibility of sunspots whose diameter is less than R' is well known.

B. SINGLE DARK LINE ON LIGHT GROUND

(*i*) Cassini's division, width about $0.''5$, was discovered with about $2\frac{1}{2}$-in aperture $(R'/3\cdot5)$.

(*ii*) W. H. Steavenson's experiments yielded about $R'/5$ as an average figure.

(*iii*) During the testing of an 11-in refractor at Harvard, a human hair was visible against a light-toned ground at a distance of nearly a quarter of a mile, where it subtended only $0.''029$ $(R'/14)$.

(*iv*) W. H. Pickering glimpsed a dark line $0.''03$ wide with his 10-in $(R'/15)$.

C. PARALLEL DARK LINES ON LIGHT GROUND

(*i*) W. H. Pickering: minimum separation for resolution with a 10-in reflector was $0.''63$ $(1\cdot4R')$.

(*ii*) A similar performance was given by the Arequipa 15-in, which resolved a pair of parallel lines when their separation was increased past $0.''42$ $(1\cdot4R')$, in good seeing; slight atmospheric deterioration immediately raised the threshold to about $2R'$. At less than $0.''42$ the lines appeared as a grey band of width about $1\frac{1}{2}$ times their separation.

(*iii*) Resolution of the lines at $12''$ with a $0\cdot4$-in OG $(1\cdot1R')$. See also sections 2.3, 24.6, 26.7, 26.9.

2.7 Practical demonstration of diffraction and interference effects

Experimental demonstrations of these modifications can be easily staged, and should convince anyone that it is aperture and not magnification that determines the separability or otherwise of a pair of stars.

Make a clean circular hole in two sheets of metal foil, using a fine needle, the diameter of the hole not exceeding about $0\cdot02$ ins. If one of the sheets is placed in front of an electric filament and viewed through the second, held close to the eye, a spurious disc and rings will be seen with a perfection seldom provided by a telescope and a star. Increasing the size of the view-hole brightens, and reduces the size of, the disc and rings, while increasing the number of the rings. The dependence of resolution on aperture can be very clearly demonstrated by piercing a close pair of holes in the object-sheet, and observing them through a graded succession of holes in the telescope-sheet, ranging in diameter from about 0·05 ins down to as small as the needle can make them.

If a single 'star', or a disc of appreciable diameter, is now viewed through the original 'double star', the action of the interferometer is simulated, the image being traversed by alternate bright bands and dark interspaces. It can be shown by trial that increasing the separation of the view-holes increases both the number and the crowding of the interference fringes.

The effect of diffraction upon extended and linear detail can also be demonstrated. Pierce a white card with two holes and observe them through a view-hole which, at a given distance, is just incapable of resolving them. Then below the two holes rule a black line such that its width is equal to the holes' separation. If the card is reobserved from the same distance it will be found that this line is clearly visible—i.e. R' is too large a figure for a linear image. Lack of resolution of surface detail through insufficient aperture is easily shown by observing any patterned surface, such as a sheet of graph paper, at different distances with the naked eye and with the pinhole.

SECTION 3

TELESCOPIC FUNCTION: MAGNIFICATION AND FIELD SIZE

Since the telescope is a compound optical instrument, in which both the objective and the ocular form images, it is necessary to distinguish between (a) the magnifying power of the objective, and (b) the magnification of this primary image by the ocular in forming the final image.

3.1 Magnifying power of objective

The size of the image at the primary focus is dependent on F alone. If θ=angular height of object (i.e. the angle subtended by the primary image at the centre of the objective), h=linear height of the primary image, in inches, and F=focal length of objective, also in inches, then

$$h=\frac{\theta F}{k} \qquad \ldots \ldots \ldots \quad (a)$$

where k=57·3 (° per radian) when θ is in °, or 206,265 (″ per radian) when θ is in ″.

D, it will be observed, does not figure in the expression for h. For example, substituting 0°53 for θ in equation (a), we have for the linear diameter of the primary image of the Sun in a telescope of 4 ft focal length (e.g. a typical 3-in refractor), 0·4 ins. Similarly, for a 150-ft telescope (e.g. the larger Mt Wilson tower telescope), diameter of primary image of Sun=16·6 ins.

3.2 Magnification of combined objective and ocular

Any textbook on elementary optics can be referred to for the simple demonstration (it can be inferred from Figure 4) that the magnification of a telescope is given by

$$M=\frac{F}{f} \qquad \ldots \ldots \ldots \ldots \quad (b)$$

52

where F and f are the focal lengths of the objective and ocular respectively.

The action of the ocular on the primary image is clearly seen from expression (b), where f and M are inversely proportional. This is confirmed by common sense: as f decreases, the distance of the eye from the image decreases, hence the angle it subtends at the eye increases, therefore the magnification of the primary image increases.

The magnification of the final image can therefore be increased in either of two ways, by increasing F or by decreasing f. Since the former

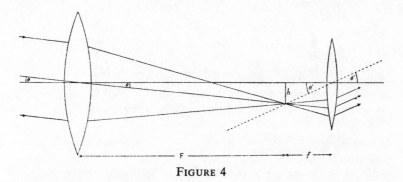

FIGURE 4

involves the use of a different objective (or of a different, e.g. Cassegrain, focus), one is in practice normally confined to the latter. This, however, cannot be carried to indefinite limits: prohibitive aberrations are introduced by the highly curved surfaces of very short focus lenses, and reduction of f also reduces the size of the field and the brightness of extended images.

3.3 Lower limit of useful magnification

The size of the exit pupil (section 3.11) also imposes limits, both upper and lower, on the range of magnification possible with a fixed value of F. Consider the passage of rays from a star situated on the optical axis:

It is clear (Figure 5) that

$$\frac{D}{d} = \frac{F}{f} = M \quad . \quad . \quad . \quad . \quad . \quad . \quad . \quad (c)$$

where D is the diameter (clear aperture) of the objective, and d is the diameter of the exit pupil. (This is the principle behind the use of the Berthon dynamometer for measuring the magnification of an ocular.)

It can be seen from equation (c) that the size of the exit pupil is inversely proportional to the magnification, or

$$d = \frac{D}{M}$$

If the exit pupil is so large as to be incompletely accepted by the pupillary aperture of the dark-adapted eye,* the full light grasp of the objective is not being utilised, and D is in effect being stopped down—not by any diaphragm in front of the objective, but by the iris of the

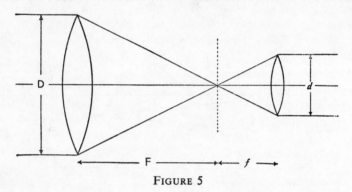

FIGURE 5

observer's eye. Thus a lower limit is set on the magnification than can be usefully employed.

If we denote this minimal value of M by M′, and take the corresponding maximal diameter of the exit pupil as 0·3 ins (equal to a mean night-time value of the diameter of the pupillary aperture), then

$$M' = \frac{D}{0·3} \qquad \cdots \qquad \cdots \qquad (d)$$

Thus the instrumental factor determining the minimum useful magnification is D, and we have

$$M' \simeq 3D$$

In other words, it is useless to employ magnifications lower than about 3 times the objective's aperture in inches. Owing to faint illumination

* The diameter of the normal pupil in subdued daylight is about 0·2 ins; in bright light it may contract to less than 0·1 ins. But with the degree of dark adaptation obtained during night-time observation of faint objects it may expand to about 0·33 ins. When brilliant objects, such as the Moon, are being observed, the pupil will contract again to 0·1 ins or less. It is interesting to note that the older observers were unaware of the extent to which dark adaptation can expand the pupil: J. Herschel, for example, always took its diameter to be 0·25 ins, and Rosse 0·1 to 0·2 ins according to the strength of the illumination.

from field stars and sky background, the diameter of the pupillary aperture may often be reduced to about 0·25 ins; hence the minimum effective magnification should be taken as lying between 3 and 4 per inch of aperture.

From (b) above (the expression for the magnification of a telescope), and putting f' for the ocular focal length giving a magnification of M',

$$f' = \frac{F}{M'} \qquad \ldots \ldots \ldots \quad (e)$$

Hence the focal length of the ocular providing the minimum useful magnification depends upon no other instrumental factor than the focal ratio, $\frac{F}{D}$ (from equation (c)).*

An entirely different consideration imposes, in certain circumstances, a lower limit on the magnification that can profitably be employed, whose value is about $4M'$. Equation (g), p. 45, gives the condition for the resolution of two points in the primary (focal plane) image of a telescope. As emphasised on p. 44, it is hopeless to try to separate two images by increasing the magnification, if they are not resolved in the focal plane. But it is true, conversely, that though they may be resolved by the objective, they will not be individually seen unless a high enough magnification is used to bring them above the resolving threshold of the human eye. No precise or universally applicable value can be assigned to this threshold (see section 24.5), but it is roughly true to say that unless two points subtend an angle of at least 1' at the eye (in the case of stellar images it will be nearer 2' or 3') they will not be separated.

Denoting by M_r the magnification necessary to separate two points, just resolved in the focal plane, to an angular distance of 60″,

$$R . M_r = 60 \text{ '' arc}$$

Also, from (g), p. 45,

$$R = \frac{4 \cdot 56}{D} \text{ '' arc}$$

(D, as always, being expressed in inches). Therefore

$$M_r = \frac{60 D}{4 \cdot 56} \text{ '' arc}$$

or

$$M_r \simeq 13 D$$

* For example, consider the case of a $f/15$ telescope of 6 ins aperture, 90 ins focal length. Then from (d), $M' = 18$, and from (e), $f' = 5$ ins. Hence for a telescope working at $f/15$ the maximum useful ocular focal length is 5 ins, irrespective of the absolute aperture or focal length of the objective.

(This is sometimes expressed in the easily remembered form that the minimum magnification fully to exploit the resolving power of an objective is equal to that objective's radius measured in millimetres.)

The less optimistic estimates of the eye's resolution threshold for stellar images (2′ and 3′) yield resolving magnifications of 24 and 36 times the aperture in inches, respectively.

With a lower magnification than M_r, there is still resolved detail buried in the focal plane image that has not yet been developed into visibility; increase of magnification beyond M_r will bring out no new detail, merely enlarging the scale of what is already visible. Thus under-magnification may waste some of the objective's resolving power as well as its light grasp.

3.4 Upper limit of useful magnification

At the other end of the scale, a decrease of the diameter of the exit pupil below about 1/30 ins leads to a progressive impairing of vision (as well as to other restricting factors) which imposes an upper limit on the magnification that can be usefully employed—irrespective of atmospheric turbulence and aberrations in oculars of very short focal length.

Using M'' to denote this upper limit, and taking the corresponding (minimal permissible) diameter of the exit pupil as 0·03 ins,

$$M'' = \frac{D}{0 \cdot 03} \quad \cdots \cdots \cdots \quad (f)$$

or
$$M'' = 30D \quad \cdots \cdots \cdots \quad (g)$$

The range of permissible magnification is also limited at its upper end by the quality of the objective, and by the fact that oculars of shorter focal length than about 0·1 ins are not practicable either to make or to use.

'Whittaker's rule', which is an empirical expression for the value of M'', states that definition suffers from diffraction effects when the magnification exceeds the number of mm in D: that is, about 25 per inch of aperture.

3.5 Discussion of theoretical limits of magnification

It should be emphasised that these 'limits', especially the upper, are not hard and fast. Vision is not affected so adversely as to prohibit observation with magnifications at least double those indicated by equation (g) above—with small instruments, anyway. For this reason, a second set of values of M'', nearer to the practical upper limit of magnification, is included in the Table and graph on p. 58.

Again, certain classes of object stand higher magnification than others; close doubles, for instance, may react to magnifications of $50D$* or $70D$, providing their angular separation is not below the telescope's resolution threshold and the atmosphere will stand it. The same is true of micrometer measures, where the highest magnification that the atmosphere will permit (seldom above $60D$) must be used.†

The use of lower magnifications than those tabulated under M' will merely involve wastage of the objective's light grasp—a matter of little importance when the object is bright. Use of a lower magnification than M_r wastes the objective's resolving power—which, again, is not vitally important in the observation of comparatively featureless and extended objects, such as some types of nebulosity, or the tails of most comets; more important, in such cases, is the loss of light.

Planets are among the most demanding of objects, since they bear a mass of minor detail, for which the maximum resolution of which the instrument is capable is required, while at the same time being for the most part not so bright that the illumination of the image can be ignored.‡ Magnifications between M_r and the theoretical value of M'' are, generally speaking, the best compromise that can be made between these conflicting demands; with such objects, to increase M far beyond M_r achieves nothing but an inferior and 'diluted' image.

At the same time it must not be forgotten that the highest magnification that can be used satisfactorily at any time is always dependent to a greater or smaller extent upon the atmosphere (see section 26), which seldom allows more than $50D$ to $60D$. The values given in the Table assume perfect atmospheric conditions. With very large apertures, such as are not found in amateur hands, the quoted values of M'' will become progressively too high, and those of M' too low. A strictly linear relationship between M'' and D is only indicative of the conditions found in practice when D is fairly small. As an example of this, Lewis has quoted the following figures showing the normal range of maximum

---

* A convenient abbreviation, of a type much used in this section and elsewhere for '50 times the aperture of the objective expressed in inches'.

† Discussed in greater detail, *O.A.A.*, section 18.3.

‡ See section 24.6 and *O.A.A.*, section 3.4.

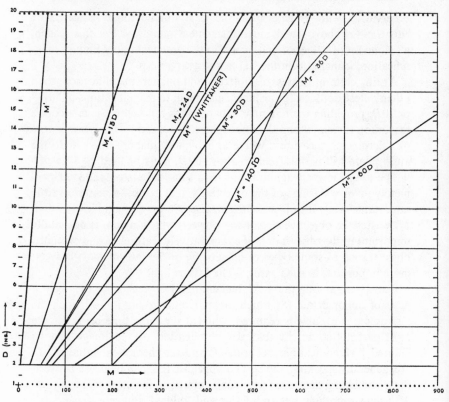

FIGURE 6

D (ins)	M″=30D (theor)	M″=60D (pract)	Whittaker's M″	Lewis's 140√D	Mr= 36D	24D	13D	M′
2	60	120	50	198	72	48	26	6
3	90	180	75	242	108	72	39	9
4	120	240	100	280	144	96	52	12
5	150	300	125	313	180	120	65	15
6	180	360	150	343	216	144	78	18
7	210	420	175	370	252	168	91	21
8	240	480	200	396	288	192	104	24
9	270	540	225	420	324	216	117	27
10	300	600	250	443	360	240	130	30
11	330	660	275	464	396	264	143	33
12	360	720	300	485	432	288	156	36
15	450	900	375	542	540	360	195	45
20	600	1200	500	626	720	480	260	60

practicable magnification for double star work with three large telescopes:

Dearborn 18½-in refractor: 390–925 (cf. M'' (pract)=1110)
Greenwich 28-in refractor: 670–1120 (cf.   „   =1680)
Lick 36-in refractor: 1000–1500 (cf.   „   =2160)

As an alternative, he has found that closer correspondence with practice is obtained from

$$M''=140\sqrt{D}$$

It will be seen from the graph that the values of this expression approximate to those of $M''=60D$ for the smaller values of D, and to those of $M''=30D$ for the larger.

Summarily, $M'' : M_r : M' : D :: 60 : 13 : 3 : 1$.

Thus although the size of the primary image is determined by F, and that of the final image by F and f, yet the highest and lowest magnifications that can be usefully employed are functions of D, since both light grasp and resolving power (section 2.4) are determined by D.

3.6 Observational characteristics of low and high magnifications

High magnifications suffer from the following drawbacks, as compared with low:

(a) smaller field;
(b) hence more frequent manual adjustment required, in the case of an altazimuth or hand-driven equatorial, with consequent increased difficulty in following, and more disjointed observation;
(c) decreased brightness of extended images;
(d) exaggeration of atmospheric defects;
(e) exaggeration of any defects in mounting, drive, etc.

Perhaps as a counterblast to the beginner's natural instinct, a good deal has been written on the advantages of low powers, with their wider fields. The pendulum may, indeed, have swung too far, for a wide field is in many spheres of work no more than a luxury—provided the mounting is steady, accurately adjusted, and suitably designed. Given these necessary conditions, wide fields are chiefly of interest to beginners 'sight-seeing' in the Milky Way, and in certain specialised fields such as variable star work (when suitable comparison stars may be some distance from the variable) or the study of telescopic meteors or large comets. As regards the 'sightseer', it should be pointed out that the type of instrument he frequently uses (a small refractor) is usually designed in such a way as to defeat its own object: the diameter of the drawtube being too small to permit the use of properly designed oculars of the low-magnification range in which the instrument particularly excels. This point is returned to in section 3.10.

3.7 True and apparent angular fields

The apparent angular field (α) is the angular diameter of the field stop in the ocular as seen through the ocular from the observing position, namely, the exit pupil.

The diameter of this stop is limited by three conditions:

(*a*) it should not be much more than about 45°, the maximum that the average eye can take in at a single glance, i.e. without movement;

(*b*) it must not be so wide that the abaxial aberrations are intolerable at its edges;

(*c*) it must not include more than the field of full illumination (section 3.8).

The true field (θ) is the angular diameter of the area of sky whose image is included in the apparent field. Hence, clearly,

$$\frac{\alpha}{\theta} = M$$

The apparent angular field of different oculars varies somewhat, but an average figure is 40°. Thus,

$$\theta = \frac{40°}{M}$$

This means that the linear diameters of the satisfactory fields of oculars, measured in their focal planes, are commonly of the order of $0 \cdot 5f$ to $0 \cdot 75f$.*

3.8 Field of full illumination

When an aperture stop is placed some distance in front of a lens, the total field visible through that lens is not fully illuminated—i.e. illuminated by rays passing through the total aperture of the lens.

In the case of a telescope, the clear aperture of the objective may be regarded as the aperture stop of the ocular, which is in fact so placed

---

* Taking the linear diameter as $0 \cdot 7f$, $\dfrac{\theta F}{57 \cdot 3} = 0 \cdot 7f$, whence $\theta F \simeq 40f$ or $\theta \simeq \dfrac{40°}{M}$.

that it reveals only the field of full illumination. Figure 7 represents this stop, *Ss*, located in the focal plane, and (for simplicity) a single-lens ocular, *O*. *S* is the image of the furthest object from the optical axis whose divergent cone is received in its entirety by the ocular lens. Thus the whole of the pencil from *S* passes through the exit pupil *ab*, while only the shaded section *ac* of the pencil from *S'*, a point in the focal plane further from the axis, is included in it. The brightness of such

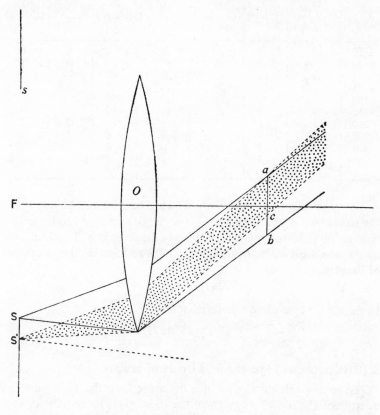

FIGURE 7

images being progressively weakened as the distance *FS'* increases, the field of vision is restricted to *Ss* by the diaphragm of the aperture stop.

The size of the fully illuminated field is a function of the aperture, the focal lengths of ocular and objective, and the focal length and aperture of the eye lens.

3.9 True field diameters of different designs of ocular

Below are tabulated the true field diameters (θ) provided by oculars of apparent fields ranging from 15° to 60°, an example of each being named, for different magnifications:

Magn	Apparent diameters						
	15° (Cemented doublet)	20° (Kepler)	25° (Cemented triplet)	30° (Ramsden)	40° (Huygh)	45° (Cooke 5-lens)	60° (Gifford)
10	1° 30′	2° 0′	2° 30′	3° 0′	4° 0′	4° 30′	6° 0′
25	36′	48′	1° 0′	1° 12′	1° 36′	1° 48′	2° 24′
50	18′	24′	30′	36′	48′	54′	1° 12′
75	12′	16′	20′	24′	32′	36′	48′
100	9′	12′	15′	18′	24′	27′	36′
150	6′	8′	10′	12′	16′	18′	24′
200	4′.5	6′	7′.5	9′	12′	13′.5	18′
300	3′	4′	5′	6′	8′	9′	12′
400	2′.25	3′	3′.75	4′.5	6′	6′.75	9′
500	1′.8	2′.4	3′	3′.6	4′.8	5′.4	7′.2

The true angular diameter of any field can be determined by clamping the telescope p a star of known Dec, so that the star will trail the field diametrically. Then if ϕ is the diameter of the field in ′ and ″ of arc, t the time of the star's diametrical transit in mins and secs of time, δ the Dec of the star,

$$\phi = 15t . \cos \delta$$

In the special case of an equatorial star

$$1^h\ 1^m\ 1^s \equiv 15°\ 15'\ 15''$$

3.10 Diameters of eye and field lenses of oculars

Just as the exit pupil must not be larger than the eye's pupillary aperture if the full LP capacity of the telescope is to be utilised, so it must not be larger than the field lens of the eyepiece. In fact, however, this is a condition that is not observed by instruments working at $f/15$ and thereabout (e.g. the typical small refractor) as normally constructed.

We have seen (section 3.7) that the angular diameter of the true field is given by

$$\theta = \frac{\alpha}{M} \quad . \quad . \quad . \quad . \quad . \quad . \quad . \quad . \quad (i)$$

From (*a*), p. 52, its linear diameter is

$$h = \frac{\alpha f}{57 \cdot 3}$$

Hence the linear diameter of the objective's image in the focal plane varies directly with f; if the ocular is to take the available field, and the relationship in (*i*) is to be observed, h may be taken as the diameter of the field lens.

Taking the case of a $f/15$, 3-in refractor, for which $M' = 9$, using an ocular giving $\alpha = 40°$,

$$h = \frac{40 \times 5}{57 \cdot 3} \backsimeq 4 \text{ ins}$$

i.e. the ocular is larger than the OG; even with a magnification of $\times 11$ they would be of roughly the same size.

There are two obvious solutions of this difficulty: (*a*) to modify the eye-end of the telescope to take a larger draw-tube, (*b*) to use an instrument of smaller focal ratio when using very low magnifications. For by reducing F, while maintaining D constant, a smaller value of f is involved for a given magnification, with consequent reduction of the size of the exit pupil. Suppose, for example, that the telescope is a 3-in, $f/5$ reflector. Then $F = 15$, and f (giving $M = 9$) is 1·6. Hence

$$h = \frac{40 \times 1 \cdot 6}{57 \cdot 3} = 1 \cdot 1 \text{ ins}$$

instead of 4 ins with the $f/15$ instrument.

This—and not, as is often supposed, increased image brightness or photographic 'speed' as such—is the reason for the small focal ratios of 'comet seekers': only thus can the full extent of the LP field be accepted by the ocular without the latter becoming prohibitively cumbersome.

The condition for full acceptance of the incident pencil by the eye lens is less obvious; nor is the matter one of such practical importance. In fact, the diameter of the eye lens must exceed

$(2 \tan \frac{\alpha}{2} \times$ distance of eye lens to exit pupil$) +$ diameter of exit pupil

In the instance of the 1·6-in ocular yielding $M = 9$ with the $f/5$, 3-in reflector the minimum diameter of the eye lens is roughly 0·7 ins.

3.11 The exit pupil and the position of the eye

The exit pupil, or Ramsden disc, is that section of the emergent pencil, perpendicular to the optical axis, through which rays from every point in the visible field pass. It has the following characteristics:

(a) rays from every point in the field pass through it;

(b) the illumination of the image is there greatest;

(c) rays from the edge of the objective pass through the boundary of the exit pupil;

(d) it is the image of the objective (the entry pupil of the system), when the telescope is pointed at the empty sky;

(e) it is coincident with the field of full illumination;

(f) it is the smallest possible cross-section of the emergent pencil;

(g) the largest field will be seen when the pupil of the eye and the exit pupil coincide.

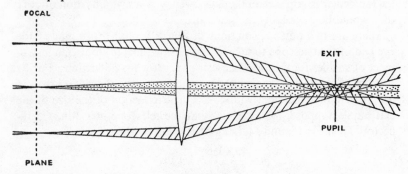

FIGURE 8

Passage of rays through the exit pupil from three points in the focal plane

To ensure that the eye and the exit pupil do coincide during observation, the cap surrounding the eye lens of the ocular is placed a few mm in front of the plane of the exit pupil. With certain types of ocular (see, e.g., section 8.5.2) the exit pupil is so near the surface of the eye lens that there may be an unconscious tendency to hold the head slightly back, thus restricting the field of view and wasting a proportion of the objective's illuminating power. The radius, r, of the field observed is given by

$$\tan r = \frac{\text{(diameter of eye pupil)} - \text{(diameter of exit pupil)}}{2 \times \text{(distance of eye from exit pupil)}}$$

This is a difficulty of which spectacle-wearers should be aware and beware, when using any normally constructed ocular.

The optical axis must also pass through the centre of the pupil of the eye, or part of the exit pupil may be intercepted by the iris and therefore contribute nothing to the retinal image.

SECTION 4

ABERRATIONS

4.1 Introduction

Elementary treatment of geometrical optics is concerned with the reflection and refraction of rays which are both close to and parallel with the optical axis.* These conditions are not observed in a telescope: the aperture of the objective is not a negligible fraction of its radius of curvature, and the incident light, though parallel, is only parallel to the axis when emanating from a source lying on the prolongation of that axis. The result of this is the breakdown of accurate correspondence between points in the object space and points in the image space. These departures are known as aberrations, and are of two species: those affecting the positions of the images, and those affecting their appearance.

The so-called Seidel errors are five in number; if we add chromatic aberration, to which refracting systems alone are subject, we have the following six errors: (1) spherical aberration, (2) coma, (3) astigmatism, (4) field curvature, (5) distortion, (6) chromatic aberration. The aberrations of a mirror are identical for all wavelengths; those of a lens vary with wavelength, the wavelength effect being known as the chromatic inequality of the aberration concerned.

(1) and (6) affect the whole field. (2) to (5) are proportional to angular distance from the centre of the field, i.e. to the inclination of the incident ray to the optical axis; of these abaxial or off-axis aberrations the most troublesome is (2) because of its asymmetrical nature. (1), (2), (3) and (6) are errors of image quality; (4) and (5) are errors of image position. The nature of the aberrations can therefore be conveniently summarised as follows:

---

* A ray can only be considered paraxial if θ, the angle it makes with the optical axis, is small enough to justify the equations $\theta = \sin \theta$, and $\cos \theta = 1$. Or again, the geometrical laws are inapplicable if either θ or $D/2f$ are of such a magnitude that they can no longer be considered negligible when raised to the third power.

Aberration	Uniform or Abaxial		Position or Quality	
Spherical	*	—	—	*
Coma	—	*	—	*
Astigmatism	—	*	—	*
Field curvature	—	*	*	—
Distortion	—	*	*	—
Chromatic	*	—	—	*

Diffraction phenomena are also ignored in the geometrical treatment of optical systems; some of the practical implications of diffraction in telescopic image formation are discussed in sections 2.2, 2.3.

4.2 Rayleigh's limit

If the pencil emerging from the objective is free from aberration, and is therefore capable of forming a point image of a point source, its wavefront is spherical and all its wave surfaces are sections of concentric spheres; such a pencil, of finite diameter and inclination to the optical axis, is termed stigmatic or homocentric.* The wavefront of an aberrant pencil, on the contrary, is not a simple spherical surface. It has been found empirically that the aberration in the pencil is in general tolerable —i.e. is too small to affect the image appreciably—if its wavefront can be contained between two spheres whose distance apart does not exceed $\lambda/4$. This limit of tolerance is known as Rayleigh's limit.

The modifications of the image attendant upon progressive increase in Δf are described in sections 4.3, 13.9. If the value of Δf at which the initial brightening of the rings just becomes perceptible is Δf_1, and that at which all light has been drained from the spurious disc is Δf_2, then it has been found by observation that

$$\Delta f_1 \simeq \frac{\Delta f_2}{4} \qquad \ldots \ldots \ldots \ldots \quad (a)$$

But the value of Δf depends solely upon the focal ratio and the wavelength:

$$\Delta f_2 = 8\left(\frac{f}{D}\right)^2 \lambda \qquad \ldots \ldots \ldots \ldots \quad (b)$$

whence from (a), $\qquad \Delta f_1 = 2\left(\frac{f}{D}\right)^2 \lambda$

* The aberrations of inferior objectives may often be greatly reduced by using oculars of similar aberration but opposite sign. This is a makeshift arrangement which cannot be recommended, since it may be required to use the objective for photography at the prime focus, or the oculars with micrometer webs. Both objective and ocular should be stigmatic.

It can be shown that a point on the axis, situated at a distance of Δf_2 from the focus, is one from which the distance to the periphery of the concave wavefront converging from the objective is 1 wavelength less than the distance to the centre of the wavefront:

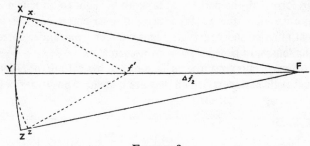

FIGURE 9

F being the focus, $\qquad FY=FX=FZ$,

but from f', situated at a distance Δf_2 from F,

$$f'Y=f'x=f'z,$$

where $Zz=Xx=\lambda$.

Therefore, from (a), the departure from a sphere in the case of Δf_1, the just perceptible aberration, is $\lambda/4$.

Taking $\lambda=5600\text{Å}$, the wavelength of the yellow region to which the eye is most sensitive,

$$\text{Rayleigh's limit}=\frac{\lambda}{4}=1400\text{Å}=0\cdot14\mu$$

To achieve this condition in the emergent beam, no local departure from the theoretical curve of the objective's surface or surfaces must exceed $\lambda/2$ in the case of a lens, or $\lambda/8$ in the case of a mirror.* This is the justification for the claim that it is easier to make a refractor than a reflector: although there are more surfaces to work, the accuracy to which they are carried need be only one-quarter that of a speculum.

4.3 Spherical aberration

Spherical surfaces are distinguished from other surfaces of revolution in that they do not yield stigmatic reflected or refracted pencils. Where

* The figure of the Mt Palomar 200-in paraboloid was worked until every part of its 209 square feet of surface was within $\lambda/10$ (two-millionths of an inch) of theoretical perfection.

the aperture is large, relative to the radius of curvature, spherical aberration occurs even in pencils parallel to the axis.

A lens or mirror suffering from spherical aberration is incapable of bringing rays to a focus in the same plane normal to the optical axis if the distances of these rays from the axis are different. If the aberration is under-corrected, the peripheral rays are brought to a focus nearer to the objective than the paraxial rays; if over-corrected, the reverse. Spherical mirrors and converging lenses suffer from under-correction; diverging lenses, in general, from over-correction.

The reflected rays from the surface of a spherical mirror (Figure 10) define a caustic surface to which they are tangent, whose apex lies at the

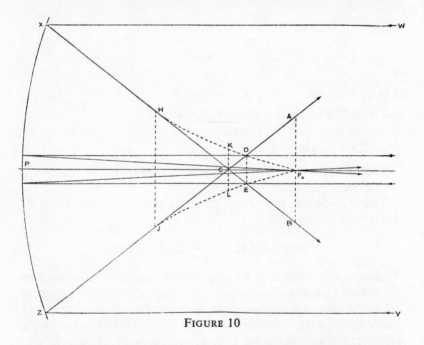

FIGURE 10

true focus of the mirror, i.e. the point at which paraxial rays cross the axis. The outer and fainter portions of the caustic are formed by peripheral rays, and hence reduction of the aperture leaves a progressively smaller and brighter portion in the vicinity of F_0, approximating to the theoretical condition.

Instead of a single point image being formed of a point object on the axis, a series of poorly defined images are strung out along the axis, their edges defining the caustic surface. If the object is a point source (and leaving out of account, for the moment, the effects of diffraction), a screen at AB will show a faint disc with a bright central point; as the screen is moved from AB towards DE the disc will decrease in size and increase in brightness, while the central bright spot will expand and fade; at DE, five-eighths of the distance from AB to KL, the size of the image is minimal: this is the so-called least circle of confusion. From this position inwards the image expands again, the edges becoming markedly brighter than the centre. Thus between C and F_0 the brightest region of the image indicates the region of the objective which is focusing long (centre of an under-corrected objective, periphery of an over-corrected one); while between C and P the brightest region of the image indicates the region of the objective which is focusing short (periphery of an under-corrected objective, central region of an over-corrected one).

In the actual case of a stellar image these effects are superimposed upon the normal diffraction pattern of spurious disc, rings and interspaces. At F_0 the disc is as bright and small as if there were no aberration, and the brightness of the rings diminishes rapidly from the centre outwards; hence spherical aberration tends not to reduce the instrument's resolving power in the case of a pair of equally bright stars. Inside KL the brightness of the rings increases from the centre outwards, the central disc being relatively weak. (This is a convenient and sensitive test for residual spherical aberration— see section 13.9; the effects of a departure from the spherical on the part of the converging wavefront which amounts to $\lambda/8$ are visible in the extrafocal image; this is only half the minimum detectable at the best focus.) The increased brightness of the rings reduces the contrast in an extended image, and hence spherical aberration adversely affects the instrument's resolving power on planetary detail.

In the case illustrated, WX and YZ being peripheral rays, AF_0 is the extent of the lateral aberration and CF_0 the longitudinal. If we write Δf for CF_0, the difference in focal length as measured by paraxial and peripheral rays, r for the distance of the peripheral ray from the axis, and take the focal length, f, as unity, then

$$\Delta f = k \cdot r^2$$

Hence the longitudinal spherical aberration is a function of the square of the distance of the incident ray from the axis.

But $r = \dfrac{D}{2}$, and hence

$$\Delta f = k \cdot \frac{D^2}{4} \text{ (when } f \text{ is unity)} \quad . \quad . \quad . \quad . \quad . \quad (c)$$

69

k is termed the coefficient of longitudinal spherical aberration, and its value varies with the design of the objective. Thus the longitudinal aberration of a $f/15$ Littrow objective, $k=+0.321$, is $+0.00036f$: from (c)

$$\Delta f = 0.321 \frac{D^2}{4} \text{ (when } f=1)$$

$$= 0.321 \frac{(\frac{1}{15})^2}{4} \text{ (since } D=\tfrac{1}{15} \text{when } f=1 \text{ and the focal ratio is } f/15)$$

$$= \frac{0.321}{900}$$

$$= +0.00036 \times f$$

We have also seen that the least circle of confusion lies inside the focus of the paraxial rays and outside that of the peripheral rays (still considering the case of an under-corrected objective) by distances of, respectively,

$$\tfrac{5}{8}.k.\frac{D^2}{4} \quad \text{and} \quad \tfrac{3}{8}.k.\frac{D^2}{4}$$

f being taken as unity in each case.

If the aberration is under-corrected, k is negative; if over-corrected, postive. In the case of concave mirrors there can only be one form obeying the condition that $f=\frac{R}{2}$, and hence one value of k for spherical mirrors generally: namely $-\tfrac{1}{8}$.

The importance of focal ratio in connexion with spherical aberration is revealed by the following expression (arrived at empirically from the examination of the focal and extrafocal images) for the limit of tolerance of longitudinal aberration:

$$\Delta f_{max} = 0.00035 \left(\frac{F}{D}\right)^2 \text{ (all quantities measured in inches).}$$

Hence the precision with which the aberration need be corrected falls off rapidly with decreasing relative aperture:

		ins
$f/1$	$\Delta f_{max}=$	0.00035
$f/3$		0.00315
$f/5$		0.00875
$f/7$		0.01715
$f/10$		0.0350
$f/15$		0.0788
$f/20$		0.140

If we no longer take f as unity, expression (c) becomes

$$\Delta f = k \cdot \frac{D^2}{4f}$$

If ω is the angle ZCP subtended between the peripheral ray ZC and the axis, and ξ the lateral spherical aberration AF_0, then in triangle CF_0A,

$$\omega = \frac{\xi}{\Delta f}$$

and in triangle ZCP,

$$\omega = \frac{r}{f} = \frac{D}{2f}$$

Combining the last three expressions, we have

$$\xi = k \cdot \frac{D^3}{8f^2} \quad \cdot \quad \cdot \quad \cdot \quad \cdot \quad \cdot \quad \cdot \quad \cdot \quad (d)$$

Correction in the case of a spherical mirror (for which, as we have seen $k = -\frac{1}{8}$) is obtained by deepening the centre of the curve, a process

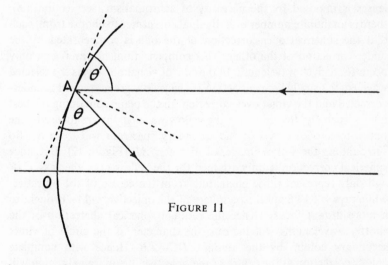

FIGURE 11

known as figuring, so as to convert the sphere into a paraboloid. For it is clear that the curve must be such that θ and θ' (Figure 11) must be equal for any position of the point of incidence, A; and the curve defined by this condition is the parabola. The spherical aberration of a parabolic mirror is zero so long as the object is at infinity. For objects nearer than this it is spherically over-corrected. If v is the distance of the image from the mirror, f the focal length, and x the distance of the point of incidence from the pole of the mirror, then the longitudinal spherical aberration is given by

$$\Delta v = \frac{x^2}{2f^2}(v-f)$$

Lenses behave analogously, converging lenses being in general under-corrected and diverging lenses over-corrected; in the latter case the apex of the caustic is towards the objective. By juggling with the radii of curvature of the two surfaces of a lens it is possible to reduce the spherical aberration to a minimum without departing from spherical surfaces. In the case of a converging lens the aberration is minimal when $r_1 = -6r_2$—i.e. one surface six times more strongly curved than the other, and turned towards the object; such a lens is termed a crossed lens. No manipulation of r_1 and r_2, however, can completely suppress the aberration in a single lens. Complete freedom from spherical aberration involves the combination of two components, one concave and the other convex, the total refraction being shared as equally as possible between the four surfaces. Although the total curvature* of the two lenses is imposed by the necessity of achromatism (see section 4.8), there is an infinite number of individual curvatures to choose from, such that the spherical over-correction of the one is compensated by the under-correction of the other. This compensation, however, can only occur for a given wavelength. In the case of visual objectives the selected wavelength usually lies near 5600Å in the yellow-green; the red is under-corrected and the violet over-corrected. In the plane containing the best image—namely, that in which the yellow wavelengths are focused—the out-of-focus violet rays of the secondary spectrum will form a halo surrounding the yellow image, of diameter AB (Figure 12). But since spherical correction is only attained for the yellow rays, the rays V, V will only represent those emanating from the centre of the objective, while rays V', V' from the periphery of the objective will be brought to a more distant focus. Hence the residual spherical aberration of the shorter wavelengths will increase the diameter of the circle of violet secondary colour by the amount $DC - AB$. Hence with complete colour correction at the centre of the objective, the marginal region will behave as though it were slightly over-corrected. This exaggeration of

* The total curvature of a lens is given by $C = c_1 - c_2$, where the individual curvatures of the two surfaces are the reciprocals of their radii of curvature, r_1 and r_2. r is positive if the surface is convex to the incident light, negative if concave, and the reciprocals take the same sign. Hence for a biconvex lens $C = c_1 + c_2$ (r_2 and hence c_2 being negative), and for a biconcave, $C = -c_1 - c_2$.

the secondary spectrum is small, however—from 3% to 9% at normal focal ratios ($f/15$ to $f/20$). It can be reduced still further by correcting the outer zone of the objective for colour, instead of the centre, which is then under-corrected. The importance of this under-correction is reduced by the smaller angles of incidence of the paraxial rays. Maximum chromatic inequality of spherical aberration then occurs in a zone

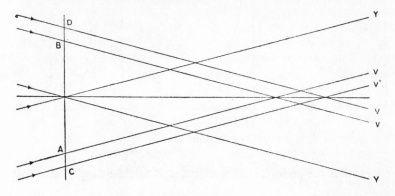

FIGURE 12

about midway between the centre and the edge of the objective, where it is less than 0·4 of the marginal inequality when the central correction is perfect. It is thus, for all practical purposes, swamped in the ordinary secondary spectrum of 'achromatic' doublets. d'Alembert showed how spherical aberration could be suppressed for a second wavelength (the so-called d'Alembert condition); the Gauss lens approximates closely to this condition.

The adequate correction of spherical aberration is of the first importance in the case of telescope objectives. Photographic lenses, on the other hand, often exhibit a certain amount of residual spherical aberration, since this is less harmful than other aberrations which cannot be removed simultaneously with it.

4.4 Zonal aberration

Zonal aberration may be described as a condition of disjointed spherical aberration. That is to say, the objective is divided into two or more concentric zones, each of which has a different focal length. If we were to plot the distance of the incident ray from the pole of the mirror against the distance of its focus from the pole, the curves for (a) a perfectly corrected objective, (b) an objective suffering from spherical aberration, and (c) one suffering from zonal aberration, would be of the general character shown in Figure 13.

Judged by its effect on the appearance of the image, and therefore upon definition, the value of Δf is more important than the number of

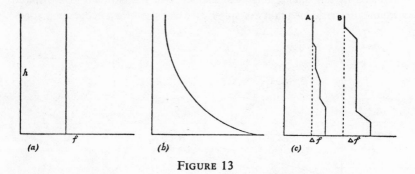

FIGURE 13

zones; thus the performance of objective A would be superior to that of B.

Zonal aberration is not a common defect of objectives that have been passed by professional makers, though the amateur who makes his own paraboloids will probably be familiar enough with it.

4.5 Coma

Coma is grouped with astigmatism, field curvature, and distortion as the abaxial aberrations: they all characterise rays which before reflection or refraction are not parallel to the optical axis (i.e. are proceeding from abaxial objects); hence the greater the angle θ between the axis and the incident ray (consequently the greater the distance of the image from F, the centre of the focal plane), the greater will be the deformation of the image.

Thus if d is the distance of the image from F (Figure 14), then the extent of the coma increases as d increases. Coma can therefore occur (a) when the aperture is large compared with the focal length—it in fact varies with the square of the relative aperture, (b) when through maladjustment the plane of the objective is not normal to the line joining its centre and the centre of the ocular, or in other words when the optical axes of objective and ocular are not coincident.

An objective corrected for spherical aberration will focus paraxial rays from a point object not on the axis to a point image; rays passing through the outer zones will be focused to disc images whose diameters will depend upon the distance from the centre of the objective of the zone forming them. The radius of each of these disc images will be half the distance separating its centre from the point image, thus producing the characteristic pear-shaped comatic image, brightest at its 60° apex (the

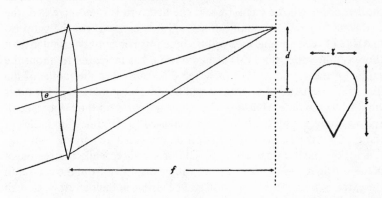

FIGURE 14

point image) and fading rapidly outward (Figure 14). The major axis of the image will, if produced, pass through F, and ξ is a function of d; if the apex of the image is directed towards F, the condition is termed external coma; if towards the edge of the field, internal coma.

The dimensions of the comatic image of an abaxial point object are given by

$$\xi = 3B\left(\frac{D}{2f}\right)^2\theta \quad . \quad . \quad . \quad . \quad . \quad . \quad (e)$$

$$\zeta = 2B\left(\frac{D}{2f}\right)^2\theta \quad . \quad . \quad . \quad . \quad . \quad . \quad (f)$$

where $\theta = \dfrac{d}{f}$

B = the aberrational coefficient of the objective.

Coma is thus a function of the distance of object and image from the optical axis (increasing as θ) and of the relative aperture (increasing as its square). It follows that the coma at a given distance from F can be reduced by stopping down the objective, thus reducing the relative aperture; and, alternatively, that a larger field with the same degree of coma can be obtained.

Coma is not usually observed in its pure form in the telescope. First, because other aberrations, notably astigmatism, are usually present;

secondly, because the axis of the ocular is also inclined at an angle θ to the incident rays, so that its own coma contributes to the total observed effect; and thirdly, because in any case the image of a point source is not a point but a diffraction pattern upon which the coma is superimposed. If an abaxial stellar image is observed with a sufficiently large focal ratio, it will be coma-free; if the aperture is then increased, the following progressive modifications will be observed: (a) the rings on one side of the spurious disc will become just perceptibly brighter than those on the other side; this occurs when ξ has increased to about the radius of the spurious disc; (b) when ξ is four times the radius of the disc, this inequality of brightness will become most marked in the inner ring, while the outer rings may be reduced to arcs on the bright side of the disc only, (c) by the time ξ has increased to ten times the radius of the disc, all the rings will be discontinuous, their remaining arcs outlining the comatic image on one side of the disc, while the region of greatest brightness of the disc itself will be displaced from the centre. In the extrafocal image the rings will still be visible in their entirety, though they will be eccentrically displaced.

Resolution is not affected by coma so long as ξ does not exceed about four times the radius of the spurious disc. Measures of position, on the other hand, are affected as soon as the first trace of asymmetry makes itself perceptible in the image. It is this asymmetrical nature of the comatic image that makes the aberration particularly troublesome (cf. astigmatism, for instance), since it is impossible to determine accurately the true position of the point image within the limits of the comatic image.

Ordinary visual observation is little, if at all, interfered with by the abaxial aberrations, owing to the comparative restriction of the usual visual field; and providing the objective is at least of average quality, and is correctly squared on, coma will be confined to the outer edges of the field, using normal magnifications. The abaxial aberrations assume great importance in photographic work, on the other hand, as also in micrometer measures; for such work rigorous correction of coma is required.

The condition for a coma-free objective is (Figure 15): B is reduced to zero for all values of x when

$$\frac{x}{\sin \theta} = \text{constant} \quad \ldots \ldots \ldots \quad (g)$$

Whence it follows that points a, b, c, $\ldots$ must lie on the surface of a sphere whose centre is the focus F: $aF = bF = cF = dF = eF$. Where the sine condition is not observed, the effect is equivalent to a non-uniform magnification across the object from the centre outwards.

It can be seen, therefore, that the spherical mirror, which has no unique axis, is coma-free; if its spherical aberration could be eliminated

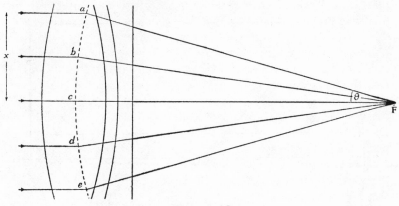

FIGURE 15

it would be an ideal objective, as was demonstrated by Schmidt (see section 9.16).

The sine condition cannot be rigidly observed by a cemented doublet, which consequently tends to suffer more from abaxial aberration than a separated-lens doublet. An objective of the latter type, corrected for spherical aberration and at the same time obeying the sine condition, is often termed aplanat even if the corrections are not perfect.

Neither can the sine condition be fulfilled by a single parabolic mirror, which is therefore subject to coma (Figure 16):

since $$aF \neq bF \neq cF$$

and $$\frac{x}{\sin \theta} = aF$$

$$\therefore \quad \frac{x}{\sin \theta} \neq \text{constant}$$

and the sine condition is not fulfilled; it is made possible, however, by the introduction of a second reflecting surface, a device employed in the two-mirror aplanatic telescopes (section 9.11). In terms of transverse magnification—which we have seen must be constant if the sine condition is obeyed—the magnification given by the zone of radius x of a

77

paraboloid is a function of aF; also, the magnification given by the axial region of the mirror is a function of PF. But $PF{\neq}aF$; on the contrary, $PF{=}aF{+}a'a$. Since $a'a{=}\dfrac{x^2}{4f}$, it may be very considerable for large values of x, and the coma correspondingly pronounced.

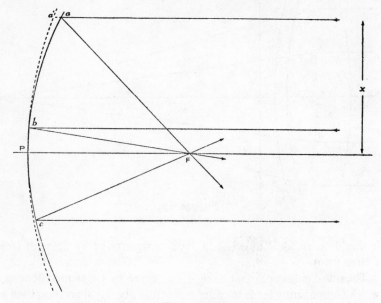

FIGURE 16

Alternatively, the coma of a parabolic mirror may be eliminated by a correcting plate—such as the Ross zero-corrector*—which is a thin lens whose power is zero and whose coma is equal to that of the mirror but of opposite sign. It is fitted in the convergent pencil in front of the ocular or photographic plate, and is impracticable at relative apertures greater than about $f/3$ owing to its uncorrected spherical aberration.

The aberrational constant, B, is $+0{\cdot}250$ in the case of a Newtonian paraboloid, the coma being external. From (e) we have, therefore,

$$\xi{=}\tfrac{3}{4}\left(\frac{D}{2f}\right)^2\theta$$

Coma becomes serious—i.e. visibly distorts images within a few minutes of arc of the axis—at about $f/3$, the practicable limit of relative aperture with a Newtonian.

* See B. 52.

4.6 Astigmatism and field curvature

An aberration characterising the refracted or reflected pencil from any surface of revolution at which the incident pencil is not normal. Unlike coma and spherical aberration, therefore, the pencils emergent from an objective of small aperture (or from the central region only of a large objective) may be astigmatic, provided only that the incident pencils were oblique. The amount of the aberration will indeed be greater when the used aperture is large than when it is small.

Astigmatism will also be inherent in an objective whose surface (or surfaces) is not a surface of revolution. If, for example, the rays from the two ends of one diameter of the objective are refracted or reflected to a shorter focus than those from the ends of the diameter perpendicular to the first, then a well-defined image will be found in no single plane.

An astigmatic ray from a small circular source comes to two foci, neither corresponding geometrically to the object, at different distances from the objective. Midway between them is located the least circle of confusion, or focal circle, but in no position is a geometrical duplicate of the object to be found. If a screen is interposed in the path of the pencil shown in Figure 17, the succession of images caught on it as it is moved away from the lens will be as follows:*

From lens to T: an ellipse of decreasing size and increasing eccentricity, its minor axis lying in the vertical plane.

At T: the ellipse has fined away to a short straight line, lying in a horizontal plane; it is known as the tangental focal line.

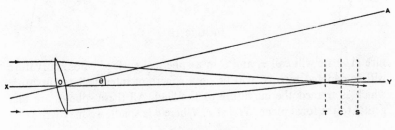

FIGURE 17

From T to C: the focal line expands into an ellipse of decreasing eccentricity.

At C: it is a small circular disc, the least circle of confusion, whose diameter is half the length of each of the focal lines (T and S).

* The appearance of astigmatic stellar (point) images will be described below.

Though blurred, the image bears the closest geometrical resemblance to the object.

From C to S: the circle elongates in a direction perpendicular to the tangental focal line.

At S: it has narrowed to a second focal line (the sagittal focal line) lying in a plane perpendicular to that containing the tangental focal line (i.e. vertical).

From S outwards: an increasingly large ellipse, whose major axis is parallel to the sagittal focal line.

The distance ST is termed the astigmatic difference, or longitudinal astigmatism of the lens. In the diagram the lens is represented as being tilted forward in a vertical plane, and the incident pencil as lying in a horizontal plane: then the sagittal focal line, the optical axis of the lens, and the incident ray are contained in a single vertical plane; the tangental focal line and the incident ray are contained in a single horizontal plane.

In Figure 18, OA is, as before, the optical axis of the lens; XY is the axis of an oblique pencil yielding a tangental focal line, a focal circle, and a sagittal focal line at T, C, and S respectively. Then T is contained in a spherical surface whose curva-

FIGURE 18

ture $(1/r)$ we will call τ, and S in a spherical surface whose curvature will be denoted by σ; C will lie on a sphere of intermediate curvature which is termed the curvature of the field. All these spheres are tangental to the focal plane, GH, at F. Where θ is small, we may write

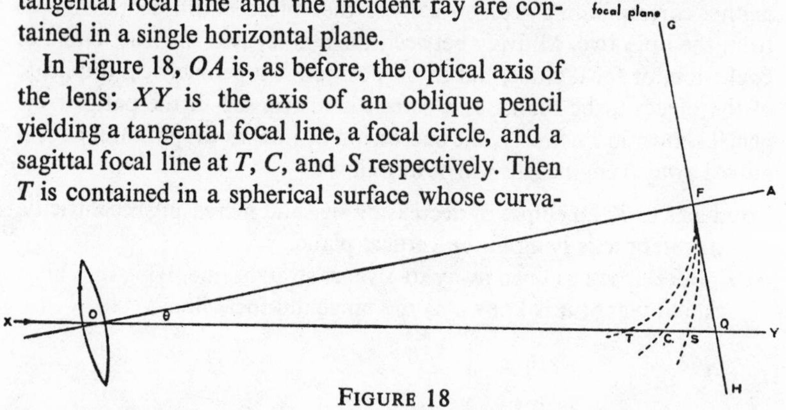

$$\left. \begin{array}{l} TQ = \dfrac{\tau}{2} . \theta^2 \\[2mm] SQ = \dfrac{\sigma}{2} . \theta^2 \end{array} \right\} \qquad \cdots \cdots \cdots \quad (h)$$

whence the longitudinal astigmatism

$$ST = \frac{\sigma - \tau}{2} . \theta^2$$

from which it can be seen that it increases very rapidly with increasing inclination of the incident ray. The displacement of the focal circle from the focal plane is

$$CQ = \frac{\sigma+\tau}{2} \cdot \frac{\theta^2}{2}$$

The radius of this circle is given by $r = \frac{D}{2f} \cdot \frac{ST}{2}$, whence its diameter is

$$d = \frac{D}{2f} \cdot \frac{\sigma-\tau}{2} \cdot \theta^2$$

The image formed in the focal plane by the rays passing through this focal circle will be an ellipse whose radial and tangential axes will be respectively (using (h))

$$d_r = \frac{D}{2f} \cdot \tau \cdot \theta^2$$

$$d_t = \frac{D}{2f} \cdot \sigma \cdot \theta^2$$

where d_r is astigmatic and d_t is a function of field curvature, $\frac{\sigma-\tau}{2}$ being termed the coefficient of astigmatism, and $\frac{\sigma+\tau}{2}$ the curvature of the field. If we denote the former by A, and the latter by C, the last five expressions give

$$\left.\begin{aligned} ST &= A.\theta^2 \\[4pt] CQ &= C.\frac{\theta^2}{2} \\[4pt] d &= A.\frac{D}{2f}.\theta^2 \\[4pt] d_r &= (C-A).\frac{D}{2f}.\theta^2 \\[4pt] d_t &= (C+A).\frac{D}{2f}.\theta^2 \end{aligned}\right\} \qquad \ldots \ldots (i)$$

The extrafocal images are distorted but (cf. coma) symmetrical, being ellipses whose major axes are parallel to the nearest focal line. It is for this reason that astigmatism is a less objectionable error than coma. In its pure form, astigmatism can only be observed in aplanatic instruments or when the aperture is heavily stopped down (spherical aberration and coma being functions of D^3 and D^2 respectively).

Below a certain value of ST the focal circle is indistinguishable from a normal spurious disc, and the astigmatism is tolerable. If xYz in Figure 9 is taken to represent an astigmatic wavefront, and the focal plane of the ocular is made to coincide with either of the focal lines, then astigmatism can be detected when $Xx=0\cdot1\lambda$: the rings will be just perceptibly brighter at the two ends of a diameter, which defines either the tangential or the sagittal axis. When the focal plane of the ocular coincides with that containing the least circle of confusion, Xx can be about $0\cdot2\lambda$ before there is perceptible distortion of the image; the rings will then be brightest at the ends of two mutually perpendicular diameters. If Xx is increased to $0\cdot5\lambda$, the image becomes definitely cruciform; by $Xx=0\cdot25\lambda$ the resolving power of the objective begins to deteriorate.

We have from (b), p. 66, that

$$ST=8\cdot\left(\frac{f}{D}\right)^2\cdot Xx$$

and taking $0\cdot25\lambda$ as the limiting value for Xx, we have for the limit of tolerance of longitudinal astigmatism

$$ST_{max}=2\cdot\left(\frac{f}{D}\right)^2\cdot\lambda$$

A range of values of this quantity, taking $\lambda=5600\text{Å}$, is tabulated below:

		ins
$f/1$	$ST_{max}=$	$0\cdot000044$
$f/3$		$0\cdot000397$
$f/5$		$0\cdot001103$
$f/7$		$0\cdot002161$
$f/10$		$0\cdot00441$
$f/15$		$0\cdot00992$
$f/20$		$0\cdot01764$

Single optical surfaces are incapable of yielding a flat image: i.e. the the surface on which the best images lie is a sphere tangent to the focal plane at the point where the optical axis intersects it. In astigmatism-free systems such as the Couder telescope (section 9.15), FT and FS are made coincident ($\sigma=\tau$), but the focal surface is still curved.

The condition for minimum field curvature in the case of two lenses in contact is

$$\mu_1 f_1=-\mu_2 f_2$$

The coefficients of astigmatism and field curvature, A and C, being independent of the form of the lens, they are the same for all achromatic doublets made of the same types of crown and flint. The values of σ and τ for a typical astronomical doublet are $-1\cdot73$ and $-3\cdot85$ ($A=+1\cdot06$ and $C=-2\cdot79$), the corresponding radii of curvature being $0\cdot578f$ and $0\cdot026f$. The radius of curvature of the surface containing the least circles of confusion is $0\cdot37f$. Hence the best images lie on a spherical surface, concave to the objective, whose radius of curvature is approximately one-third of the focal length.

In the case of a Newtonian reflector (paraboloid), the values of A and C are respectively $+1$ and -1.

Comparing expressions (e) and (f) with (i) it can be seen that the relative importance of astigmatism and coma depends upon the relative aperture of the instrument and the distance of the image from the axis:

(a) As the focal ratio is increased, the distortion of the image due to astigmatism decreases more slowly than that due to coma; hence at large focal ratios coma is less important relative to astigmatism than at small focal ratios, and vice versa.

(b) As the distance of the image from the centre of the field increases, the element of its total distortion that is due to astigmatism will gradually swamp that due to coma.

These relations are illustrated in the following Table (B. 54):

$\theta =$	15'			30'			1°			2°		
	Astig	Coma	A+C	A	C	A+C	A	C	A+C	A	C	A+C
$f/3$	1″3	18″7	19″3	5″0	37″4	39″9	20″0	74″9	84″9	83″8	140″8	191″7
$f/10$	0·4	1·7	1·9	1·6	3·4	4·2	6·3	6·8	10·0	25·2	13·6	26·2

If astigmatism predominates, definition at the centre of the field can be improved by increasing the magnification; if coma predominates, magnification is no help. But in normal astronomical instruments the two are of roughly equal importance, and the image distorted by both astigmatism and coma is not markedly different from one exhibiting either aberration in the pure form. In Figure 19 are plotted the values of the longitudinal axes (which is never less than $1\frac{1}{2}$ times the minor axis) of the combined astigmatic and comatic images formed by parabolic mirrors of different focal ratios at different distances from their optical axes.

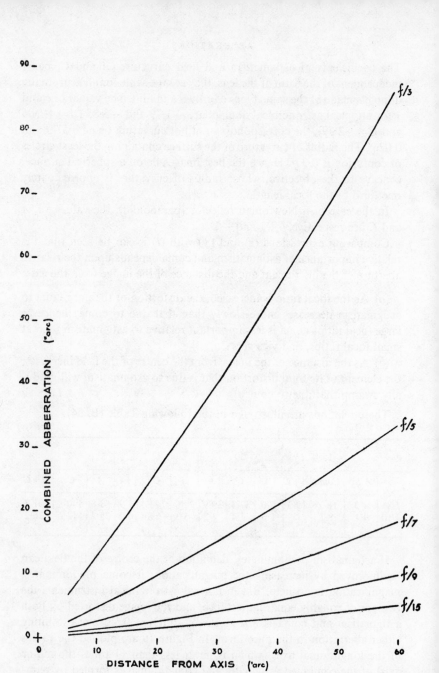

FIGURE 19
Combined astigmatism and coma

Neither astigmatism nor field curvature plays an important part in the performance of visual objectives, owing to the comparatively restricted fields that are used; as with coma, however, they are vitally important in the case of wide-field photographic objectives, or those designed for accurate measures of position.

4.7 Distortion

If all the foregoing aberrations are suppressed, the image will indeed be well defined. But unless the so-called tan condition is satisfied it will be distorted. In the discussion of magnification in terms of geometrical optics (section 3.2) it was assumed that if for one ray the magnification is given by $\dfrac{\theta_1'}{\theta_1} = M$, then for a second ray from the same object $\dfrac{\theta_2'}{\theta_2} = M$; i.e. it was assumed that all regions of the field would be magnified equally This is only true, however, if the condition

$$\frac{x}{\tan \theta} = \text{constant} \quad \ldots \ldots \ldots \quad (j)$$

is satisfied (see Figure 20); the constant is, of course, the magnification, M. Any optical system obeying this condition is termed orthoscopic (cf. equation (g), the sine condition for freedom from coma).

If the orthoscopic condition is not satisfied, we have some such state of affairs as

$$\frac{\tan \theta'}{\tan \theta (1 + E \tan^3 \theta)} = M$$

where E is the coefficient of distortion.

Alternatively, if ξ is the distance of the object from the optical axis, and ξ' the distance of the image from the axis, as given by $\xi' = M$, and if, further, the actual distance of the image from the axis is ξ'', where $\xi'' = \xi' \pm \Delta \xi$, then

$$\xi'' = \xi' \pm E(\xi')^3$$

and hence

$$\Delta \xi = E(\xi')^3$$

$$\Delta \xi = E\theta^3$$

If E is positive, M will vary directly with distance from the centre of the field, and the image of a square will be distorted as shown on the right of Figure 21. If E is negative, M will vary inversely with distance from the field centre, and the image of a square object will be distorted in the opposite sense.

Since the distortion is radial (i.e. the images are displaced towards or away from the point where the axis intersects the focal plane) a set of

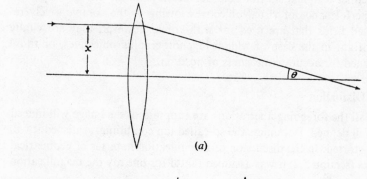

(a)

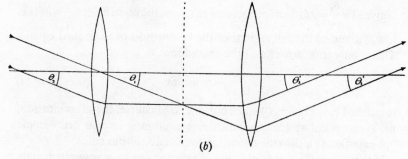

(b)

FIGURE 20

(a). Orthoscopic objective: $\dfrac{x}{\tan \theta} = M$

(b). Orthoscopic combination: $\dfrac{\tan \theta_1'}{\tan \theta_1} = \dfrac{\tan \theta_2'}{\tan \theta_2} = M$

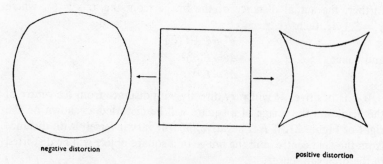

negative distortion

positive distortion

FIGURE 21

concentric circles, their centre on the optical axis, will be undistorted as regards shape though their relative sizes will be falsified.

It will be clear from a comparison of equations (g) and (j) that an objective cannot at the same time be orthoscopic and free from coma. In practice, attention is paid to the correction of coma, since distortion due to the objective is negligible, θ being small and E close to zero.

This does not mean that an ordinary astronomical refractor will necessarily and always provide an undistorted field. It may be orthoscopic, and uncorrected spherical aberration in the ocular still produce distortion. If it is under-corrected it will introduce negative distortion; if over-corrected, positive distortion. As has already been emphasised, both elements in a compound optical instrument must be homocentric.

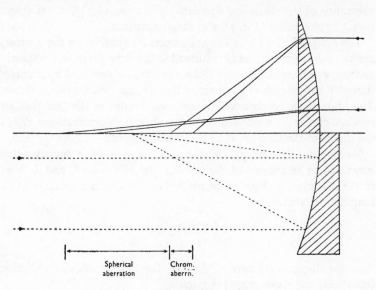

Spherical aberration   Chrom. aberrn.

FIGURE 22

Lens and paraboloid compared in respect of spherical and chromatic aberration (after F. G. Pease)

4.8 Chromatic aberration

Chromatic aberration being an effect of refraction, mirrors and all optical systems consisting wholly of mirrors are free from it. In the case of lenses it can be effectively reduced by combining components in such a way that their individual aberrations tend to cancel one another out.

It can in practice never be wholly removed, however, the residual aberration of a technically 'achromatic' two-lens combination being known as secondary spectrum.

Since the refractive index of glass varies inversely with the wavelength of the light with which it is measured, images in different wavelengths are formed at different positions along the axis of the objective.* Since

$$f=k.\lambda$$

a converging lens (Figure 23) brings the violet rays to a shorter focus than the red; and a white object will produce a series of sharply focused monochromatic images strung out along the axis. Thus two dimensions of chromatic aberration can be distinguished (analogous with longitudinal and lateral spherical aberration)—aberration of position, and aberration of size; these are alternatively known as longitudinal aberration, and chromatic inequality of magnification.

The images formed by a fully achromatic objective, on the contrary, are all of the same size and are formed in the same position, irrespective of their wavelength. This condition can only be provided by a mirror, though it can be closely approached by three-component combinations. The importance of chromatic aberration resides in the fact that any residual chromatism sets a limit to the objective's performance even at the centre of the field (cf. the abaxial aberrations).

Writing f_R, f_B, and f_Y for the focal lengths of a biconvex lens measured in radiations of the wavelengths of the C, F, and D lines† in the red, blue, and yellow respectively, we have as a measure of the longitudinal aberration

$$\frac{f_R-f_B}{f_Y}=\frac{\Delta f}{f} \quad . \quad . \quad . \quad . \quad . \quad . \quad . \quad . \quad (k)$$

And the dispersive power of the lens for these particular radiations (effectively the visual range) is given by

$$\frac{\Delta f}{f}=\frac{\mu_B-\mu_R}{\mu_Y-1}=\frac{1}{\nu} \quad . \quad . \quad . \quad . \quad . \quad . \quad . \quad (l)$$

where $\nu=\dfrac{f}{\Delta f}$ is known as the constringence of the lens. Its value for crown glasses averages about 60; for flint, about 36.

* The difference of size of the monochromatic images is a matter of less importance than their formation in different planes.

† 6563Å, 4861Å, 5890Å.

If C is the total curvature $(c_1 - c_2)$ of the lens, then

$$C = \frac{1}{f} \cdot \frac{1}{\mu - 1}$$

whence

$$f = \frac{1}{C(\mu - 1)} \quad \cdot \quad \cdot \quad \cdot \quad \cdot \quad \cdot \quad \cdot \quad \cdot \quad (m)$$

Taking f_Y as unity,

$$C = \frac{1}{\mu_Y - 1}$$

whence, taking μ_Y for crown glass as $1 \cdot 589$,

$$C = 1 \cdot 93084 \quad \cdot \quad \cdot \quad \cdot \quad \cdot \quad \cdot \quad \cdot \quad (n)$$

Taking the values of μ_R and μ_B as $1 \cdot 5153$ and $1 \cdot 5241$, we have for the longitudinal chromatic aberration of a simple biconvex lens, from (k), (m) and (n)

$$f_R - f_B = \frac{1}{C(\mu_R - 1)} - \frac{1}{C(\mu_B - 1)} \cdot f_Y$$

$$= \frac{1}{1 \cdot 93084(1 \cdot 5153 - 1)} - \frac{1}{1 \cdot 93084(1 \cdot 5241 - 1)} \cdot f_Y$$

$$= 0 \cdot 017 f_Y{}^*$$

and from (l), $\nu = 58 \cdot 8$

If a screen is moved along the axis, the succession of images of a polychromatic point source will be as follows (Figure 23):

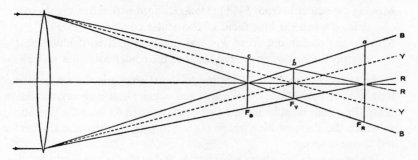

FIGURE 23

At a: The red focus; hence the image will include a central red point. But the out-of-focus blue rays will form a circular blue disc in which it is centrally placed.

$$\left. \begin{array}{l} f_R : +0 \cdot 0051 \\ f_B : -0 \cdot 0119 \end{array} \right\} \times f_Y$$

89

From a to b: The whole disc shrinks, position b being the least circle of confusion; at the same time the central red point expands and fades as the screen moves away from F_R.

At b: The red and blue discs are here coincident, since F_Y is the focus of neither the red nor the blue rays. The diameter (minimal) of the image is

$$d = \frac{D}{2\nu}$$

where D is the aperture of the lens, and ν the constringence.

From b to c: The disc expands again, the blue component at the same time shrinking, so that the whole image in expanding grows a red border.

At c: The blue rays being at their focus, the image will consist of a circular disc with a red outer fringe and a central blue point. It is twice as large as the least circle of confusion, or

$$d = \frac{D}{\nu}$$

In the case of a compound optical instrument consisting of an objective and an ocular, the eye automatically selects (by directing the hand focusing the instrument) that image in the yellow to which it is most sensitive; i.e. the ocular is adjusted so that its yellow focus coincides with the yellow focus of the objective. This yellow image will be blurred by the out-of-focus images in all the other colours.

Moving the ocular in from F_Y: blue image, edged with red, most sharply defined when the blue focus of the ocular coincides with F_B.

Moving the ocular out from F_Y: red image, edged with blue, most sharply defined when the red focus of the ocular coincides with F_R.

Hence the longitudinal aberration of the telescope is the sum of the longitudinal aberrations of objective and ocular: that is to say, the distance the ocular must be moved along the optical axis from the position at which the sharpest blue image is seen to that at which the sharpest red image is seen.

The relative contributions of objective and ocular to the total aberration are a function of their focal lengths (measured at the yellow focus, the mean position): writing f_R' and f_B' for the ocular, we have

$$\frac{f_R - f_B}{f_R' - f_B'} = \frac{F}{f}$$

Hence it is that the aberration of the ocular becomes increasingly troublesome when very low magnifications are used.

The possibility of constructing a compound lens that approximates to achromatism springs from the fact that flint and crown glasses are unequally dispersive ($\nu \backsimeq 60$ and 36 respectively), while their powers of refraction (1·6 and 1·5 respectively) are approximately the same. Hence a single concave surface of flint is capable of neutralising to a large extent the colour dispersion of two convex surfaces of crown glass (its aberration being of the opposite sign) without similarly neutralising their power of refracting the incident light to a focus (Figure 24). Complete neutralisation of the colour error is not possible with only two

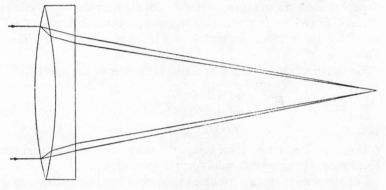

FIGURE 24

components since it would involve the use either of insufficiently durable types of glass or of curvatures of such strength as to make their manufacture very difficult and therefore prohibitively expensive.

Given that the combined dispersions are to be zero, and that the combined focal length is unity, it is possible to calculate the values of C_1 and C_2, and the powers of the two lenses, in terms of their constringencies and refractive indices. For most astronomical achromatic objectives, C_1 is about $2\frac{1}{2}$ times that of a single lens of the same focal length as the doublet. Also the focal length of the concave component is about twice that of the convex, since to secure achromatism the condition must be satisfied that

$$\frac{\nu_1}{\nu_2} = \frac{f_1}{f_2}$$

The achromatic doublet does not provide complete colour correction; the dispersion of the combination only approximates to zero. What it does is to bring two colours to the same focus, and the remainder nearly so (Figure 25). The residual colour (about 5%) visible at the best focus is the so-called secondary spectrum. The doublet achieves this by doubling the primary spectrum back on itself, and by correct selection of the coinciding wavelengths it can be arranged that the wavelength selected for use occurs at the point of minimum focus where the spectrum turns over, and where there is consequently a minimum displacement per wavelength along the axis.

With the correction most commonly employed for visual objectives nowadays, the radiations of the F line in the blue and of the B line in the deep red are brought to the same focus, leaving the yellow rays to which the eye is most sensitive (about 5600Å) close to the minimum focus at 5550Å (Figure 25b). In practice all degrees of correction between $C=F$ and $B=F$ are to be encountered among the best objectives of various makers.

The longitudinal aberration of a typical doublet of this sort is

$$\left.\begin{array}{l} f_R: \ +0\cdot00050 \\ f_B: \ +0\cdot00051 \end{array}\right\} \times f_Y$$

and $$\Delta f = 0\cdot0005f$$

A slightly different correction brings the red focus nearer to, and the blue further from, the minimum focus (Figure 25c).

For a given aperture the trouble experienced from secondary colour increases as the focal ratio is reduced. Taking four times the Rayleigh limit as the greatest tolerable discrepancy between the combined foci and the minimum focus (since the former are red and blue, which are little visible compared with the latter, yellow), the minimum permissible focal length of a visual doublet of aperture D is quickly derived:

From (b), p. 66,

$$\Delta f = 8\left(\frac{f}{D}\right)^2 \lambda$$

Also we have just seen that

$$\Delta f = 0\cdot0005f$$

Hence $$f = 16,000\left(\frac{f}{D}\right)^2 \lambda$$

Taking $\lambda = 5500\text{Å} = 2\cdot17 \times 10^{-5}$ ins,

$$f \nless 2\cdot88D^2$$

or, roughly, $$f \nless 3D^2 *$$

---

* For a single converging lens, primary colour becomes intolerable at about $f = 100D^2$, which was one reason for the enormous length and unwieldiness of the seventeenth-century refractors.

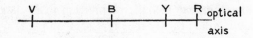

(*a*) Arrangement of foci along the axis of an
uncorrected biconvex lens

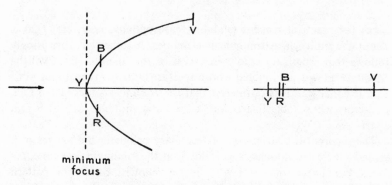

(*b*) Arrangement of foci of a visual doublet.
(Scale 5 times that of (*a*))

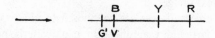

(*c*) Alternative colour correction of an achromatic doublet.
(Scale 5 times that of (*a*))

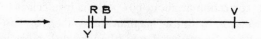

(*d*) Photographic correction.
(Scale 5 times that of (*a*))

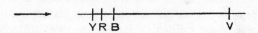

(*e*) Apochromatic correction.
(Scale 25 times that of (*a*))

FIGURE 25

93

Aperture (ins)	Minimum permissible f	Maximum permissible relative aperture
2	12	f/6
7	147	f/21
10	300	f/30

Since for practical reasons (stability, ease of movement, etc) f/20 is about the limit for astronomical telescopes, it somewhat surprisingly follows that secondary colour in excess of the tolerable limit will be encountered with all doublets whose aperture is greater than about 7 ins. In this connexion it is of interest that the $H\alpha$ and D_2 foci of the Lick refractor, whose objective is an 'achromatic' doublet, are 81·5 mm apart.

The theoretical distribution of secondary colour in the images of a visually corrected doublet is as follows: at the minimum focus (i.e. *the* focus, visually) a white star will show a yellowish diffraction pattern with little surrounding colour; the image of a very bright object will be surrounded by a halo of out-of-focus secondary colour, bluish or possibly with a tinge of purple; inside the focus a red fringe will develop as the rings proliferate; just outside the focus a minute red point will develop at the centre of the diffraction pattern; further out this vanishes again, and the whole diffraction pattern will be suffused with blue 'flare'. If the objective is under-corrected, the image at the focus will be surrounded by a red fringe which becomes increasingly conspicuous as the ocular is racked in; if it is over-corrected, no red fringe will develop round the intrafocal image.

The curves of object glasses are calculated to give optimum colour correction with a particular magnification, usually in the neighbourhood of $60D$ to $70D$ with instruments of moderate aperture. Since the natural under-correction of the ocular and the eye is exaggerated when the diameter of the pencil with which they are dealing is increased, it follows that the degree of over-correction in the objective necessary to counteract this under-correction at about $60D$ will be insufficient to do so at lower magnifications. With low magnifications, therefore, an objective will appear to be under-corrected, and (though this is much less noticeable or important) with very high magnifications, over-corrected.

The observational effects of secondary colour are important, and not always appreciated. It is not merely a question of falsification of colour, but of light grasp and resolution also:

(a) Light grasp is reduced by the secondary spectrum:* light which should be contributing to the intensity of the image on the retina is spread over a wider area. One can easily demonstrate the effect of such dilution of the image by selecting a star in the telescopic field which is just on the threshold of vision, and then throwing it out of focus, when it will disappear altogether.

(b) The contrast of an extended image is reduced. Dark detail on a bright ground, for example, is tinted blue by the overlapping unfocused fringes of the images of all the bright adjacent points, and tends to be lost. If there is plenty of illumination, so that a slight light loss is of no importance, contrast can be improved by a pale red filter, which absorbs the out-of-focus blue and violet radiations. The freedom of reflectors and photovisuals from this 'image haze' gives them a decided advantage over achromatics in planetary work.

The only way a visually corrected objective can be used for photography is to employ a panchromatic plate at the visual focus, and eliminate the out-of-focus rays with a filter. These rays, however, are the very ones which are most actinic. There is therefore a great loss of sensitivity, with consequently lengthened exposures.

To overcome this difficulty, objectives for photographic use are differently corrected: f_B and f_V (the latter about 4050Å) are brought into coincidence, leaving the G' line, in the region of greatest sensitivity of ordinary plates (about 4350Å), at the minimum focus (Figure 25d):

$$\left. \begin{array}{l} f_R: +0.00371 \\ f_Y: +0.00235 \\ f_B: +0.00048 \\ f_V: +0.00048 \end{array} \right\} \times f_{G'}$$

The highest degree of colour correction possible with doublets, without the curvatures becoming impracticably strong, is achieved by the so-called semi-apochromats. Their secondary spectrum is reduced to about 30% that of an ordinary achromatic objective.

Triple apochromats, consisting of three components, each of a different kind of glass, approach very close to the perfectly achromatic object glass, though secondary colour is never completely eliminated. By means of a double twist in the primary spectrum they can bring three different wavelengths to the same focus, with enormous reduction of the overall longitudinal aberration (Figure 25e):

$$\left. \begin{array}{l} f_R: +0.00002 \\ f_B: +0.00013 \end{array} \right\} \times f_Y$$

* See further B. 58.

The minimum focus lies midway between the shorter of the three selected wavelengths. Ideally, to give the best visual achromatism, wavelengths in the green, yellow, and red should be chosen; in practice, since this involves exaggerated curvatures and difficulty of manufacture, wavelengths from the blue, orange, and infrared are normally used. Even so, the curvatures are stronger than those of less completely corrected objectives.

Objectives corrected to bring the regions of maximum visual and actinic sensitivity to the minimum focus, so that they can be focused by eye for photographic use, are termed photovisuals.

Apochromatic objectives are extremely expensive, and a good doublet will perform better—both regarding its light grasp and resolving power —than the smaller apochromat that the same money will buy, notwithstanding the secondary colour. Secondary colour, indeed, only becomes obtrusive in a doublet when its aperture exceeds about 8 ins. For practical reasons, therefore, and whatever the theoretical optician may have to say on the subject of colour correction in apochromats, the reflector offers the only effective escape from chromatic aberration.

SECTION 5

TYPES OF OBJECT GLASS

5.1 Introduction

Among the many types of achromatic objective in use, the following are described briefly in order to indicate the dependence of optical characteristics upon the curvatures of the component surfaces. It is convenient to group them as follows:

A. Doublets with lenses in contact, or with very small air spaces:
   (a) producing internal coma,
   (b) producing no coma,
   (c) producing external coma.
B. Doublets with separated lenses.
C. Triplets.
D. Anastigmats (photographic objectives).

In the majority of objectives, the crown component is nearer the source of the incident light. Flint outside is theoretically possible, but the advantages of the former arrangement are that the more durable type of glass is exposed to the external air and to the necessity of more frequent cleaning, and that weaker curves are involved when the converging component is placed first, thus facilitating manufacture and reducing costs.

5.2 Contact doublets with internal coma

A series of objectives with lenses in contact, crown component outside, which runs from

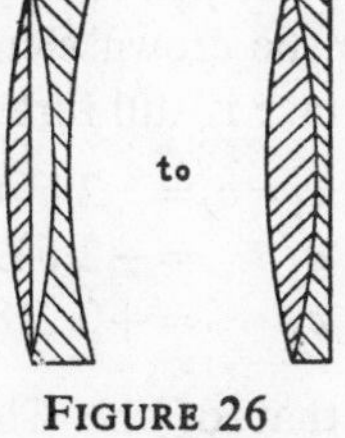

to

FIGURE 26

97

by increasing c_2 and decreasing c_4, the curvatures of the second and fourth surfaces. Along this series the field of good definition increases in size with decreasing coma, and sensitivity to errors of squaring-on decreases.

The advantages and disadvantages attendant upon cementing the components may be summarised as follows:

Advantages:

(*i*) 8% gain in transmission (see section 6.4).

(*ii*) Increased contrast resulting from the elimination of internal reflections.

(*iii*) Centring, once fixed by the makers, cannot get out of adjustment, thus allowing the use of a simpler type of cell.

(*iv*) The effects of local imperfections of figure are minimised.

(*v*) The production of ghosts is avoided; these (usually about 6 mags fainter than the parent image) result from internal reflection in uncemented combinations.

Disadvantages:

The aperture is limited to a few inches. Differential expansion of the lenses, owing to their different coefficients of expansion (9·5 and $8·0 \times 10^{-6}$ for crown and flint respectively), threatens to crack the balsam; also the different expansions of two cemented lenses sets up strains within them which may cause double refraction.

The Clairaut objective is an example of this class. It is commonly found in Galilean telescopes, with curvatures:

$$c_1 \quad = +2·694$$
$$c_2 = c_3 = -2·628$$
$$c_4 \quad = +0·247$$

It suffers from relatively strong coma ($B = -1·047$), which limits the usable field. Spherically under-corrected for divergent incident pencils (e.g. from terrestrial objects) when stigmatic in the focal plane; in fact, Clairaut objectives are frequently far from stigmatic in the focus.

In the Littrow objective the crown element is equiconvex, and the curvature of the fourth surface is still further reduced:

$$c_1 = c_2 = \quad 2·661$$
$$c_3 \quad = -2·649$$
$$c_4 \quad = +0·218$$

Coma, though less than that of the Clairaut, is still considerable

($B=-1\cdot008$). Coefficient of longitudinal spherical aberration$=+0\cdot321$; it is under-corrected for divergent pencils.

The Littrow objective offers the following advantages: (*i*) minimum departure from the d'Alembert condition, among the objectives of this series, (*ii*) relative insensitivity to errors of parallelism between the planes of the two lenses, and to distortion near the supporting points (the latter negligible with small objectives, in any case). Its chief disadvantage is that coma is still strong, compared with objectives of types more usually encountered in astronomical telescopes—four times that of the paraboloid ($B=+0\cdot25$), for example—which limits its usefulness for photographic work.

5.3 Coma-free contact doublets

The Fraunhofer aplanat exemplifies this type. The equiconvex character of the crown lens is departed from, the second surface being about three times as strongly curved as the first; the flint component has become a concave meniscus, though the fourth surface is only weakly curved (in one form, the flint component is plano-concave). The sine condition is satisfied when

$c_1=+1\cdot631$
$c_2=-3\cdot670$
$c_3=-3\cdot580$
$c_4=-0\cdot737$

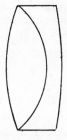

FIGURE 27

The lenses are not cemented, the airspace being narrowest at their centre; relatively thick peripheral spacers are therefore required when the objective is mounted in its cell. Slight modification of the curves allows the objective to be cemented, whilst retaining its aplanatic character.

$B=0$; hence a wide field of good definition, limited only by astigmatism; also the objective is relatively insensitive to errors of squaring-on (astigmatism only, in the oblique image). If $c_1>1\cdot631$, internal coma is reintroduced; if less than $1\cdot631$, external coma. The spherical correction is good, the coefficient of longitudinal aberration being $+0\cdot442$.

The Fraunhofer objective is the best of the contact types.

5.4 Contact doublets with external coma

By strengthening the convexity of the fourth surface, coma is reintroduced, though now it is positive (external).

The Herschel objective, though not used in astronomical objectives, is of interest in that

$c_1 = +1.460$
$c_2 = -3.842$
$c_3 = -3.754$
$c_4 = -0.894$

FIGURE 28

satisfies what is known as Herschel's condition for virtual stigmatism in more than one plane. Whereas the objectives so far considered, if rendered stigmatic for parallel incident light, are either over- or under-corrected for spherical aberration if the incident pencil is divergent, the spherical aberration of the Herschel objective is a function of Δf^2, where Δf is the difference between the distances of the image from the objective when focused for divergent and parallel pencils. It is therefore negligible for small values of Δf.

The sine and Herschel conditions are mutually incompatible; but in fact the spherical aberration of the aplanat and the coma of the Herschel objective, when focused on terrestrial objects, are almost negligible.

$B = +0.125$, about half that of the paraboloid. Coefficient of longitudinal spherical aberration $= +0.449$.

5.5 Separated doublets

The Clark objective—similar to the Littrow and aplanatic types, but with the components separated by a distance equal to about 1.5% of the focal length—is an example of this class. A large degree of suppression of the normal spherical over-correction of the short wavelengths is obtained.

Owing to the separation of the lenses, the distances of the yellow and the blue rays from the axis are materially different by the time they reach the diverging component. From (d), p. 71, it follows that ξ is a function of this distance raised to the third power; hence the relative over-correction of the blue ray is decreased. This advantage, however,

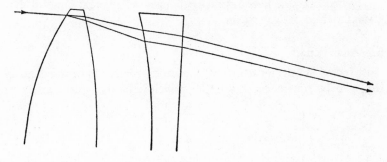

FIGURE 29

is bought at the cost of reintroduced chromatic inequality of magnification, the violet image being about 0·2% larger than the red.

$B = -0.504$; hence the Clark objective stands about midway between the Littrow and the Fraunhofer as regards coma.

5.6 Achromatic triplets

The Cooke photovisual, exemplifying this class of objective, is a 3-lens apochromat of as nearly perfect chromatic correction as is

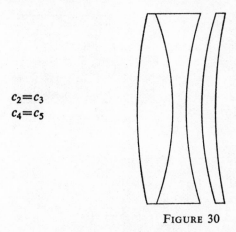

$c_2 = c_3$
$c_4 = c_5$

FIGURE 30

feasible with a system of lenses; its residual colour is from one-eighth to one-tenth that of an ordinary achromatic doublet. Freedom from coma is comparable with that of the aplanats. Differential spherical aberration with wavelength is reduced, as in the Clark objective. It is comparatively insensitive to flexure and to errors of squaring-on.

The first and second components are made of different kinds of flint glass, the third of light crown.

5.7 Anastigmats

By increasing the separation of the lenses, correction of astigmatism can also be obtained. This is the principle of the great variety of

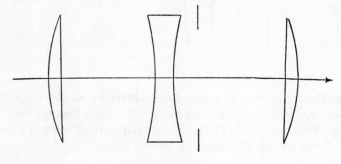

FIGURE 31

anastigmats put out today by manufacturers of photographic lenses. Together with suppression of field curvature, this gives a focal plane

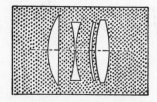

Wray Lustrar f/2.8

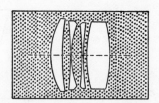

Wray Lustrar f/4.5

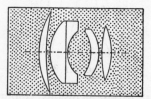

Wray Unilite f/2.0

FIGURE 32

(instead of a spherical focal surface, with F at its pole), to the edges of which the definition is determined by spherical aberration and coma alone.

One of the simplest forms is the triple anastigmat, illustrated in Figure 31. The first and third components are dense barium crown biconvexes, almost plano-convex; the diverging lens is a dense flint concave. There is some residual spherical aberration, but the separation of the astigmatic foci from the focal plane is kept within $0.01f$ to angular distances of nearly 30° from the axis.

The Zeiss Tessar is a slight improvement on the above, which it resembles except that the single third component is replaced by a cemented doublet. The complexity and variety in design of modern anastigmats is illustrated by Figure 32, showing three of Wray's series.

OPTICAL MATERIALS

6.1 Glass: composition and optical properties

Glass consists essentially of fused silicates. Originally, lead and calcium silicates, with sodium or potassium carbonate, were alone used; the former gave the range of flint glasses, the latter of crown. Modern optical instruments owe much to the work of Abbe and Schott, started at Jena about a century ago; as a result of this a very wide range of glasses is nowadays available for the lens designer, with refractive indices varying from 1·45 to 1·96 (μ_D), and constringences (ν) from 19·7 to 70. The most important modern varieties are:

light phosphate crown	baryta flint
barium crown	antimony flint
zinc silicate crown	boro-silicate flint
boro-silicate crown	silicate flint
dense baryta crown	dense silicate flint
silicate crown	borate flint

The composition and optical qualities of average crown and average dense flint, together with those of Pyrex glass and silica for comparison, are summarised below:

	Average crown	Average dense flint	Pyrex	Silica
Composition (%)	SiO_2: 70 K_2O: 20 CaO: 10	SiO_2: 45 PbO: 46 K_2O: 9	SiO_2: 81 B_2O_3: 12 Na_2O: 7	SiO_2: 100
Density	2·56	3·60	2·3	2·2
Refractive index (μ_D)	1·516	1·618	1·5	1·46
Constringence (ν)	59	36	60	67
Linear coefficient of expansion	$9·5 \times 10^{-6}$	$8·0 \times 10^{-6}$	3×10^{-6}	5×10^{-7}

The mean refractive index of a glass is taken as its value of μ in light of the wavelength (5896Å) of the D_1 line of sodium, μ_D. If μ_F and μ_C are the values of its refractive index at the wavelengths of the F and C lines of hydrogen (4861Å, 6563Å) respectively, then $\mu_F - \mu_C$ is termed its mean dispersion, or partial dispersion for the interval F to C.

The dispersive power of a glass is commonly expressed in the form of its reciprocal ν, known as the constringence:

$$\nu = \frac{\mu_D - 1}{\mu_F - \mu_C}$$

It is in general true that for crown, both mean dispersion and mean refractive index are relatively small; for flint, relatively large. For average crown and dense flint, we have:

	μ_D	$\mu_F - \mu_C$
crown	1·516	1·523−1·515=0·008
flint	1·618	1·634−1·616=0·018

The distinction is not absolute, however; barium crown, for example, has a large mean refractive index but small mean dispersion.

6.2 Plastics

Plastic optical apparatus is now being produced satisfactorily. Since it is moulded, its manufacture eliminates grinding, figuring and polishing. Polymerisation takes place in the mould at moderate temperatures and without pressure being applied. Cellulose acetate, methyl methacrylate, styrene, and polyvinyl acetate have been in use for a number of years, though only recently have plastics suitable for precision optics been developed.

Two at least are now suitable in this field:

(a) Polycyclohexyl methacrylate (CHM): satisfactory in all respects except low softening temperature (70°C) and the ease with which it is scratched. Optically similar to crown glass.

(b) Polystyrene: chief weakness is its rather large shrinkage, causing loss of figure in the mould; also rather easily scratched. Optically similar to flint glass.

These materials are lighter than glass, and allow of higher curvatures than are practicable with glass. Their main drawbacks are softness and mould shrinkage, causing optical imprecision. When these are overcome it is likely that they will have a much wider employment for astronomical components than at the present day.

6.3 Light transmission of glass

The loss by absorption in a lens depends both upon its refractive index and the wavelength of the radiation, and is directly proportional to thickness. Percentage transmissions of a selection of optical glasses, per inch thickness, are given below:

	3600Å ultraviolet	5200Å green	7000Å red
Boro-silicate crown $\mu=1\cdot5081$	93·75	98·57	97·00
Extra-light flint $\mu=1\cdot5298$	79·66	97·86	96·63
Barium light flint $\mu=1\cdot5512$	91·54	98·42	97·89
Light flint $\mu=1\cdot5795$	86·94	98·18	96·99
Barium dense crown $\mu=1\cdot6150$	76·02	96·61	95·62

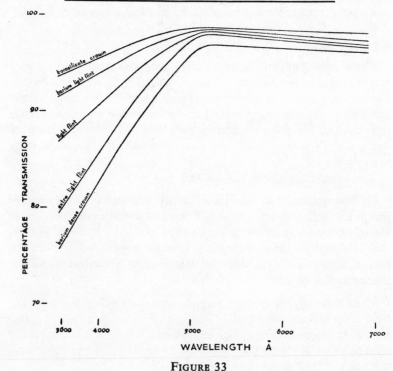

FIGURE 33

Taking the light loss resulting from absorption in 1 inch of average optical glass to be 4% in the visual range, we have:

Thickness of lens	Transmission factor
1 in	96%
2 ins	92%
3 ins	89%
4 ins	85%
5 ins	82%
6 ins	78%

It will be noticed that crown and flint are, roughly, equally transparent in the visual range (95% to 98% per inch), and both of them considerably less transparent in the ultraviolet. For optical glasses generally it may be said that a very rapid fall-off in transmission sets in at about 5000Å, and below 3000Å they are virtually opaque.

Various materials are transparent to shorter wavelengths, allowing investigation of the ultraviolet. Pettit, for example, has traced the solar spectrum as far as 2900Å, using a grating spectroscope with a quartz objective. Schott's 'Uviol' carries a transmission of about 90% per inch thickness to 2970Å; Schott's UBK.5 and Vitaglass are also much superior to ordinary optical glass in ultraviolet transparency.

Glass also becomes increasingly opaque in the far infrared. Transmission drops sharply at about

25,000Å with glass
40,000Å with quartz
110,000Å with fluorite (calcium fluoride)
180,000Å with sylvin (rock salt)

The last-named, being hygroscopic, requires to be looked after carefully, and kept in a desiccator when not in use, if frequent polishings are to be avoided. (See Figure 183.)

The transmission factors of materials in wavelengths below about 3000Å are of little interest to astronomers, since this region of stellar and other external radiations is removed by telluric absorption.* Above this threshold, lenses and silver mirrors make a poor showing compared with aluminium films (see below, section 6.6); flint glass generally is less transparent than crown, some flints being opaque almost up to the visible region; even so, 0·08 ins of average crown will completely absorb wavelengths shorter than about 3200Å. The gelatin coating of a photographic plate is strongly absorbent below about 2200Å, while Iceland

---

* 1 mm of air at 760 mm pressure will absorb all radiations below 1700Å. By means of vacuum spectrographs employing a Rowland grating, laboratory spectra have been obtained as far into the ultraviolet as 136Å.

spar is opaque below about 2150Å, quartz below about 1850Å, and fluorite or fluorspar below about 1000Å.

A further source of loss, which becomes appreciable with lenses of large aperture, is secondary spectrum, into which goes an appreciable fraction of the light transmitted by the objective.* The light loss from this cause is about doubled when the aperture and focal ratio are doubled; trebled when the aperture is doubled without changing the focal ratio.

6.4 Reflection at lens surfaces

A proportion of the light projected at a lens is not refracted through it but is lost by reflection. This occurs at every glass/air and air/glass surface, and causes an integrated light loss of some 4% to 5%.

Where the incident ray is not far from the normal to the surface of the lens we have, putting π = transmission factor,

$$\frac{\pi}{100} = 1 - \left(\frac{\mu-1}{\mu+1}\right)^2 \quad \cdots \cdots \cdots \quad (a)$$

Putting $\mu = 1 \cdot 5$ for the average refractive index of optical glass,

$$\frac{\pi}{100} = 1 - \tfrac{1}{25} = 0 \cdot 96$$

or $\pi = 96\%$

The precise value of π, the percentage of the incident light of different wavelengths not lost by reflection at the surface of an average crown and flint combination, is given below:

Wavelength (Å)	2800	2900	3000	3500	4000	4500	5000	6000	7000
Crown	0	90·6	90·7	91·1	91·3	91·4	91·6	91·7	91·9
Flint	0	86·2	86·6	87·8	88·4	88·7	88·8	89·1	89·4

It may be noted in passing that the combined transmission of a crown and flint doublet is increased by about 8% if the two lenses are cemented. For optical cementing, Canada balsam is generally used,† since it is

---

* 42% of the light transmitted by a 24-in object glass, working at $f/15$, is thus lost to the image.

† It has been reported that methylene iodide, dissolved in alcohol, gives a spectacular decrease in scattered light, with corresponding improvement of definition and resolution. A single drop will spread across the surface of a small lens or prism, the film being dry in from 4 to 12 hours, according to temperature.

easily worked, is of satisfactory transparency and stability, and its refractive index is virtually identical with that of crown glass. Thus, if we write $\mu_c = 1\cdot5$, $\mu_f = 1\cdot6$, $\mu_b = 1\cdot5$ for the respective refractive indices of crown, flint, and balsam, we have from (a) above:

for air-spaced combination:

$$\frac{\pi}{100} = \left[1 - \left(\frac{\mu_c - 1}{\mu_c + 1}\right)^2\right]^2 \left[1 - \left(\frac{\mu_f - 1}{\mu_f + 1}\right)^2\right]^2$$
$$= 0\cdot826$$
$$\therefore \quad \pi \simeq 83\%$$

for cemented combination:

$$\frac{\pi}{100} = \left[1 - \left(\frac{\mu_c - 1}{\mu_c + 1}\right)^2\right] \left[1 - \left(\frac{\mu_c - \mu_b}{\mu_c + \mu_b}\right)^2\right] \left[1 - \left(\frac{\mu_f - \mu_b}{\mu_f + \mu_b}\right)^2\right] \left[1 - \left(\frac{\mu_f - 1}{\mu_f + 1}\right)^2\right]$$
$$= 0\cdot908$$
$$\therefore \quad \pi \simeq 91\%$$

showing a saving of 8%.

Various non-reflecting films for coating lenses by the vacuum evaporation method have been devised (see B. 63–5, 68, 71), the best of which virtually eliminate light-loss by reflection at lens surfaces. The principle involved is the suppression of the reflected 5% of the incident light by destructive interference, balanced by constructive interference in the transmitted pencil arising from reflections within the film.

The condition for this to occur is that the film's μ^2 should be equal to the glass's μ, the optical thickness of the film being $\frac{\lambda}{4}$. Herein lies the first difficulty, for $\sqrt{\mu}$ for crown and flint are respectively $1\cdot231$ and $1\cdot271$, and no solids have refractive indices as low as this. Among the nearest are lithium fluoride ($\mu = 1\cdot39$, reducing the loss from 4% or 5% to about $0\cdot1\%$ over the whole visible range), calcium fluoride ($\mu = 1\cdot43$), sodium aluminium fluoride ($\mu = 1\cdot34$), sodium and magnesium fluorides, magnesium chloride, certain fatty acids, silicon dioxide ($\mu = 1\cdot46$) and cryolite.

None of these is perfect: lithium fluoride is fragile and soluble in water, hence useless for unprotected lenses, outer surfaces of objectives, etc; the magnesium salts, though more durable, are less efficient; the fatty acids are also only suitable for protected surfaces; cryolite is resistant to most corrosive agents but not to water;* silicon dioxide, more stable than cryolite in the presence of water, is less resistant to

* Coating the internal surfaces of a Tessar camera objective with cryolite resulted in a 20% improvement in total transmission, and nearly a doubling of the contrast.

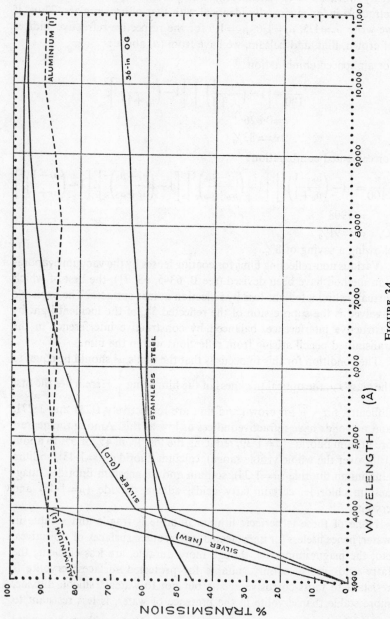

FIGURE 34

abrasion. Of the materials at present available, it is probably the most generally satisfactory.

6.5 Percentage reflection at a silver surface

A freshly deposited silver film reflects about 93% of the incident radiation at the wavelength to which the eye is most sensitive, though this can only be an approximate figure since individual films vary considerably among themselves.

Percentage reflection is also a function of the age of the film and of the wavelength of the incident light. The first two entries in the Table below give figures for silver-film reflectivity at stated wavelengths for a fresh and for an old film; the latter are somewhat arbitrarily chosen, but indicate the sort of effect produced by tarnishing. All the figures, however, must be regarded as to some extent approximate: individual tests have yielded results differing slightly from the average figures quoted here; in particular, the short-wave reflectivity, especially of the old film, may be too optimistic.

Wavelength Å	3000	4000	5000	6000	7000	8000	9000	10,000	11,000	12,000
Equivalent colour	ultra-violet	violet	blue/green	orange	red	infrared				
Fresh silver	4	89	93	95	96	97	97	97	96	96
Old silver	0	63	77	85	90	93	95	96	96	96
Aluminium	88	89	88	87	87	86	88	91	91	90
Stainless steel	46	56	59	60	60	—	—	—	—	—
36-in OG (Lick)	0	53	60	62	63	65	66	68	69	70

These figures, plus a second set for aluminium and for fresh silver, are represented graphically in Figure 34.

Beyond the minimum (4% to 6%) at about 3160Å, the reflectivity of silver rises again to about 34% in the region of 2150Å; at the other end of the spectrum it maintains a tolerably constant value of 96% to 98% at least as far as 15,000Å. The progressive loss of the shorter wavelengths as a result of tarnishing should be noted by photographers.

6.6 Percentage reflection at an aluminium surface

It can be seen that compared with silver films, aluminium is markedly superior in the ultraviolet below about 4350Å,* and inferior to a fresh

* The photographic threshold of the 100-in Mt Wilson reflector was lowered about 0·5 mag when the present aluminium film was substituted for the original silver film.

silver film by 5% to 6% over the visible range, and by 7% to 8% over the infrared as far as 12,000Å.

Its outstanding advantages over silver are its high transmission of the wavelengths to which the photographic plate is sensitive, and its relatively very great freedom from tarnishing.

6.7 Percentage reflection at other metallic surfaces

The reflectivity of stainless steel is considerably lower than that of even a badly tarnished silver film in all wavelengths, varying with the nature of the alloy and the lapse of time since polished. Where it scores over a silver- or aluminium-on-glass mirror is in its freedom from tarnish, and a high coefficient of expansion which enables it to reach thermal equilibrium after any temperature change more rapidly. During a temperature change, for the same reason, definition will be poorer with a steel than with a glass mirror.

The transmission factor of speculum metal, also a comparatively inefficient reflector, is a direct function of wavelength, giving the image a colour bias of a yellowish-pink tint.

6.8 Telescopic transmission factors

The actual brightness of the telescopic image is always less than the theoretical, owing to light-loss in the optical train. This loss, which varies with wavelength, we have seen to be primarily due to

(a) partial reflection at air/glass and glass/air surfaces in object glass, ocular, and possibly Newtonian prism;

(b) absorption in object glass;

(c) imperfect reflectivity of speculum, and possibly flat;

(d) secondary spectrum of object glass.

Of these, (b) and (d) are negligible in small refractors.

Defining the transmission factor, π, of the instrument as

$$\pi = \frac{\text{actual brightness of image}}{\text{theoretical brightness of image}}$$

we can now reach the following conclusions regarding refractors and reflectors:

Refractor: Compound OG (two separated lenses) transmits about 80%
Compound ocular transmits about 80%
(Neglecting (b) and (d) above)

$\left. \begin{array}{l} \\ \\ \\ \\ \end{array} \right\}$ $\pi \simeq 64\%$ (80% without ocular)

Reflector: Newly silvered mirror transmits about 93%
  Flat transmits about 93%
  Ocular transmits about 80%
  Silhouetting $\left(\text{diameter of flat} \simeq \dfrac{D}{3}\right)$, say
  10% loss

$\pi \simeq 62\%$
(78% without ocular)

For small instruments, therefore, there is little to choose between a refractor and a Newtonian reflector, providing the latter is resilvered as soon as the reflectivity starts to fall off. But whereas with increasing aperture the light-loss due to absorption in the increasingly thick components of the OG further reduces the transmission factor of the refractor, that of the reflector is independent of aperture.

A comparison of (i) a silvered, two-mirror reflector, s, (ii) an aluminised two-mirror reflector, a, and (iii) a refractor of about 5 ins aperture (OG rather less than 1 in thickness of glass) r, would be somewhat thus:

Ultraviolet	$a > s > r$:	62	0·2	0
Photographic	$a = s > r$:	63	63	61
Visual	$s > r > a$:	71	63	62

When it is borne in mind that these figures for s are based upon a brand-new film, and that tarnishing occurs fairly rapidly, the photographic supremacy of an aluminised reflector is obvious.

SECTION 7

METAL-ON-GLASS FILMS

7.1 First-surface mirrors

Astronomical mirrors belong without exception to the type known as first-surface reflectors: i.e. the reflecting film is deposited on the front surface of the glass. The principal advantage of second-surface construction is that the film is protected on one side by the glass, and can be protected on the other by any opaque medium such as an electrolytically deposited film of metal, paint, etc. Such mirrors, however, are useless for any purpose demanding the degree of optical perfection required by an astronomical instrument: the light has to pass twice through the glass in front of the reflecting surface, it undergoes refraction and secondary reflection each time it encounters the glass/air surface, it will be made aberrant if the first surface is not as accurately worked as the second, and the glass itself would have to be of a high grade. The multiple reflections occurring when a ray is projected at a second-surface mirror are clearly demonstrated in Figure 35.

With first-surface mirrors, a single surface has to be worked, and the optical quality of the glass is immaterial. The disadvantage of an unprotected film is met in various ways: by the periodical replacement of the film, by the application of a protective layer of some transparent and resistant material, or by the use of a resistant material for the reflecting film itself.

Metals suitable for depositing on glass to form first-surface mirrors include:

Silver: Deposited chemically; its affinity for sulphur causes rather rapid tarnishing; easily scratched; removable by friction when wet; for optical qualities, see section 6.5.

Aluminium: Deposited by evaporation in a vacuum; unaffected by sulphur; on contact with air forms a very thin surface layer of alumina (aluminium oxide, Al_2O_3), which is extremely hard and impervious to water, though it produces a slight diffusing 'bloom';

114

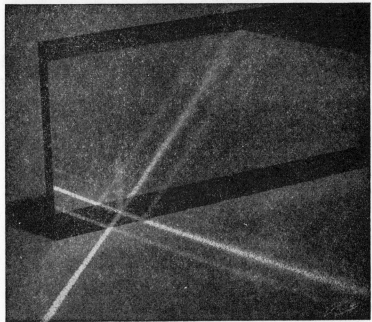

FIGURE 35

Reflection at first-surface and second-surface mirrors

visual reflectivity about 88%, but superior to silver in the photographic range; for optical qualities generally, see section 6.6.

Rhodium: A method of depositing rhodium by vacuum evaporation was developed in Germany in 1936; even more resistant to weather and physical maltreatment than aluminium; reflectivity only slightly inferior to that of aluminium (about 85%), but only about 70% when deposited electrolytically; free from the surface 'bloom' of aluminium. It is so tough that the glass backing, and not the film, is the weakest part of the mirror.

7.2 Length of life

The frequency with which the old film has to be replaced depends upon the local incidence of those substances that attack it:

Aluminium: Sodium chloride; other things being equal, therefore, expectation of life can be related to the distance of the telescope from the sea.

Silver: (i) Sulphurous acid (HSO_3) and hyposulphurous acid ($H_2S_2O_4$), resulting from the dissolving of sulphur dioxide—released into the atmosphere by combustion—in water droplets, are the chief tarnishing agents. The sulphur reacts with the silver to form compounds which reduce the reflectivity more especially in the shorter (photographically active) wavelengths. A few weeks' exposure to air containing a high enough sulphur content to reduce the visual reflectivity by 10% to 20% will reduce the reflection of the actinic wavelengths by as much as 50%. The length of life of a silver film is, generally speaking, proportional to its distance from the nearest industrial centre or large town, though a single smoke-stack in the immediate vicinity can cause as much trouble as a city twenty miles away.

(ii) Dew: if a mirror is allowed to remain dewed for long periods it will gradually develop a matt surface. Industrial impurities in the dew may cause the film to blister.

The life of aluminium films is increased by periodical (say six-monthly) washings with a strong detergent, such as Dreft or Aerosol, and distilled water (see also B. 67, 69). That of silver films, by more frequent (according to local conditions—about six-weekly is an average figure) burnishings with the minimum of finely divided rouge and a chamois leather pad stuffed with cottonwool. A thoroughly dry film of normal thickness ($0 \cdot 1 \mu$ to $0 \cdot 3 \mu$) will withstand 30 or 40 such treatments,

116

but it is absolutely essential that the surface of the mirror first be cleared of every trace of the gritty dust that will have accumulated on it; for this purpose a loose, dry wad of cottonwool should be used as a sweeper. If this precaution is not carefully and thoroughly carried out, there will be a serious risk of scratching both film and glass, and burnishing can do more harm than good.

No rule can be given for the period during which a film will remain serviceable, since too much depends on local conditions. All that can be said is that an aluminium film should last for years, and a silver film should be replaced when its reflectivity falls to about 70%, which may take several months or several years. At the Lowell Observatory—whose location at Flagstaff, Arizona, is virtually free from all atmospheric pollution—the silver films are only replaced every four or five years; even after five years there is no visible deterioration of the films, though reduced reflectivity in the shorter wavelengths is beginning to make itself felt by the necessity of longer photographic exposures. The silver films at Lowell Observatory are periodically burnished between silverings.

When, in 1947, the aluminium films of the 100-in and 60-in Mt Wilson reflectors were renewed for the first time since before the war, it was found that their reflectivities had dropped by 1% and 3% respectively—an astonishing figure, especially as the observatory is only fifteen or so miles from the sea. Both these mirrors had been given six-monthly washdowns with Dreft; the reflectivity of another mirror which had not been so treated had fallen off by 11% during the eight years.

7.3 Care of metallic films

The reflectivity of a silver film could be maintained indefinitely if it were kept permanently in an airtight container. The best that can be done for a mirror mounted in a telescope is to provide it with some sort of protection against atmospheric sulphur and damp when not actually in use.

Cottonwool next the mirror, backed by a thick layer of blotting paper which has been impregnated with lead acetate from a concentrated solution, has been recommended. Its function is to absorb the sulphur dioxide in the air.

Better, perhaps, is an airtight cap containing a circular disc of cork which has been warped to the exact curve of the mirror by heat and pressure. Before insertion in the cap it is coated with shellac, and when the shellac is dry a sheet of chamois leather is ironed on to it. Close contact should be made by the leather with all parts of the mirror. It is essential that while the telescope is in use, the cap should be kept in a desiccator (calcium chloride), or before a fire; also that the silver film be perfectly dry before the cap is replaced.

Even a well-fitting metal cover, enclosing a very shallow airspace above the silver film, will materially retard tarnishing.

7.4 Protective films

7.4.1 Celluloid: Make up a 0·1% solution of celluloid in amyl acetate. Stir frequently (a mechanical stirrer can easily be rigged up) for a period of two to three days. Filter the solution, and pour enough of it over the dust-free mirror to swill the whole surface when the mirror is rocked from side to side. Allow to dry (a) with the mirror standing upside-down on several thicknesses of blotting paper (not on its edge, or the thickness of the film will not be uniform across the mirror), or (b) on a rotating gramophone turntable, which throws off the excess solution; the slight rim left at the edge is of no consequence. It is important that the rotation should not be stopped until the film is perfectly dry. Speed of rotation should be, roughly, inversely proportional to the area of the mirror: about 8 r.p.s. for a 1·5-in, or 3 r.p.s. for an 8·5-in, has been found satisfactory. Drying should be carried out at a room temperature not much less than about 60°F.

The thickness of the film should be comparable with that of the silver —roughly between $0·1\mu$ and $0·3\mu$. If much less than this, interference between the light reflected from the two surfaces will cause an apparent dulling of the silver. Looked at obliquely, the dry film should be of uniform colour (greenish or pinkish) and free from interference rings.

Though impervious to water, the celluloid film scratches very easily. Remove dust, as infrequently as possible, with a tuft of swan's down or equally soft material. Also it becomes opaque and brittle with age, losing its protective qualities. It can be removed with amyl acetate and a cottonwool pad; this may not injure the silver film, but nine times out of ten it is partially removed along with the celluloid.

7.4.2 Perspex: Make up a 0·5% solution of Perspex (e.g. a Perspex article, reduced to a fine powder with a clean file) in butyl acetate. Stir at intervals for two to three days. Filter, and apply as described in section 7.4.1. Of the commoner solvents of Perspex, butyl acetate produces a good, even film on drying; amyl acetate gives a gelatinous solution at concentrations of even 0·5%, while benzene tends to dry leaving a film of uneven thickness.

The Perspex film is slightly tougher than celluloid, but still scratches easily; it is colourless. It becomes opaque if allowed to remain wet for any length of time; dew must therefore be removed at once (see section 11.1). Old films can be removed by benzene, usually without harming the underlying silver.

7.4.3 Collodion: A lacquer can be made up of collodion in redistilled ether, 1%. The lacquering should be carried out in a fairly cool room, as too rapid evaporation tends to produce an uneven film.

7.4.4 Mirrolac: A proprietary lacquer (see section 40) which has been very highly recommended by B.A.A. users. The film is very thin, hard (withstands vigorous rubbing with a soft cloth), and is unaffected by condensation and salt.

7.4.5 Varnish: Now that Mirrolac is on the market, ordinary white varnish need hardly be used, although it gives a reasonable degree of protection. It should be of the highest quality, and should be diluted with about eight times its own volume of the special diluter that is supplied with it. After filtering, it is used as described in 7.4.1.

7.4.6 Silicon dioxide: Much more satisfactory, for aluminium films, is the silicon dioxide coating, since it is permanent, gives complete protection, and is virtually unscratchable. Unfortunately the amateur cannot apply the film himself, but must rely upon a laboratory or commercial firm which has facilities for aluminising (see section 40).

Experiments in the production of these films for large first-surface mirrors were carried out during the war by Heraeus, Steinheil, and Siemens-Schuckert in Germany. Essentially the process consists of the evaporation of silicon monoxide* by means of an 8-kw heating element immediately after the evaporation of the aluminium, and at the same reduced pressure. According to Heraeus, the small amount of oxygen remaining in the chamber under an evacuation of 10^{-6} mm combines with the vaporised silicon monoxide to form the dioxide, which is deposited as a continuous film of quartz. The Perkin-Elmer Corporation (who since the war have treated a 36-in mirror for the Link Observatory, and another mirror belonging to Harvard) state that this process is only completed after about a week's exposure to the air. The film so produced is much more durable than that resulting from the direct evaporation of the dioxide.

The correct thickness is given by $0·25n\lambda$, where n is an odd number (so as to eliminate reflections from the surface of the quartz), and λ is the wavelength of light. The thickness is controlled by observing the colour of a beam of light reflected from the surface of the mirror as the silica is being deposited; as its thickness increases, the colour of the reflected beam passes through all the colours of the spectrum in turn,

* Apparently titanium dioxide was also used successfully.

and the thickness is found to be most satisfactory when it becomes yellow for the second time.

Perkin-Elmer state that the reflectivity of the mirror in both the visual and actinic ranges is slightly increased by the quartz film. Steinheil, on the other hand, were satisfied with films that did not reduce the reflectivity by more than 3%.

7.4.7 Chroluminium: Chroluminium films, such as that given to the 82-in McDonald mirror in 1939, consist of aluminium on a chromium base (chromium adhering to glass more satisfactorily than aluminium), protected by a film of silica. The combined thickness of the three films is about 0.1μ to 0.2μ. Reflectivity over the visual range and the near ultraviolet better than 85% is guaranteed.* Its rate of deterioration when exposed to the amount of water and sodium chloride that an astronomical mirror normally encounters is very slow indeed. It can be cleaned with an ordinary soft cloth, if carefully dusted first, and washed with soap or detergent and water.

7.5 Semi-transparent films

With a perfect semi-transparent mirror, the sum of the reflected and transmitted portions of the incident ray would be 100% of the incident light. In fact a certain percentage is always lost by absorption. A compromise has to be found between small absorption and practicable durability of the material of the film.

The following have been used:

(*a*) Thin coating of silver, aluminium, or rhodium. Any degree of transmission relative to reflection can be obtained by varying the thickness of the film.

(*b*) Thin coating of ferrous oxide (the Steinheil 1 mirror, B. 71). Maximum reflectivity 52%, minimum absorption about 7%; resistant to abrasion and water, and does not corrode in air. The transmitted light is strongly orange.

(*c*) Thin coating of titanium dioxide (the Steinheil 2 mirror). Maximum reflectivity 42%, minimum absorption 1% to 2%; transmitted light practically colourless; stability similar to that of the Steinheil 1. Both resist vigorous scrubbing with a mixture consisting of equal parts alcohol, water, and 10% ammonia.

(*d*) Duolux.* Absorption higher than with silver or aluminium films (up to about 40%), but more resistant to chemical erosion and physical

* Evaporated Metal Films Corporation, Ithaca, N.Y.

abrasion. Can be cleaned with detergents, or even nitric acid, and will not abrade appreciably under hard rubbing with a cloth.

7.6 Aluminising

Most metals can be evaporated in a vacuum on to glass, forming a thin bright layer. The process is not one that the amateur can undertake unless he has the run of a well-equipped physical laboratory; commercial firms which have facilities for aluminising are listed in section 40.

The aluminising process may be summarised as follows:

(1) thorough cleaning of the glass surface by physical and chemical methods;
(2) loading the vacuum chamber with the mirror and the aluminium pellets;
(3) evacuation of the chamber to 10^{-3} mm;
(4) further cleaning of the glass surface by glow discharge at from 3000 to 5000 volts;
(5) further evacuation to 10^{-5} or 10^{-6} mm;
(6) evaporation of the aluminium (heating elements rated at about 3 kw);
(7) cooling for 30 to 50 minutes; destruction of the vacuum; removal and examination of the mirror.

Individual techniques vary slightly in detail; that described is the Siemens-Schuckert procedure. (See further B. 71, 67, 69.)

7.7 Silvering

Silver films are more conveniently deposited by one of several chemical methods than by evaporation.

7.7.1 Theory of the precipitation of metallic silver from the solution of a silver salt: All chemical methods of precipitating a film of metallic silver are fundamentally similar, differing among themselves chiefly in the nature of the reducing agent and in the general manipulation of the process.

Two solutions are prepared: one of a silver salt—commonly silver nitrate in ammoniacal solution—and the other an alkaline solution of some reducing agent. The latter, when mixed with the silver solution, reduces the oxygen-containing silver compound, with the precipitation of metallic silver.

Thus silver nitrate plus sodium or potassium hydroxide produces a dark precipitate of silver oxide:

$$2AgNO_3 + 2NaOH = 2NaNO_3 + Ag_2O \downarrow + H_2O$$

This is redissolved by the addition of ammonia, the solution being extremely unstable. The addition of the reducer precipitates metallic silver on any surface that is in contact with (i.e. wetted by—hence the necessity for complete freedom from grease) the solution:

$$Ag_2O + H_2 = 2Ag \downarrow + H_2O$$

The speed with which the silver is precipitated is a function of temperature, and although different methods work best at different temperatures silvering should never be carried out in a room, or with solutions, whose temperature is below about 55° F.

7.7.2 Preliminary cleaning of the glass surface: Absolute cleanliness of mirror, silvering bath, and materials is essential. The chief enemy is grease. By the same token, only chemicals of the highest purity (of the grade known as A.R.) should be used; it is advisable to get them from a wholesale firm, like B.D.H., rather than from the local chemist.

A carelessly cleaned mirror, or one which has been allowed to become contaminated with grease or chlorides from the fingers after cleaning, is among the commonest sources of failure to obtain a uniform and firm film. A variety of cleaning routines are advocated by different workers. If the glass surface can be cleaned subsequently by a H.T. discharge, it is enough to give it a hard swabbing with linen or cottonwool and commercial detergent, followed either by absolute alcohol or a mixture of equal parts of absolute alcohol, distilled water, and 10% ammonia.

A satisfactory routine for a glass surface on which silver is to be deposited chemically is as follows:

(*i*) Wash thoroughly with a strong detergent, such as Dreft. This is a sulphonated organic compound which is preferable to soap, having a neutral reaction; it forms soluble compounds with calcium and magnesium ions.

(*ii*) Rinse thoroughly with tap water, followed by distilled water.

(*iii*) Swab the surface with concentrated nitric acid, which oxidises any soluble impurities. Since, however, some are insoluble the mirror must be scrubbed thoroughly with a cottonwool pad over the end of a glass rod (being careful that the rod does not slip through the pad) until all regions of the mirror produce a slightly grating feeling of resistance to the swab. Any part of the mirror over which the swab slides smoothly and with little resistance is not yet clean. The back and sides of the mirror must be similarly treated.

(*iv*) Rinse thoroughly in distilled water. (An intermediate rinse in dilute alkali is unnecessary if this water rinsing is thorough.) A rubbing down of the surface with stannous chloride, between the acid bath and the final rinse, is an article of faith with some workers; at least it does no harm.

(*v*) Keep the mirror under distilled water, at the temperature at which the silvering is to be carried out, until ready for the silvering bath. At no stage during the cleaning, or subsequently, must any part of the surface be allowed to dry. If it does, the routine should be repeated from the beginning.

(*vi*) Treat the silvering bath to the same cleaning.

If the surface to be cleaned is not glass, but an old film, omit stages (*i*) and (*ii*), and start with the acid treatment, changing swabs and acid when all the old film has been removed. Handle throughout with clean rubber gloves.

As a rough check that the surface is clean, lift it from the distilled water bath, and run off the excess; if the film of water covers the whole surface uniformly, and does not avoid certain areas, the condition may be regarded as satisfactory. Care must be taken that the mirror is not kept out of the bath long enough for the film, or any part of it, to evaporate.

When the silver solution is ready to be poured on the mirror, the distilled water that has been covering it can be poured off, or the solution can be added to it.

7.7.3 Brashear's method: This and Martin's method (employing sugar as the reducing agent) and Lundin's (employing formaldehyde) are the most commonly used, and, although each has its faithful adherents, there is very little to choose between them.

Brashear set himself to devise a method of producing a bright film, rapidly precipitated, and of a toughness that would withstand vigorous rubbing while still wet from the silvering bath. Brashear films also, and for the same reason, stand up well to the periodic burnishings that are essential (especially if the mirror is used for photography) to remove tarnish, thus considerably lengthening the life of the film.

Four stock solutions are made up, and may be kept indefinitely:

I	Silver nitrate	60 gms
	Distilled water	to 1000 c.c.
II	Ammonium hydroxide (conc.)	1000 c.c.
III	Potassium hydroxide	60 gms
	Distilled water	to 1000 c.c.
or		
	Sodium hydroxide	42 gms
	Distilled water	to 1000 c.c.

IV White table sugar   100 gms
    Distilled water   150 c.c.
    ⎰Tartaric acid (cryst.)   5 gms
    ⎱or Nitric acid (conc.)   5 c.c.

Boil for 15 mins to convert the sugar to invertase (a mixture of dextrose and laevulose). Add

    Distilled water   500 c.c.
    Alcohol (100%)   175 c.c.
    Distilled water   to 1000 c.c.

Solution IV should be made up several days before it is required.

or

Dextrose may be used with advantage instead of invertase; since it is a rather more powerful reducing agent it accelerates the formation of the film, allows the process to be carried out at a lower temperature, and reduces the proportion of sludge:

    Dextrose   100 gms
    Alcohol (100%)   175 c.c.
    Distilled water   to 1000 c.c.

This solution can be used at once, and must be made up afresh on each occasion; it may ferment if kept.

or

A third reducer, which employs ordinary table sugar but may be used immediately, is as follows:

    Table sugar   100 gms
    Nitric acid (conc.)   40 c.c.
    Distilled water   to 1000 c.c.

Bring to the boil, and allow to cool before using. If it is to be stored, add 175 c.c. alcohol.

For a mirror of area x square inches, the following quantities will be required (great precision in measuring out the relative quantities is not necessary): of I, that containing $x/5$ gms silver nitrate; of II, as required to clear the precipitate (see below); of III, half that of I;* of IV, that containing a weight of sugar equal to about half that of the silver nitrate used.† Thus for a 10-in mirror:

* Some workers use half this quantity of III, apparently with good results.

† If less reducing agent than this is used, the precipitation of the silver will be slower, though the film tends to be tougher. If dextrose is used, a quantity of IV containing as little as one-fifth or one-tenth of this silver nitrate by weight will suffice.

Area	78·5 sq ins
Weight of silver nitrate required	16 gms
∴ Volume of I required	266 c.c.
Volume of III required	133 c.c.
Weight of sugar required	8 gms
∴ Volume of IV required	80 c.c.

The preparation of the silver solution by the mixing of the appropriate quantities of I, II, and III is the tricky part of the process, since it is essential that there should be a residual excess of the nitrate, not of ammonia, but at the same time too long must not be spent achieving the correct balance between them, since the precipitates are unstable. It is here that practice and familiarity with the behaviour of the solutions count for a great deal, and the amateur who has done no silvering is advised to try his hand several times with small mirrors, or even with odd pieces of window or picture-frame glass, before attempting to silver his speculum.

The procedure is as follows:

(*i*) The required quantity of I is poured into the silvering bath. This should be of glass, and of such a size that when the mirror is placed in it, the surface of the combined silver and reducing solutions comes about halfway up its rim if it is being silvered face down (see section 7.7.8).

(*ii*) Add II slowly, stirring all the time. At first a brown precipitate of silver oxide will be thrown down; further addition of II will redissolve this. Towards the end, add II drop by drop with a pipette, stirring between drops, till the precipitate is *just* redissolved.

(*iii*) Add III slowly. This will again produce a dark-brown precipitate.

(*iv*) Add enough of II just *not* to clear the solution. Adding it finally drop by drop, stop when the solution is still slightly cloudy.

(*v*) If too much of II has been added, so that the solution clears completely, the excess ammonia must be removed by adding more of I, drop by drop, until the liquid again has a slight brown cloudiness due to free oxide.

(*vi*) Finally, the solution may be filtered, though this is not essential.

(*vii*) The solution is now ready, and will precipitate metallic silver immediately the reducer is added. Pour in IV, stir vigorously for a few seconds, and lower the mirror into the bath.

Throughout the silvering, keep the solution in motion across the surface of the mirror by gently tilting the bath from side to side and by rotary rocking. As soon as the film has formed over the whole surface of the mirror, and black specks are beginning to settle on it (it is being assumed that the mirror is face upwards—see section 7.7.8), start swabbing gently with a cottonwool wad. The end of the reduction of the silver oxide is indicated by the precipitate being black and granular, and the liquid itself clearing. Pour off the liquid and rinse the mirror in distilled water.

Five to 10 minutes is the average time for the precipitation of a Brashear film, though it may take as long as 20 minutes, according to the temperature of the bath. Brashear recommended 70°F, but experience has shown that a slightly lower temperature is preferable. About 60°F is probably most satisfactory; higher temperatures, though they accelerate the process, tend to produce a softer film* and also traces of silver fulminate, an extremely sensitive explosive.

If the film is found to be too thin (see section 7.7.10), the mirror is placed under distilled water and a second silvering bath prepared.

7.7.4 Martin's, or Petitjean's, method:

I	Silver nitrate	60 gms
	Distilled water	to 1000 c.c.
II	Ammonium nitrate	90 gms
	Distilled water	to 1000 c.c.
III	Potassium hydroxide	150 gms
	Distilled water	to 1000 c.c.
or		
	Sodium hydroxide	105 gms
	Distilled water	to 1000 c.c.
IV	Reducer as in Brashear's method.	

To the appropriate volume of I (determined as explained on p. 124) add an equal volume of II, stirring all the time. Then either (a) add an equal volume of III, immerse the mirror, add the appropriate quantity of IV, or (b) add the volumes of III and IV together, place the mirror in I–II, and add III–IV.

Martin's process works best at the same range of temperature (60°F to 70°F) as does Brashear's, which it closely resembles.

7.7.5 Lundin's method: Also simpler than Brashear's, yet nine times out of ten equally satisfactory in its results. A Lundin film perhaps tends to be slightly less bright than a Brashear film, but the difference is not much better than negligible.

* Not invariably true, satisfactorily tough films having been produced at temperatures as high as 80° F. At such a temperature, however, a higher proportion of failures may be expected.

I	Silver nitrate	22 gms
	Distilled water	to 1000 c.c.
II	Formaldehyde	200 c.c.*
	Distilled water	800 c.c.

Any cloudiness in solution I can be cleared up by the addition of a few drops of ammonium hydroxide. The two solutions are mixed together and the mirror immediately immersed. The end-point of the silvering is indicated by the clearing of the solution, apart from numerous black grains, and usually occurs in from 3 to 5 minutes. The temperature of the bath should be about 65°F. It has been stated that if the temperature of the mirror is a few degrees higher than that of the bath, a larger proportion of the precipitated silver will go to the formation of the film, and less to sludge.

7.7.6 Liebig's method: Of great historical interest, Liebig's method is little used today. About the middle of last century the construction of the reflecting telescope was revolutionised, and the way opened to the giant reflectors of today, by the substitution of silvered glass for the old polished speculum metal. This was made possible by Liebig's discovery, in 1856, of a reliable method of precipitating metallic silver on glass in the form of a thin, highly reflecting film.

I	Silver nitrate	6 gms
	Distilled water	114 c.c.
II	Potassium hydroxide	28 gms
	Distilled water	710 c.c.
III	Lactose	14 gms
	Distilled water	142 c.c.

These quantities are sufficient for silvering about 50 square inches of glass; for other sizes, different amounts in the same proportion. III must be freshly prepared.

Half the volume of I is poured into the silvering bath. Ammonium hydroxide is added, with constant stirring, a drop at a time towards the end, until the precipitate is *just* cleared.

To this solution 114 c.c. of II are added; and again the precipitate is just redissolved by the addition of ammonium hydroxide.

The I–II mixture is then diluted to 425 c.c. with distilled water, and enough of the remainder of I dropped in, with constant stirring, just to form a faint cloudy precipitate which is not redissolved after a few minutes' stirring.

* Subsequent users of this formula have found that satisfactory results can be obtained with as little as 10% of this quantity of formaldehyde.

Finally, the mixture is diluted with a further 425 c.c. distilled water, and 60 c.c. of III added; after stirring, the mirror is immersed.

Common's and Draper's processes are slight modifications of Liebig's, and need not be described in detail.

7.7.7 Böttger's (Rochelle salt) method: The characteristic feature of this process is the use of Rochelle salt (potassium sodium tartrate) as the reducer. Quantities for a surface of 30 square inches are as follows:

I	Silver nitrate	5 gms
	Distilled water	300 c.c.
II	Rochelle salt	0·8 gms
	Distilled water	10 c.c.
III	Silver nitrate	1 gm
	Distilled water	500 c.c.

Add ammonium hydroxide to I until the precipitate is nearly redissolved. Filter if necessary, and make up to 500 c.c. with distilled water.

Boil III, and to the boiling solution add II. Continue boiling until a grey precipitate is thrown down. Filter, and dilute to 500 c.c.

Pour equal quantities of I and II–III into the silvering bath, and immediately immerse the mirror. The process is rather slow—for which reason it is often used in the production of semi-transparent films of specified transmission/reflection. At 115°F to 120°F the formation of the film may take anything from 20 to 80 minutes, the end-point being indicated by the clearing of the solution.

7.7.8 Position of mirror in bath: Mirrors may be silvered either face-up or face-down in the bath. Generally speaking, small mirrors (less than 6 to 8 ins) are more conveniently silvered face-down, larger mirrors face-up.

(*i*) Face-down: Fix a narrow plank of wood across the back of the mirror with pitch; this must be long enough for the ends overlapping the mirror to rest on the sides of the silvering bath. Throughout the silvering the solution must be kept in motion across the surface of the mirror by gently and continuously rocking the bath. The mirror must be lowered into the solution at a slant to prevent air being trapped in its concavity. The progress of the film can be observed through the back of the mirror. It is for this reason that it is inadvisable to allow the solution to submerge the mirror completely (apart from the necessity it involves of cleaning both surfaces): inspection then necessitates removing the mirror from the bath, and on its return it is inevitable that some of the floating flakes of silver will be trapped against the film.

When it is judged that the film is thick enough, or the reduction is complete, remove the mirror; quickly check the thickness of the film against an electric-light filament (see section 7.7.10). If it is too thin, place the mirror in a dish of distilled water containing a few drops of ammonium hydroxide, and prepare for a second silvering. If satisfactory, it is rinsed thoroughly with distilled water.

(*ii*) Face-up: Floating flakes of silver settling on the mirror cause innumerable small holes in the film, but these are of little consequence (providing the flakes are not allowed to adhere), since the subsequent polishing will in any case produce tiny pores.

The silvering can either be carried out in an ordinary bath (in which case the mirror should be submerged to a depth of at least an inch), or the mirror itself can be made the base for a bath. Take a strip of tough brown paper, five times the circumference of the mirror in length, and the thickness of the mirror plus one-third its diameter in width. Dip it in melted paraffin wax and hang up to cool. Smear the median line of the edge of the mirror with a little vaseline, taking every precaution against the grease getting spread around the place. Wrap the paper strip round the mirror, so that one edge is flush with its back; fix it in place with several turns of light string. The acid-cleaning can be carried out in this improvised bath, but rinsing must be very thorough.* When the solutions are prepared, tip out the distilled water and pour in first the silver solution and then the reducer. Rock the mirror gently with rotary and transverse motions throughout the precipitation. The progress of the film can be judged by periodically tilting the mirror far enough for the edge of its surface to be uncovered. If there is any sign of the flakes sticking to the film, a long wad of wetted cottonwool must be trailed across the silver surface in all directions It is essential that no pressure be exerted, and that rubber gloves be worn.

At the end of the process, tip out the solution, immediately replace it with distilled water containing a few drops of ammonia, and rough-check the thickness of the film. If satisfactory, rinse, remove the paper, clean off the vaseline with cottonwool while the mirror is still under running water; change over to distilled water for the final rinse. If the film is too thin, leave the mirror under ammoniated distilled water while the new solution is being prepared.

Films deposited with the mirror in the face-up position tend for some

---

* Many prefer to clean the surface before fixing the waxed band. In this case the mirror must be carefully levelled so that it will remain covered with water whilst the band is being put in position.

reason to be rather easier to polish than those deposited in the reverse position.

7.7.9 Post-treatment of the film: After rinsing with distilled water, prop the mirror at an angle and rinse several times with strong alcohol. Stand it on edge on a good thickness of blotting paper and dry with a fan or in front of an open window. As it dries, the yellowish tint that it may have while wet is replaced by a bluish bloom, which has to be removed by burnishing and polishing; it should nevertheless be highly reflective at this stage.

Once thoroughly dry, it is ready for burnishing. This consolidates the film and prepares the way for the polishing. It should be done with a chamois leather pad stuffed with cottonwool, roughly one-third to one-quarter of the mirror in diameter. Exerting a gentle pressure, cover the whole surface with a circular motion. After several minutes of burnishing, scatter a little finely sifted rouge* on the pad, and continue polishing; gradually increase the pressure, which initially must be very light. Best results are obtained during the rouge polishing if the surface of the polisher has first been made smooth and rather shiny by a preliminary rubbing on a piece of perfectly clean glass.

If after several minutes' polishing the quality and reflectivity of the film are unsatisfactory, resilvering is unavoidable.

7.7.10 Determining the thickness of the film: Upon the thickness of the film depends the number of polishings it will stand, and therefore its length of life. 0.1μ to 0.2μ is quite normal; a film of this thickness is not completely opaque, transmitting about 0.01% of the incident light. The filament of an electric-light bulb will therefore be visible through it. If, however, a window frame silhouetting the day sky is visible through it, it is on the thin side. Even if the film is so thick that the Sun cannot be seen through it, there is no cause for worry. By means of its transparency, or otherwise, to light from a window, an electric filament, and the Sun, the thickness of the film can be gauged quickly and precisely enough to decide whether or not a second film must be applied.

The thickness can be determined easily and with great accuracy by means of Nobili's rings, once the mirror is dry. A single minute crystal of iodine is placed on the film. Over this is placed an inverted beaker, to exclude draughts. In a few minutes it will be seen that the crystal is surrounded by a number of concentric yellow rings. These consist of silver iodide, and the ring formation stops when the film immediately beneath the crystal is completely converted to the iodide.

* E.g. the dried scum taken from the surface of a jarful of water in which rouge has been stirred up.

The number of rings formed depends directly upon the thickness of the film:

Number of rings	Thickness of film (μ)	
2	0·04	thin
3	0·07	
4	0·11	
5	0·15	moderate
6	0·18	
7	0·22	
8	0·26	thick
9	0·29	
10	0·33	

7.7.11 Reclamation of silver residues: Unreduced silver solution should never be left standing about; it is unstable, and silver fulminate, an extremely sensitive detonator, may be formed. It can either be thrown away, or the silver precipitated as silver chloride by the addition of sodium chloride solution.

The greyish or black sludge left in the silvering bath is rich in silver—indeed, from 90% upward of the silver that is precipitated is left in this form. Metallic silver can be recovered by washing the precipitate through a filter paper, drying it, and grinding it up with borax; if it is then heated to a bright red heat in a crucible, allowed to cool, and swilled around in a beaker of water to dissolve out the matrix, a bead of metallic silver will be left.

SECTION 8

OCULARS

Oculars are rather frequently overlooked as possible sources of inferior performance, which is then blamed on the objective. The general function of the ocular is described in section 3.2; maximum and minimum useful magnification in sections 3.3, 3.4; and field size in sections 3.7–3.9. Here the optical characteristics of oculars in general, and of individual types of ocular in particular, will be dealt with in greater detail.

8.1 Characteristics of the perfect ocular

The perfect ocular exists only in the astronomer's wistful imagination. The following are the main features that are required of an ocular, but no ocular possesses them all, the strength of one type being the weakness of another:

(a) imperceptible aberrations, if not intrinsically, at least with the appropriate objective;

(b) a wide field, with good definition over at least the greater part of it;

(c) a flat field;

(d) a dark field, i.e. freedom from internal reflections which either produce ghosts, or lessen contrast, or both;

(e) bright images, i.e. the reduction to a minimum of light lost by internal reflection and (though this is negligible in most types of ocular) by absorption;

(f) sufficient clearance between the eye lens and the exit pupil to ensure comfortable observing, even when glasses are worn.

8.2 Ocular aberrations

The importance of the aberrations increases with the diameter of the pencil that the ocular has to deal with, and is therefore a function both of the relative aperture of the objective and of its own focal length.

132

(*a*) The aberrations increase rapidly with increase of the relative aperture, since the latter increases the diameter of the pencil delivered to the field lens of the ocular. It is explained in section 4.3 that spherical aberration is a function of D^2: hence the spherical aberration exhibited by a given ocular will be nine times as great at $f/5$ (e.g. Newtonian) as at $f/15$ (e.g. typical refractor). An ocular which gives perfectly satisfactory images when used with the latter objective will nevertheless be intolerable with the former, possibly leading the observer to the unjustified conclusion that his speculum is faulty, or that the refractor has some inherent advantage over the reflector.

(*b*) In general the size of the aberrations increases in direct proportion with the focal length of the ocular, and they are therefore more troublesome with low magnifications (when the exit pencil is comparatively large) than with high. The superiority of highly corrected oculars over the commoner types is only apparent with low magnifications, except at small focal ratios when they are essential even with high magnifications.

8.2.1 Spherical aberration: All types of ocular in common use suffer from spherical aberration to a greater or smaller degree. At relatively large focal ratios (around $f/15$) this is not noticeable with high powers, but may become so with low. Since the residual spherical aberration of both the Ramsden and Huyghenian ocular (the two types in commonest use) is positive, these work better with slightly over-corrected than with perfectly corrected objectives.

We have seen that to increase the relative aperture of an objective by x is to increase the illuminated aperture of the ocular in the same proportion, and therefore to increase the spherical aberration by x^2. Hence uncorrected spherical aberration in the ocular only becomes a real menace at small focal ratios, when highly corrected (and therefore expensive) oculars are an absolute necessity if the potential qualities of the objective are to be realised.

8.2.2 Chromatic aberration: Uncorrected oculars suffer from both the longitudinal aberration and chromatic inequality of magnification. The longitudinal aberration introduced by an ocular consisting of simple crown lenses is negligible so long as $f < \dfrac{1}{190} \cdot \dfrac{F^2}{D}$ approximately. If f increases past this limit, a progressively strong modification of the secondary spectrum of the objective will be observed, becoming really serious when $f > \dfrac{1}{50} \cdot \dfrac{F^2}{D}$ approximately. Achromatic oculars are therefore a necessity when the objective has a fairly large relative aperture, though they are less essential with refractors of normal construction.

A $f/6$ Newtonian used with, say, a Huyghenian ocular cannot be described as an achromatic instrument.

An ocular may be termed achromatic if it preserves the colour correction of the objective. Hence the term 'achromatic' carries somewhat different connotations when applied to objectives and to oculars. Whereas in the former context it signifies the bringing together of two wavelengths at the same distance from the objective, an ocular is said to be achromatic if it is free from chromatic inequality of magnification, producing an image which is the same size in all colours although the individual monochromatic images are formed in different positions. Images are therefore affected by fringes of out-of-focus colour, though this is uniform over the whole field.

The residual under-correction of the ocular and the human eye in combination is compensated by making the objective slightly over-corrected. This compensation, however, is only strictly possible for a single magnification (see section 4.8). Magnifications higher than this will produce residual over-correction, lower magnifications under-correction. The colour correction of the OG is therefore to be improved by employing slightly over-corrected oculars of long f and under-corrected oculars of short f.

Even the condition of chromatic equality of magnification is progressively departed from when the diameter of the pencil becomes large compared with the diameter of the ocular lens through which it is passing. When it becomes marked, its effect is to draw out stellar images in the outer part of the field into short radial spectra, red towards the centre of the field and blue towards the edge, whose lengths are proportional to their distances from the field centre. It is objectionable, therefore, in that it limits the usable field.

The suppression of chromatic inequality of magnification requires that the emergent rays AF_r and BF_v, of different wavelengths, shall be parallel (Figure 36), i.e. that $\theta=\theta'$. The expression for the equivalent focal length of the combination is

$$\mathcal{F}=\frac{f_1 f_2}{f_1+f_2-d} \qquad \cdots \cdots \cdots \quad (a)$$

Since both lenses have the same refractive index, for a given $\Delta\lambda$ their focal lengths will both be altered by Δf. Hence

$$\frac{f_1 f_2}{f_1+f_2-d}=\frac{f_1 f_2(1+\Delta f)^2}{f_1(1+\Delta f)+f_2(1+\Delta f)-d}$$

Neglecting powers of Δf and simplifying, we have

$$2d=f_1+f_2$$

which is the condition for achromatism in a pair of separated converging lenses made of the same material. It is observed by the Huyghenian and, approximately, by the Ramsden ocular.

Reflectors and apochromats deserve oculars whose high colour correction approaches their own perfect achromatism. With ordinary

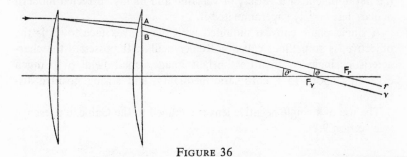

FIGURE 36

'achromatics' it is less important, since there is inevitably some secondary colour, and an imperfectly corrected ocular may even be discovered that reduces the residual chromatism.

8.2.3 Other aberrations: Coma and astigmatism are normally negligible, though both appear if oculars are used at grossly inappropriate focal ratios. Field curvature is generally present to some extent, its complete suppression being coupled with intolerable astigmatism towards the edges of the field. Distortion, when present, takes the positive form, and is negligible in the central region of the field.

8.3 Single-lens oculars

These have the highest transmissions of any, but also the most pronounced aberrations; hence their fields of tolerable definition are very restricted.

The field of full illumination of a single-lens ocular is given by

$$\tan R = \frac{r - \frac{1}{2}d}{f}$$

where R = radius of the field, defined by the ocular diaphragm,
r = radius of the lens,
f = its focal length,
d = diameter of the exit pupil.

The aberrations become intolerable when $2R > 20°$, approximately. The exit pupil lies at a distance from the lens of $f + \left(1 + \frac{1}{M}\right)$, or, for practical purposes, f. Good eye clearance is therefore one of the strong points of single-lens oculars.

While the single-lens ocular was recommended by Herschel for all purposes not demanding a large field, and even today has its advocates for such specialised tasks as the delineation of minute planetary detail, the development of a variety of versatile and highly corrected modern oculars has largely superannuated it.

A single plano-convex, mounted with the convex face towards the objective, is sometimes called a Kepler ocular. It possesses the characteristic single-lens features: bright image, small field (not much exceeding about 15°), chromatic inequality, and strong positive distortion.

The use of a single negative lens is confined to the Galilean telescope (see section 9.2).

8.4 Two-lens oculars

These (notably the Huyghenian and Ramsden) were developed in the early days of the telescope to provide an escape from the prohibitive aberrations and restricted field of the single-lens ocular. In particular, the fact that the total refraction is distributed over four surfaces, instead of only two, allows chromatic aberration to be reduced very considerably. As against the Kepler ocular, however, the loss from internal reflections, with resultant loss of contrast and the occurrence of ghosts,* is doubled.

The function of the field lens is to collect a wider cone of rays than would otherwise be received by the eye lens, and it contributes little to the magnification. The eye lens is responsible for the magnification† and for delivering a parallel emergent beam which will come to a focus on the retina of an eye placed at the exit pupil and focused at infinity. A diaphragm in the focal plane of the eye lens defines the field of full illumination.

The exit pupil coincides with the resultant focal plane of the two lenses. This is located between the eye lens and its own focal plane;

---

* Single reflections brighten the field indirectly, by illuminating the back of the diaphragm. Pencils resulting from two reflections are travelling in the same direction as the incident light, and if their section in the focal plane is small they will be seen as faint ghosts; such ghosts lie on the diameter of the field passing through the parent image, and are at least 7 mags fainter than it. Those too faint or too diffuse to be seen as ghost images merely reduce the general contrast and help to destroy the darkness of the field. (See further, B. 72.)

† The addition of a field lens actually reduces the magnification that would be yielded by a given eye lens, since it reduces the effective focal length of the objective.

hence the eye clearance of a single lens is decreased by the addition of a field lens, and the field of full illumination correspondingly increased.

8.4.1 The Ramsden: The so-called 'positive' ocular, thus named for the not very obvious reason that the focal plane does not lie between the two lenses. It consists of two plano-convex crown lenses of equal radius of curvature, mounted with their convex surfaces facing one another.

If the achromatic condition of $2d=f_1+f_2$ is exactly obeyed, the focal plane of the eye lens will coincide with the field lens, and the exit pupil with the eye lens; disadvantages of which are, respectively, that any dust particles or defects on the surface of the field lens are visible, magnified, in the field; and that since the eye cannot be placed at the exit pupil, the visible field is decreased and some of the illuminating power of the objective is lost.

Hence d is usually reduced by from 25% to 40%: both focal plane and exit pupil are then external to the system, though at the cost of introducing a sensible degree of chromatic inequality of magnification. Trouble is also encountered from out-of-focus ghosts due to double reflection in the field lens. Since the focal plane is external, a Ramsden can be used as a magnifying glass.

The Ramsden has considerably less spherical aberration (positive) than the Huyghenian, and less field curvature. It is therefore preferable to the Huyghenian for use with objectives of small focal ratio. It has more colour due to chromatic inequality of magnification, but smaller longitudinal aberration. Slight astigmatism is not troublesome, and there is also slight positive distortion. Useful field in the region of 35° apparent diameter. In the interests of achromatism, the clearance of the exit pupil from the eye lens is not very great—usually about $0.25f$.

8.4.2 The Kellner and achromatic Ramsden: The former may be regarded as a semi-achromatic Ramsden, the simple eye lens of which has been replaced by an achromatic doublet. The doublet consists of a crown biconvex element, nearest the objective, backed by a flint plano-concave. In the achromatic Ramsden, both field and eye lens are replaced by achromatic doublets, consisting of a crown biconvex and flint meniscus.*

The Kellner is one of the finest of the older oculars for LP work, giving wide, flat fields (up to about 40° or 45° apparent diameter), and an eye clearance of about $0.45f$; colour correction and orthoscopy are also excellent. It is, however, one of the most haunted of oculars (if one

---

* Strictly speaking, both these oculars are achromatised Ramsdens, the true Kellner having a crossed biconvex field lens (and doublet eye lens). The usage employed in the text seems, however, to be the settled practice nowadays.

may use the expression for an ocular which is afflicted with ghosts); the eye clearance, though better than that of the Ramsden, is still rather small; and spherical aberration is introduced by the eye lens if ordinary crown and flint are used (the requirements are large relative dispersions combined with small $\Delta\mu$).

An improved form of the Kellner, with a triplet eye lens, extends the usable field to about 50°.

The Ramsden ghost is much less obtrusive in the achromatic form of the eyepiece, which also has a wider field and a closer approach to orthoscopy than the simple Ramsden. Colour correction is, as would be expected, excellent.

Kellners and achromatised Ramsdens are commonly found in prismatic binoculars, finders, and micrometers.

8.4.3 Aplanatics: A family of oculars developed from the Kellner and possessing its general features while surpassing it in width and flatness of field. Freedom from scattered light also darkens the field.

8.4.4 Orthoscopics: These also share the Kellner characteristics of abnormally wide, flat fields, but in an even more marked degree. Numerous orthoscopics exist,* variants on the original design by Mittenzwey. The eye lens is a plano-convex, and the main function of the field lens—which is a combination such as a pair of separated equiconvex lenses, or a cemented triple achromat—is to supply aberrations opposite to those of the eye lens. Orthoscopics all share the following characteristics: excellent correction, both spherical and chromatic, permitting their use at relative apertures of $f/5$ or even larger; wide, flat fields—25° even at $f/6$; freedom from ghosts and scattered light,† giving darker fields than the simple two-lens types; greater eye clearance than the Kellner $(0\cdot8f)$.

8.4.5 The Cooke 5-lens ocular: Though not a two-lens ocular, the Cooke eyepiece is conveniently treated here, since it was designed as an improvement on the orthoscopics; it consists of a cemented triplet field lens and an achromatic doublet. Its excellent correction allows it to be used at $f/4$. Perfect achromatism and good definition over a wide, flat field (up to about 65°); comfortable eye clearance $(0\cdot7f)$. Transmission and freedom from internal reflections not significantly inferior to the simple two-lens oculars.

8.4.6 The Gifford: The 'Orthochromats' of C. Baker are of this type.

---

* See, further, Taylor's very useful summary of 27 types of inverting ocular in B. 74.

† It has been said that orthoscopics have a tendency to produce ghosts, but these are not inherent in the design, since the makers will always correct the trouble.

Generally similar to the orthoscopics. Flat, dark fields with apparent diameters of nearly 60° can be obtained, though some types suffer from rather marked field curvature.

Orthochromats have been recommended for planetary work, for which, like highly corrected oculars generally, they are specially suited.

8.4.7 Euryscopics: Reported to give good definition over wide, flat fields. Made by Hensoldt of Wetzlar.

8.4.8 The dialsight eyepiece: This development of the Ramsden is generally superior to the two-lens types, and only slightly inferior to the more expensive oculars such as monocentrics and orthoscopics. Two varieties are obtainable. The first consists of two achromatic and aplanatic doublets, crown components facing one another and almost in contact.* It is orthoscopic, provides a field of 35° to 40°, and has good eye clearance ($0 \cdot 77f$). A development of this type by Zeiss gives a slightly larger field at the cost of reducing the eye clearance to $0 \cdot 64f$.

Alternatively, the crown components may both be faced towards the objective, and be used with or without a field lens. The Zeiss model gives a 70° field with an eye clearance of $0 \cdot 32f$.

8.4.9 The Huyghenian: Termed 'negative', in contradistinction to the Ramsden and its derivatives. In effect it is a double Kepler, consisting of two plano-convex or convex meniscus lenses, of unequal curvatures, convex towards the objective.

Since the separation of the lenses and their focal lengths are related by $2d=f_1+f_2$ to secure achromatism, and $f_1=3f_2$ is the condition of minimum spherical aberration (in fact f_2 is often as large as $\frac{1}{2}f_1$), it follows that the focal plane lies between the two lenses, since $d=\frac{2}{3}f_1$ and (from (a), p. 134) $f=\frac{1}{2}f_1$ only. Hence the Huyghenian cannot be used as a magnifying glass. For the same reason, micrometer webs located in the focal plane will be flanked by fringes of uncorrected colour, since they are being observed with the eye lens alone, and this is not achromatic; the webs and a stellar image, in other words, are observed with different optical systems and are therefore affected by different aberrations.

Though the chromatic aberration of size is thus corrected, that of the field lens being cancelled by that of the eye lens, the longitudinal aberration of the two lenses is additive; there is therefore residual colour, which becomes very noticeable when the ocular is working at small focal ratios. At the same time, strong positive spherical aberration and slight internal coma appear; there is also some distortion towards the edges of the field. For these reasons the Huyghenian is quite unsuitable

* The once common 'Browning' achromatic was of this type.

for use with Newtonians of normal construction, its working limit being in the region of $f/12$.

The field is rather larger than that of the Ramsden, though slight positive distortion, astigmatism, and curvature begin to obtrude at more than about 20° to 25° from the centre of the field. The field of full illumination is twice that of a single-lens ocular.

Like the Ramsden, the negative ocular has rather small eye clearance (about $0·3f$). It and its derivatives are, however, less troubled by ghosts than the positive eyepieces.

8.4.10 The Airy ocular: A development of the Huyghenian, but not a notable improvement apart from a slight gain in spherical correction.

It consists of a convergent meniscus field lens and a crossed biconvex eye lens. It has been reported to be usable at relative apertures where the spherical aberration of an ordinary Huyghenian would be prohibitive.

8.4.11 The Mittenzwey ocular: Similar to Airy's ocular, the biconvex eye lens of the latter being replaced by a plano-convex and the meniscus being less strongly curved. Good definition is obtainable over abnormally wide fields (50° or so).

8.5 Solid and cemented oculars

In theory these combine the best features of both single-lens and separated-lens oculars: high transmission and a high degree of correction. In practice their superiority over the best oculars with separated lenses is often not spectacular; though between the best of the cemented types, such as the monocentric, and the more primitive types of ocular there is a gulf fixed.

Generally, they are characterised by dark fields, bright images, and virtually complete freedom from both spherical and chromatic aberration. The lack of scattered light gives them the advantage over separated-lens types in cases where (as in planetary observation) the detection of contrast near the threshold of visibility is a factor of importance.

8.5.1 Coddington's eyepiece: Not much used, since it is badly chromatic towards the edge of the field. It is cut from a sphere, and the field is limited by a central groove which is blackened or frosted.

8.5.2 The Tolles eyepiece: A development of the Coddington lens which may be regarded as a solid Huyghenian: it is cut from a rod of crown glass with spherical ends (Figure 37). The groove defining the image of the objective is cut one-third of the distance from the objective end towards the eye end.

Chromatic inequality of magnification is suppressed when the overall length of the lens d is related to the refractive index of the glass and the radii of curvature of the surfaces by

$$d = \frac{r_1 - r_2}{1 - \frac{1}{\mu^2}}$$

and in its usual form, $r_1 = 1\cdot5r_2$.

Its spherical aberration is very considerably less than that of the Huyghenian, and its transmission is higher provided d does not exceed

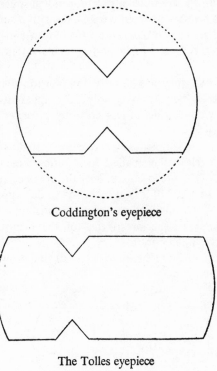

Coddington's eyepiece

The Tolles eyepiece

FIGURE 37

about 1 inch. For darkness of field and freedom from ghosts the Tolles rivals the monocentric, and its general correction is such that it works well at $f/7$. Like the Mittenzwey, it has a very wide field.

The exit pupil coinciding with the rear surface, the Tolles is rather inconvenient in use. Since the image is formed inside the glass, it cannot be used with webs.

8.5.3 Cemented doublets: These are in effect achromatic object glasses of small aperture and very short focal length. The elimination of two air/glass surfaces gives a gain in transmission of from 8% to 10%. Very flat, but rather small, field (the apparent diameter of the Steinheil doublet is only 15°). Some distortion, but good eye clearance.

8.5.4 Cemented triplets: Consist of a central crown equiconvex element flanked by identical flint meniscuses, both focal plane and exit pupil lying well outside the system, They give wider fields than the doublets (25° or so), are free from ghosts, and have negligible distortion and field curvature. There is usually slight residual spherical aberration.

8.5.5 Monocentrics: Cemented combinations, the surfaces of whose components are sections of concentric (or almost concentric) spheres. Without doubt the most nearly perfect oculars that have yet been designed.

Highly corrected, they are suitable for Newtonians of small focal ratio; indeed, their perfection is only fully apparent at small focal ratios, when the aberrations of less perfectly corrected oculars are all too painfully visible.

Their achromatism is practically perfect; very slight spherical aberration makes itself felt at the smallest focal ratios. Fine definition, even better than that of the orthoscopics, over the greater part of a flat, very dark field—for, say, 20° of a field from 25° to 30° apparent diameter. Freedom from scattered light and ghosts; satisfactory eye clearance $(0.85f)$; transmission superior to that of the orthoscopics, despite greater loss through absorption.

Steinheil's monocentric (Figure 38) consists of a crown biconvex and

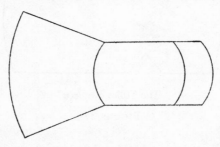

FIGURE 38

two flint meniscuses; all six spherical surfaces are concentric. Spherical aberration restricts its use to objectives of larger focal ratio than $f/6$ approximately. Field about 25°.

The Zeiss monocentric employs slightly larger lenses and a thinner convex component than the Steinheil, giving a rather wider field. Spherical correction is also somewhat better, permitting its use at about $f/5$.

8.6 Terrestrial ocular

Consists of four separated simple lenses, biconvex or plano-convex, of very different curvatures (Figure 39). The rear pair constitutes an ordinary Huyghenian ocular; the other pair constitutes the erector, as shown in the diagram.

The exit pupil of the lens furthest from the eye is exactly defined by a stop between it and the second lens. The field is limited by a diaphragm between the lenses of the Huyghenian pair. The magnification of the pair furthest from the eye—which are together equivalent to a field lens —is inconsiderable, usually ×2 or ×3.

The usual field is about 35° to 40°, but can be further widened by inserting an equiconvex between the field stop and the eye lens of the Huyghenian pair.

8.7 Dual oculars

The only advantage to be gained by using a dual ocular is reduced eye fatigue during long spells of observation. Neither light grasp nor the resolution of point or extended images is affected when the pencil is divided between the objective and the eye; any parallactic effect that may appear is, likewise, illusory.

Binocular microscopes, complete, have been tried as astronomical oculars, and advocated. Their use, however, is exposed to the following disadvantages:

(*a*) the multiplication of air/glass surfaces;
(*b*) the fact that if the beam is split by a Wenham prism covering half the microscope objective (as is the common practice), half the pencil being received by each eye, then asymmetrical star images will be seen under high magnifications, and the definition of extended images will suffer similarly.

A more satisfactory arrangement would be to split the beam by a half-silvered prism in the neighbourhood of the focal plane of the tele-scope's objective (Figure 40). The base of the right-angled prism *V*, which is cemented to the end of the parallelopiped *W*, is lightly silvered, a film about 0·2λ thick transmitting and reflecting equally. The function of the second parallelopiped, *Z*, is to provide for equal light paths in glass for the two rays, without which provision the two images would be

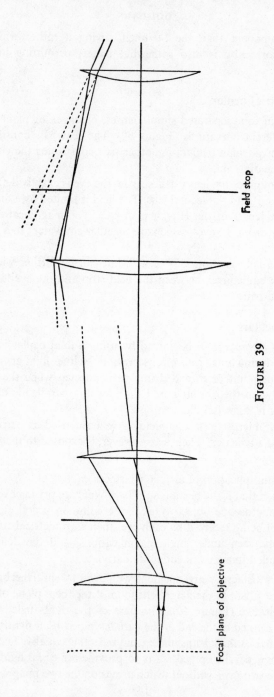

Field stop

Focal plane of objective

FIGURE 39

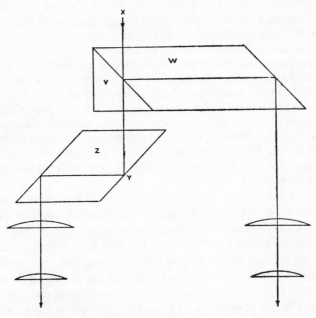

FIGURE 40

unequally bright. Adjustment of the inter-ocular distance is obtained by rotating Z about XY as axis.

8.8 The Barlow lens

The name Barlow lens is given to any negative (diverging) lens placed in the convergent cone from the objective, where, by decreasing this convergence, it increases the objective's effective focal length and focal ratio. It usually takes the form of a cemented plano-concave doublet, corrected for incidence at its plane surface. Its secondary spectrum and residual spherical aberration are, or can be, negligible.

Ordinary Barlow lenses of 8 or 9 ins focal length are less satisfactory than the shorter focus telenegative lenses, since the amplification factors possible with the latter are much less limited. Telenegatives were used before the development of the telephoto lens to increase the plate-scale of camera objectives, and are still to be found on the second-hand market. Those of Goerz have been particularly recommended.

Several notable advantages are offered by these extremely useful accessory lenses. By increasing the effective focal length of the objective,

145

thus reducing its relative aperture, it permits the use of relatively poorly corrected oculars—such as the Huyghenian—with Newtonians and similar instruments whose large relative apertures would otherwise render the uncorrected aberrations of the ocular intolerable. Astonishing improvement of definition is from time to time reported (though this should be no matter for surprise) by the addition of a Barlow lens to an optical train consisting of a paraboloid of ordinary Newtonian focal ratio and a Huyghenian ocular. There is a limit to which a Barlow can increase the focal ratio of a given objective, however, since, as will be seen shortly, increase of effective focal length is obtained by moving the lens further from the primary focus, i.e. increasing its own illuminated aperture and therefore bringing its own aberrations into increasing prominence.

A Barlow lens likewise increases the range of magnifications provided by a given set of oculars, and permits high magnifications without employing oculars of impossibly short focal lengths.

It can be seen from equation (a), p. 158, that for a given aperture the size of the flat can be reduced as the focal length of the objective is increased; we have also seen that although the Newtonian's central obstruction may improve some types of stellar resolution, it is a disadvantage in the observation of extended detail owing to the reduced contrast which it causes. Reduction of the size of the diagonal therefore tends to improve, for example, planetary definition. This can be effected by a Barlow, with whose assistance the final image plane can still be brought to the eyepiece position just outside the tube when the flat is moved in nearer to the prime focus, i.e. to a narrower part of the convergent beam.

The operation of the Barlow lens is explained by Figure 41 and the three equations below:

If F=focal length of the objective,

    f_b=focal length of the Barlow,

    $\mathcal{F}$=effective focal length of the objective and Barlow combined,

   M_b=the Barlow's amplification factor,

    d=distance of the Barlow inside the original focal plane,

    d'=distance of Barlow inside the new focal plane,

then

$$\mathcal{F}=\frac{F.f_b}{f_b-d}$$

$$M_b=\frac{\mathcal{F}}{F}=\frac{f_b}{f_b-d}$$

$$d'=f_b(M_b-1)$$

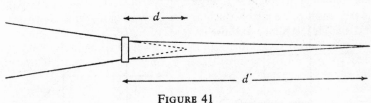

FIGURE 41

Hence when $d=0$, the Barlow has no effect on F, and $M_b=1$; at the other extreme, when $d=f_b$, $M_b=\infty$, the Barlow having converted the convergent beam into a parallel beam. The range of possible positions for the Barlow therefore stretches from the primary focus toward the objective for a distance equal to its own focal length.

Take, for example, the case of a $f/6$ 12-in Newtonian ($F=72$) and a 3-in focus Barlow set 2 ins inside the primary focus. Then from the first of the above equations, the effective focal length of the objective is now increased to 216 ins and its focal ratio from $f/6$ to $f/18$; from the second equation, the Barlow's amplification factor in this position is 3; and from the third, the new focal plane lies 6 ins outside the Barlow.

It will be noticed that d' is extremely sensitive to changes in d. In the case just quoted, where $f_b=3$, we have

d (ins)	1	$1\frac{1}{2}$	2	$2\frac{1}{2}$
M_b	$1\frac{1}{2}$	2	3	6
d' (ins)	$1\frac{1}{2}$	3	6	15

If the Barlow is mounted outside the diagonal (i.e. in the drawtube between the diagonal and the ocular) the inconvenience of widely differing focal extensions is therefore encountered. More conveniently the lens may be cemented to the diagonal (prism); or a combined prism and Barlow could probably be constructed, saving one cemented surface (Figure 42). B. 75 describes in detail an arrangement where the drawtube and prism-lens are mounted in one solid piece, adjustable in parallel guides on the telescope tube and held in the required position by a simple locking device.

8.9 Polarising eyepiece

A polarising eyepiece can be easily made. A piece of Polaroid, held between microscope slips and mounted between two thin card rings, is fixed against the eye lens of an ordinary ocular. A second piece, similarly mounted, is fixed in a screw-on suncap from which the dark glass has

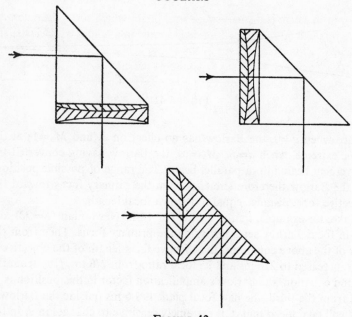

FIGURE 42

been removed; it can thus be rotated relative to the first. Variations of image brightness can be quickly and conveniently made, but there is usually trouble from flare and scattered light. Incomplete polarisation at the extremities of the spectrum produces a purplish tint when the sheets are crossed. Polaroid at the exit pupil is not suitable for solar work owing to the intense concentration of heat at and near this point.

8.10 Diagonals

These may employ total internal reflection in a right-angled prism, a silvered flat, or a prism with hypotenuse face silvered to turn the emergent beam through a right angle. Solar diagonals are unsilvered.*

The reversal of the image is awkward when working with charts, unless the chart is viewed in a mirror; in some cases it is more convenient to take a tracing and use this back to front.

With diagonals of all types it is important that the reflecting surface be kept absolutely clean, or scattered light will cause deterioration and dilution of the image.

* Sun diagonals, as well as other types of eyepiece employed in the direct observation of the Sun, are described in *O.A.A.*, section 1.6.

A zenith prism is a small equilateral prism which turns the emergent beam through 60° (Figure 43). It is mounted in a holder which is usually attached to the ocular by means of a simple clip fitting. Being so close to the exit pupil it has the advantage that it can be quite small.

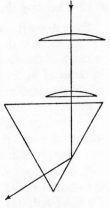

FIGURE 43

8.11 Image rotators

Eyepiece attachments which, when rotated about the optical axis, have the effect of rotating the image, are of use in solar work, variable star observation, and other fields. The type shown in Figure 44 consists of a right-angled prism which is mounted in a short tube behind the eye-piece, so that its hypotenuse face is parallel to the optical axis; when, now, this tube is rotated through θ, the image will rotate through 2θ. This type of prism is also sometimes called an erecting prism, since, like a diagonal, it introduces one-plane reversal into the image—i.e. left to right, or up and down. As such it can be used, for example, to give the projected solar image the

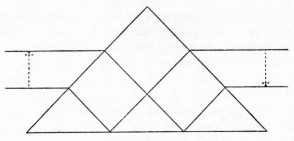

FIGURE 44

same orientation as the image viewed direct. This can also be done by the use of an ordinary star diagonal containing a right-angled prism.

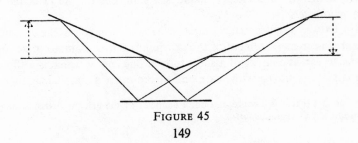

FIGURE 45

Figure 45 illustrates a similar rotator, but employing reflection at plane mirrors instead of a prism.

8.12 Checking the magnification of an ocular

Since $M = F/f$, there can be no such thing as *the* magnification of an ocular, since M depends in part upon the focal length of the objective with which the ocular is used. The only valid description of an ocular is in terms of its focal length. For this reason the magnification of each newly acquired ocular should be determined with the objective with which it is to be used. Available methods include:

(*a*) Berthon dynamometer: A graduated (to 0·001 ins) V-shaped scale, with which the linear diameter of the exit pupil may be measured. The telescope is directed at any light-toned surface—not at a bright one, such as the day sky, or irradiation will falsify the measurement—and the dynamometer held against the eye-guard of the ocular in the plane of the exit pupil; it is then viewed from a distance of about a foot with the aid of a magnifying glass. The magnification is then (equation (*c*), p. 53) given by

$$M = \frac{D}{d}$$

where D and d are the diameters of the objective and exit pupil respectively. Care must be taken that the tube stops in a refractor, which define the convergent pencil, are of the correct size and spacing; otherwise the exit pupil may not be an image of the whole aperture of the object glass. Although this method is not highly accurate—owing primarily to the difficulty of reading the scale with sufficient precision*—it is adequate for ordinary purposes. The effect of irradiation can also be eliminated by fixing across, and in contact with, the objective a card in which two V-shaped notches have been cut at a distance apart equal to about three-quarters of the aperture. The separation of the notches is then measured with the dynamometer, this quantity replacing d in the formula; their linear separation is substituted for D.

(*b*) Banks' dynamometer: Employs a split lens on the principle of the heliometer.

(*c*) Ottway's Scaleometer: Employs the same principle as the dynamometer (linear measurement of the exit pupil) but is more accurate and easier to use. It consists of a short tube at one end of which is a slip of plane glass engraved with scale, and at the other a simple magnifying

---

* The fact that it is a chord and not a diameter of the exit pupil that is measured introduces negligible error.

glass for viewing the scale. The scaleometer is held behind the ocular so that the scale lies in the plane of the exit pupil and the diameter is read with the aid of the lens.

(*d*) A scale erected at a distance from the telescope is viewed simultaneously with one eye through the telescope and with the other naked; in the case of a Newtonian a prism or plane mirror will have to be used so as to bring the naked-eye image to a position where it can be observed. The distance of the scale is varied until a position is found such that one division of the telescopic image coincides with a convenient number of divisions of the naked-eye image: this number is the magnification.

(*e*) Theoretically, the magnification can be derived from (*b*) on p. 52, by direct measurement of *F* and *f*; in practice this is awkward to carry out with reasonable accuracy, and it is more satisfactory to derive the quantities indirectly:

(*i*) Focal length of ocular: Rule two fine parallel lines on a sheet of cardboard, their distance apart, *u*, being accurately measured. Observe these with the ocular from a distance of several yards, *d*, and measure the linear separation of the images of the two lines, *v*, with a scaleometer. Then

$$f = d \cdot \frac{v}{u}$$

(*ii*) Focal length of objective: (1) Expose a plate in the focal plane of the objective with the telescope clamped, so as to photograph the trails of two stars of known Decs. Measure the linear separation of the trails, *d*, as accurately as possible. Then

$$F = d \cot \varDelta\delta$$

where *F* and *d* are, as before, measured in the same units. (2) Selecting a star of known Dec, δ, and using a Ramsden or other positive ocular (since the field lens of a Huyghenian cooperates with the objective in the formation of the image in the plane of the ocular diaphragm), time its transit across the field diametrically, *t*. Measure the linear diameter, *d*, of the ocular diaphragm with the greatest accuracy possible. Then

$$F = \tfrac{1}{2}d \cot \tfrac{1}{2}(15t \cos \delta).$$

8.13 Care and maintenance

Oculars are often subjected to a roughness of handling that would never be meted out to an objective. But it is worth remembering that from the point of view of the final image, the strength of the optical chain is that of its weakest link, not its strongest; and a faulty ocular can ruin the performance of a superlative objective with no difficulty at all.

The precautions to prevent the formation of dew on the lenses, described in section 11.1, should be observed. If condensation does form, it should never be wiped off. Oculars when not in use should at all times be kept in a dust-proof and airtight container of some sort. Lenses should be polished as infrequently as possible, and only when they are perfectly dry. They should first be brushed thoroughly with a fine sable brush (as sold to water colourists) to ensure that all dust is removed; they can then be polished with clean, soft, dry silk and the minimum of pressure. Flint is more liable to scratch than crown, and hence those oculars having a flint component outermost (e.g. mono-centrics) should be especially carefully treated.

Faulty oculars should be handed over to an optician, preferably the maker, for repair.

See also section 22.9.

SECTION 9

COMPOUND OPTICAL INSTRUMENTS

The images formed at the primary foci of short-focus concave mirrors and convex lenses are too small for effective examination. A telescope is simply an instrument employing what is in effect a magnifying glass to increase the scale of the image formed in the focal plane of the objective, so that resolved points in the focal plane image are separated to an extent which carries them above the eye's resolution threshold. By virtue of the objective the telescope performs two further functions: to reveal fainter objects, and to separate closer objects, than would be possible without it.

9.1 Astronomical refractor

For the purposes of demonstrating the action of a combined image-forming lens (objective) and examining lens (ocular), each may be taken as a simple convex lens. In fact the former is an achromatic combination, residually convex, and the latter usually a combination of two spaced lenses, also residually convex, although the single-lens ocular has its uses in astronomical observation.

Owing to the great distance of astronomical objects, the incident rays are sensibly parallel (Figure 46). Hence they are brought to a focus in the focal plane of the objective, F_{OG}. It is this real, inverted image which is the ocular's object.* The ocular may be regarded as a simple magnifying glass whose position on the optical axis is adjusted so that F_{OG} lies just inside its own focus F_{OC}. Since the eye is focused for infinity when its ciliary muscles are relaxed, it is likely that this adjustment will be unconsciously directed to approximating F_{OG} and F_{OC}. For when the two focal planes coincide, the emergent beam will be parallel and the final image, A, at infinity (Figure 47).

The ocular thus forms a virtual, inverted image which is examined by the eye; in effect the ocular permits the comfortable examination of

* Or it may be studied at the prime focus with the aid of a photographic plate.

153

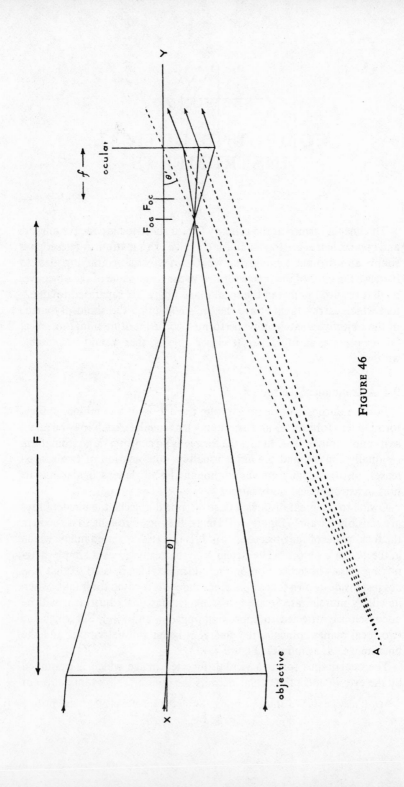

FIGURE 46

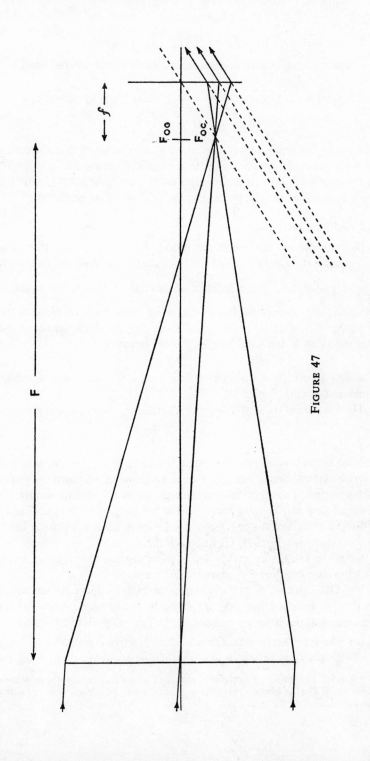

FIGURE 47

the primary image from a distance which is much smaller than the distance of distinct vision.

The final magnification, as we have seen, is given by

$$M = \frac{F}{f}$$

Secondary spectrum and the difficulty and expense of manufacturing strongly curved lenses set a limit to the useful focal ratio of about $f/10$ or $f/12$; astronomical refractors are usually built to about $f/15$, and for some types of work (e.g. planetary) even $f/20$ is to be preferred.

9.2 Galilean refractor

Differs from the modern astronomical refractor in using a diverging (negative) system as an ocular. The image is therefore virtual and erect. The magnification, $\frac{F}{f}$, is in practice seldom more than ×5. The ocular is adjusted along the optical axis so that it lies inside the focal plane of the objective by a distance about equal to its own focal length, i.e. to the algebraic sum of the focal lengths, f being negative:

$$d = F + f$$

The OG is usually a Clairaut achromatic, and the ocular a crown equiconcave. (See Figure 48.)

The distance of the virtual exit pupil in front of the eye lens is given by

$$a = \frac{df}{d-f} = \frac{d}{M}$$

It being impossible to place the eye at the exit pupil, the whole field can never be viewed simultaneously from a single position. Since the pupillary aperture is smaller than the emergent pencil the magnification is always below M'. The size of the field being, for given values of F and f a direct function of the aperture,* the Galilean telescope usually has a large relative aperture—$f/3$ is quite normal.

Since the ocular lies inside F_{OG} it is impossible to use cross wires, which would interfere with the observer's eye.

The Galilean system is chiefly employed today in opera glasses, where its compactness and lightness are valuable. It has been superseded for astronomical purposes on account of the following disadvantages:

(a) The full illuminating power of the objective is not used.

(b) Any defects in the objective—such as scratches or dust on the

---

* Whence, incidentally, a stop placed over the OG not only reduces the brightness of the image and the telescope's resolving power (as with an astronomical refractor), but also the size of the field.

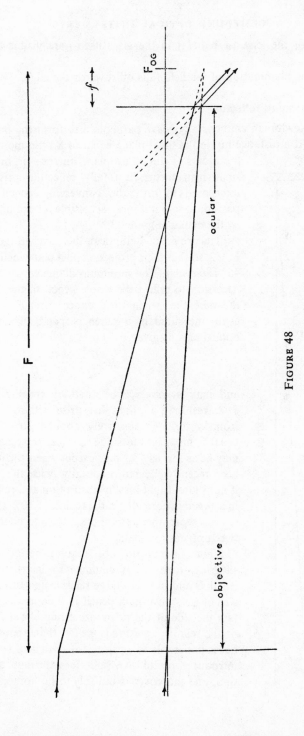

FIGURE 48

surface of the lens, or bubbles in the glass itself—are visible at the eyepiece.

(c) The illumination of the field falls off towards the edge.

9.3 Newtonian reflector

The Newtonian employs a concave paraboloid to form the primary image. At a distance inside the focal plane equal to a little more than the radius of the tube a plane mirror—or, in some small instruments, a totally reflecting prism—is interposed to turn the converging pencil to a position where it is more accessible for examination by the ocular (Figure 49).

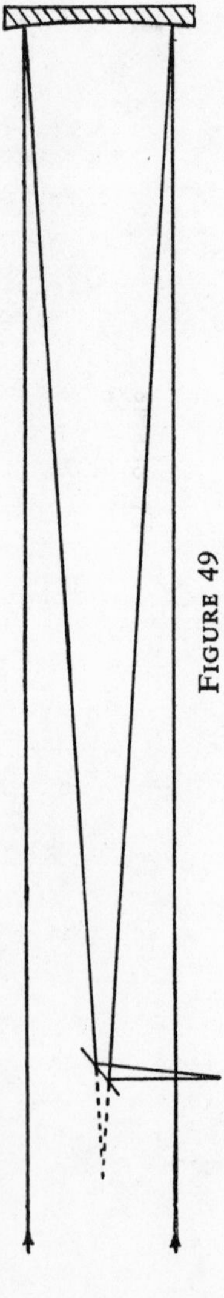

Flats are ground elliptical, their axes in the ratio $1:\sqrt{2}$, so that their projection on a plane inclined at $45°$ is circular. The minimum diameter of the flat necessary to reflect the whole pencil to the side of the tube (i.e. when its distance from F is $D/2$), again measured in a plane perpendicular to the optical axis, is given by

$$d = \frac{D^2}{2f} \quad \cdots \cdots \cdot (a)$$

and may in practice be anything from $D/10$ to $D/2$, resulting in light loss from silhouetting of from 1% to 25% respectively; about 4% to 6% is the usual figure. Light loss due to imperfect reflectivity may be as low as 7%, but, except when the flat has been recently silvered, is usually twice this figure, if not more. Light loss by absorption and reflection in a prism is normally in the region of 10% to 15%. On this score, therefore, there is little to choose between prisms and flats.

While prisms do not require periodical re-silvering, their use is confined to instruments of small D and not very large relative aperture. If the size of the converging pencil is such as to necessitate the side of the prism exceeding about 2·5 ins, or the relative aperture is greater than about $f/6$, a flat is preferable to a prism: (a) a prism of such size introduces prohibitive light loss through absorption, and moreover is unlikely to be homogeneous

FIGURE 49

158

unless very expensive; (*b*) the incident rays will not strike the prism face normally; (*c*) the prism introduces spherical over-correction, and some chromatic aberration, in proportion to the length of its side;* for a given prism it is proportional to the relative aperture. Hence a prism should not be used with an over-corrected mirror, though it may improve the performance of an under-corrected one.

The modifications of point and extended images caused by diffraction at the diagonal are discussed in section 2.2; they only attain importance when the diameter of the flat exceeds about $D/6$.

Inferior quality or faulty adjustment of the secondary reflector, especially if it is a prism, are among the commoner causes of disappointing performance by Newtonians. The reflecting surface should at no point depart from planeness by more than $0 \cdot 1\lambda$; nor should the flat be too thick, or the prism too massive, if trouble from differential cooling is to be avoided. The error introduced by a faulty flat is proportional to the linear error of the illuminated area of its surface; hence for a flat of uniform linear error performance will be improved by moving it near to the focus of the primary, where the diameter of the cone is smaller.

In general prisms, unless of the first quality, are likely to give worse performance than a flat, since three optical surfaces are involved instead of one. Small departures from the isosceles figure will prevent the entering and exit rays from being strictly normal to the surfaces, resulting in the distortion of star images into tiny spectra and the ruination of the definition of extended images. Inaccuracy of collimation can introduce abaxial aberrations—astigmatism, and even coma.

The aberrations of the Newtonian consist of coma ($B = +0 \cdot 25$), astigmatism ($A = +1 \cdot 0$), and field curvature ($C = -1 \cdot 0$). Except for such colour as is contributed by the ocular (and the eye) it is perfectly achromatic; and if correctly figured is free of spherical aberration with parallel incident light. The image at the centre of the field can therefore be as nearly perfect as makes no difference; the field is limited by coma, which sets a limit of about $f/3$ on the relative aperture, above which the field of tolerable definition is hopelessly restricted.

The linear distance from the axis at which coma limits the field, for various focal ratios, is easily derived. Given that $B = +0 \cdot 25$, then from (*e*), p. 75, we have

$$\xi = \tfrac{3}{4}\left(\frac{D}{2f}\right)^2 \theta$$

---

* Many oculars suffer from the opposite errors, and in such cases the aberrations introduced by ocular and diagonal will partially or completely counteract one another, with consequent improvement of definition.

When f is taken as unity, $\theta = d$ (see Figure 14), and

$$\xi = 3d\frac{D^2}{16f^2} \quad \cdot \quad \cdot \quad \cdot \quad \cdot \quad \cdot \quad \cdot \quad \cdot \quad (b)$$

The linear radius of the spurious disc, from equation (b), p. 38, is given by

$$r = 1 \cdot 22\lambda\frac{f}{D}$$

whence, taking $\lambda = 5600$Å,

$$r = (2 \cdot 69 \times 10^{-5})\frac{f}{D} \text{ ins} \quad \cdot \quad \cdot \quad \cdot \quad \cdot \quad \cdot \quad \cdot \quad (c)$$

The distortion of the image becomes prohibitive when ξ exceeds r by a factor of about 3. We may therefore write for the maximum permissible value of ξ

$$\xi'' = 3r$$

Hence from equation (c) above,

$$\xi'' = (8 \cdot 07 \times 10^{-5})\frac{f}{D}$$

Substituting this value in (b), we have

$$3d = (8 \cdot 07 \times 10^{-5})\frac{f}{D} \cdot \frac{16f^2}{D^2}$$

whence $\qquad\qquad d = 0 \cdot 00043\left(\frac{f}{D}\right)^3$

d being the distance from the axis, measured in inches, at which coma becomes intolerable. Tabulating d for various focal ratios, we have:

$\frac{f}{D}$	3	5	7	9	11	13	15
d (ins)	0·012	0·054	0·147	0·31	0·57	0·95	1·45

The particular *advantages* of the Newtonian may be summarised as:

(*a*) perfect achromatism;

(*b*) ease of construction;

(*c*) cheapness, compared with equivalent refractor;

(*d*) superior light grasp, compared with refractors, when D is large;

(e) comfort and convenience of a horizontal eye position (though this is dependent on the provision of a rotating tube or section);

(f) shorter tube than refractor of same D.

And its *disadvantages*:

(a) definition impaired by tube currents (see section 11.2);

(b) definition impaired by diffraction effects at central obstruction (see section 2.2);

(c) observing positions tend to become awkward with some types of mounting when the focal length exceeds 5 ft or so, even with rotating tube or upper section (without the latter, some observing positions may be almost impossible).

On balance, the disadvantages are completely eclipsed by the advantages.

9.4 Cassegrain reflector

A design of telescope whose peculiar advantages make its comparative neglect by amateurs a matter for surprise. As normally constructed, it consists of a paraboloidal primary mirror and a coaxially mounted hyperboloidal secondary, located inside the focal plane of the former. The focal ratio of the large mirror commonly lies between $f/4$ and $f/7$: these normal limits are dependent on, as regards the former, practical difficulties connected with the manufacture of the mirrors and their alignment in the instrument, and, as regards the latter, the impossibility of obtaining low enough magnifications since their fields of full illumination would be impossibly restricted.

With small Cassegrains focusing is normally effected by the movement of the secondary mirror, rather than of the ocular, along the optical axis; the focus is extremely sensitive to changes in the distance of the secondary from the focal plane, a small adjustment of the latter making a considerable difference to the former (Figure 50).

If we write:

f_1 for the focal length of the primary (PF),

f_2 for the focal length of the secondary,

s for the distance of the secondary inside the primary focus (ZF),

s' for the distance of the secondary from the final focal plane (ZF'),

then
$$\frac{1}{s'} = \frac{1}{s} + \frac{1}{f_2}$$

Values of these quantities are chosen so that a real image is formed at F', just behind the primary mirror; then $s' > s$. The equivalent focal

161

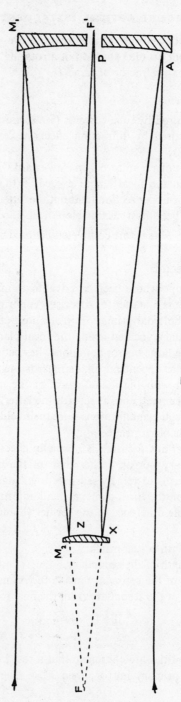

FIGURE 50

length of the combination (i.e. the focal length of a single objective giving the same image scale) is given by

$$\mathcal{J} = \frac{s'}{s} \cdot f_1$$

where s'/s is the magnifying power of the secondary mirror, or $\mathcal{J}/f_1$; in practice this is usually between 2 and 5.

The focal ratio of the primary being given by f_1/D_1, that of the combination is

$$\frac{\mathcal{J}}{D_1} = \frac{s'}{s} \cdot \frac{f_1}{D_1}$$

If the secondary is to be large enough to reflect the whole of the convergent pencil from the primary, the following condition must be satisfied:

$$\frac{D_2}{D_1} > \frac{s}{f_1}$$

The *advantages* of the Cassegrain can be summarised thus:

(*a*) It is the most compact design of visual telescope there is. This, a convenience at any time, particularly commends the Cassegrain to observers engaged on work where a large focal ratio or large linear focal length is desirable.

(*b*) The long equivalent focal length gives a large image-scale.

(*c*) Greatly reduced coma. It is sometimes stated that the Cassegrain is a coma-free system, but this is not so. The coma in the final image is as large as it would be with a single parabolic mirror of aperture D and focal length $\mathcal{J}$. The Cassegrain's freedom from coma, as compared with a Newtonian of the same aperture, derives solely from the fact that $\mathcal{J} > f$, where f is the focal length of a normal Newtonian; the extreme feebleness of the coma in a typical Cassegrain is a direct result of the comparative largeness of its focal ratio $\mathcal{J}/D_1$. The coma in any paraboloidal system (whether Newtonian or Cassegrain) is very weak at $f/15$, and virtually non-existent at $f/20$, over the restricted fields used in visual observation; but is negligible in neither at $f/8$. That the Cassegrain is not an aplanat, whilst being a near approach to one, is demonstrated by the fact that the scale of the image in the focal plane is a function of $\frac{XF' \cdot AF}{XF}$ which is not a constant for different positions of A, though it is very nearly so. To render a Cassegrain aplanatic the paraboloid would have to be deepened into a hyperbola, and the curvature of the secondary hyperboloid strengthened even further.

(*d*) Freedom from spherical aberration. Similarly, the Cassegrain's freedom (compared with the average Newtonian) from spherical aberration is dependent solely upon its greatly reduced relative aperture, and to no miraculous property in the combined action of the two mirrors. In fact the image formed by M_1 is as free from spherical aberration as that of a similarly figured Newtonian paraboloid: to keep the pencil stigmatic after reflection by M_2 it is necessary that the increase in the length of the optical path, due to this reflection, should be a constant for all rays, whether paraxial or peripheral. The curve of M_2 which will maintain $XF' - XF$ a constant is a hyperbola with foci at F and F'—which is why a hyperboloid is chosen for the secondary mirror.

Owing to its comparative freedom from spherical aberration, oculars which are not highly corrected, and which are therefore useless with Newtonians of large relative aperture, perform as well with a Cassegrain as with a refractor. In this respect the Cassegrain combines the advantages of refractor and Newtonian reflector.

Disadvantages of the Cassegrain:

(*a*) Contrast in the image is apt to be seriously diminished by direct light from the sky falling on it. This, however, can be completely obviated by enclosing the convergent pencil from the secondary mirror in a tube mounted in the aperture of the primary mirror, concentric with the optical axis. When the tube is correctly placed, an eye situated on the optical axis in the focal plane should see nothing but the small mirror and the circular reflection of the primary in it. Thus all light not contributing to the formation of the image is excluded. Alternatively, and more commonly in small instruments, an ocular diaphragm exactly limiting the exit pencil may be employed, though it is rather tricky to locate with precision.

(*b*) Small field, the corollary of large image-scale. This should be taken into account when considering the type of instrument best suited to a particular branch of observation—it represents no drawback in planetary work, for example.

(*c*) Astigmatism and field curvature are more marked than with an equivalent Newtonian, though this is to some extent mitigated by the virtual freedom from coma.

(*d*) Unless precautions are taken to screen the observer from the instrument, the effect of this bodily warmth on the primary mirror will impair definition. This is easily done, however.

(*e*) The hyperbolic secondary is comparatively difficult to figure, a fact which in the past has tended to restrict the Cassegrain design to fairly large instruments.

(*f*) The size of the secondary (usually $D/3$ to $D/2$) introduces diffraction effects (see section 2.2) which tend to impair the definition of critical detail in extended images, and under certain circumstances to interfere with the resolution of close point images.

(*g*) Inconvenient observing positions at high altitudes, involving the use of a diagonal.

9.5 Cassegrain-Newtonian

A modification of the Newtonian, embodying some of the features of the Cassegrain, that is worthy of consideration. Like the Cassegrain it employs a coaxial hyperboloid mounted just inside the primary focus; but instead of the converging pencil passing through the centre of the primary, it is reflected to the side of the tube by a diagonal mounted a short distance from the centre of the paraboloid (Figure 51).

Advantages:

(*a*) Proximity of the ocular to the Dec axis provides a nearly stationary observing position; it is also nearer to the ground than with the orthodox Newtonian.

(*b*) By increasing the effective focal length it permits of higher magnifications than would be possible without the hyperboloid.

(*c*) It avoids the Cassegrain trouble of sky-light falling on the image.

Disadvantages:

(*a*) Reversal of the image.

(*b*) Smaller field of full illumination than with a normal Cassegrain.

(*c*) Small light loss at the additional reflecting surface.

9.6 Combined Newtonian and Cassegrain

Instead of some of the features of each arrangement being combined, as in the last type described, the two arrangements may be incorporated in a single mounting for alternative use. The Cassegrain hyperboloid is interchangeable with a Newtonian flat, and the paraboloid is bored to give passage to the rays converging to the Cassegrain focus. Thus a single speculum can be used at two different focal ratios.

The Newtonian focus would be used when low magnifications and wide fields are required, the Cassegrain when large image-scale is more important.

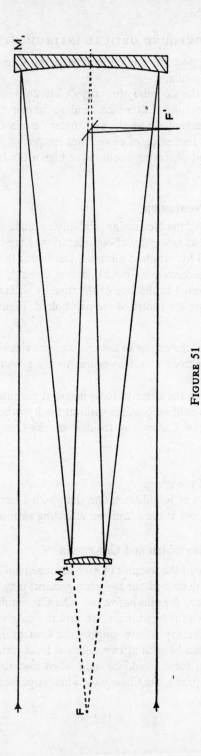

FIGURE 51

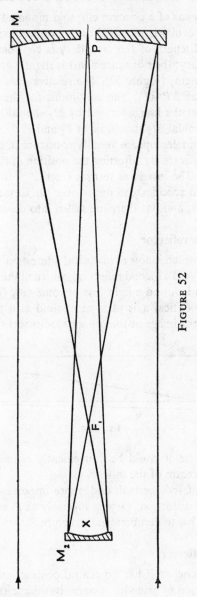

FIGURE 52

9.7 Gregorian reflector

The objective is a concave paraboloid, as in the Newtonian, but instead of a secondary flat or hyperboloid, the convergent pencil is brought

167

to a focus by means of a concave ellipsoid mounted coaxially with the paraboloid; the ocular is mounted on the optical axis, as in the Cassegrain. The focal length of the secondary is considerably shorter than that of the primary; their distance apart is slightly greater than the sum of their focal lengths (Figure 52). The relative sizes of the images at P and F_1 will be as $XP:XF_1$. The conditions for the final image to be stigmatic are that the image formed by M_1 should be so, and that the foci of the ellipsoidal M_2 should lie at F_1 and P.

The Gregorian telescope, as usually constructed, had a fixed ocular, focusing being effected by adjusting the position of the secondary along the optical axis. The image, as seen, is erect.

It possesses no specific advantages over the Cassegrain, while being some 30% longer, and has therefore fallen into disuse.

9.8 Herschelian reflector

Another type which is now of historical interest only. It was developed by Herschel to avoid a second reflection, which in the days before silver-on-glass mirrors involved a light loss of some 40% (Figure 53).

By tilting the optical axis of a paraboloid at a small angle to the incident pencil, the image at the primary focus was directed to the side

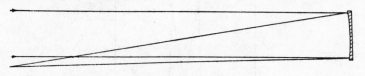

FIGURE 53

of the tube, where it could be conveniently examined by an ocular pointing to the centre of the mirror.

The image suffered reversal, and, more important, was affected by astigmatism and distortion, varying inversely as F. Avoidance of these errors therefore led to cumbersome structure.

9.9 Off-axis reflector

Like the Herschelian it has no central obstruction, with consequent improved definition of extended images. But the mirror being a section of a paraboloid mounted with its axis parallel to the incident pencil, the astigmatism and distortion of the Herschelian are avoided. The practical disadvantage of the system is the greatly increased difficulty of figuring the off-axis section of a paraboloid (Figure 54).

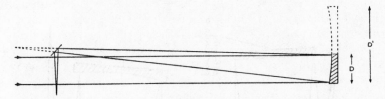

FIGURE 54

From the viewpoint of diffraction effects in the image, exactly the same result would be achieved by the use of an eccentric stop of diameter D, over a speculum of diameter D'.

A development of the off-axis design (Figure 55; B. 116) offers the the additional advantage of small tube-length/focal-length ratio.

9.10 Herschel-Cassegrain reflector

A natural development of the Herschelian reflector, which like it has been long neglected; recently, however, Pressman has constructed one and has stressed its advantages (B. 112), particularly for planetary work. It consists (Figure 56) of a concave spherical primary and an approximately oblate spheroid for a secondary.

Advantages:

(*a*) Elimination of diffraction effects arising from central obstruction (cf. Newtonian or Cassegrain).

(*b*) Increase of light grasp compared with Newtonian or Cassegrain of the same aperture.

(*c*) Avoidance of distortion due to tilting of the Herschelian mirror.

(*d*) Avoidance of trouble from thermal currents originating from the observer and passing directly across the light path to the mirror.

(*e*) Possibility of accommodating a system of large equivalent focal length in a comparatively short tube (e.g. a $f/70$ $8\frac{1}{2}$-in reflector would have a tube length of little more than 6 ft).

(*f*) Comfort attendant upon the use of relatively long-focus oculars, while still providing high magnifications, which results from focal ratios of $f/40$ to $f/70$.

Disadvantages:

(*a*) Operation is more complex, and adjustments more sensitive, than with the more orthodox designs.

(*b*) Small fields, resulting from the recommended large focal ratios; this, however, is of no importance in many fields of visual work.

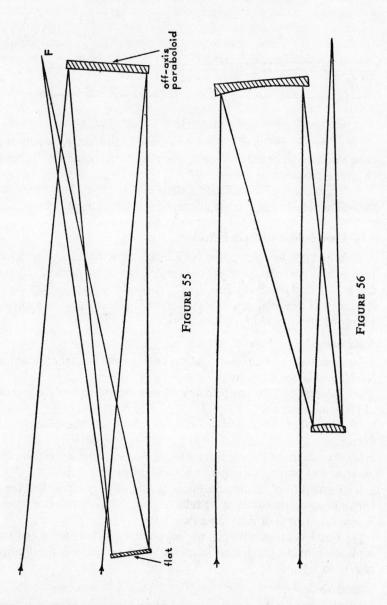

FIGURE 55

FIGURE 56

9.11 Aplanatic telescopes

We have seen that the single curved reflecting surface of the New-tonian cannot be rid of coma whilst at the same time escaping spherical aberration; but that by increasing the focal ratio, as in the Cassegrain, it can be reduced to negligible proportions over a restricted field such as is adequate for visual observation. But increasing the focal ratio by increasing F means, with paraboloids, restricted fields; on the other hand, achieving it by decreasing D means reduced resolving power; and increasing it by any method whatever means reduced photographic speed— if it is greater than about $f/8$, prohibitively long exposures are needed for faint objects. These conflicting demands have led to the development, in fairly recent years, of various two-mirror systems and lens-and-mirror systems which combine large relative aperture (and therefore photographic speed) with freedom from spherical aberration, whilst obeying the sine condition giving freedom from coma over wide fields.

Of the remaining aberrations, the effects of field curvature are overcome by distorting the photographic plate to the same curve as that of the focal surface; while the sting is taken out of astigmatism when it is pure, and not combined with coma, since the images are then symmetrical and uniformly illuminated and are therefore as accurately measurable as perfectly circular images. (See further B. 106.)

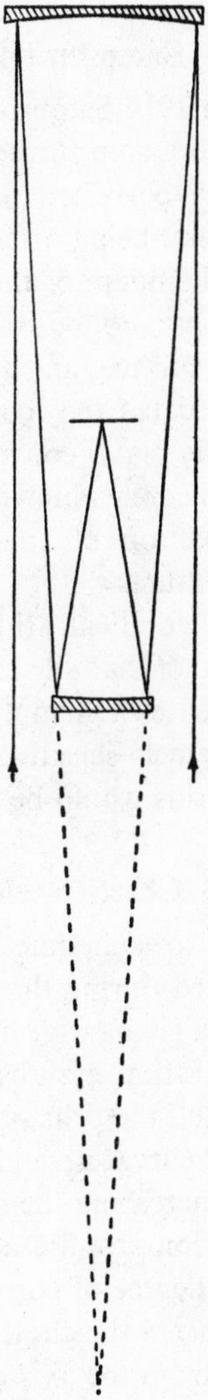

FIGURE 57

9.12 Schwartzschild reflector

Invented as long ago as 1905, the Schwartzschild is a two-mirror aplanat with aspherical mirrors, giving freedom from coma and spherical aberration over a wide, uncurved field (3° diameter at $f/3\cdot5$, 1°5 at $f/1\cdot2$). Astigmatism is retained.

Hence it combines the photographic desiderata of symmetrical images over a wide field, and extremely short exposures. The focal plane (Figure 57) is located between the two mirrors and nearer to the secondary; the latter diminishes, not magnifies, the primary image, its only function being to correct the coma of the large-D/f primary.

The focus is therefore inaccessible for visual inspection, which is

171

disadvantageous since necessitating a separate guide telescope. The Schwartzschild may therefore be considered as having crossed the boundary, ill-defined though it is, between telescopes and cameras.

The heavy light loss through silhouetting (about 25%) is more than compensated for by the phenomenally large relative aperture.

9.13 Sampson telescope

In 1914 Sampson proposed a system (B. 115) which was an attempt to go a step further than Schwartzschild in the development of a perfectly corrected telescope. Sampson's telescope is a lens-and-mirror system, being essentially a Cassegrain with added correcting lenses.

The figure of the concave primary is intermediate between spherical and paraboloid; the secondary is a convexo-concave lens silvered on its rear surface; and about two-thirds of the distance from it to the primary is situated the 'corrector', consisting of a double concave lens and a nearly plano-convex lens mounted coaxially with the mirrors. This corrects coma while introducing negligible chromatic aberration; complete correction of spherical aberration is obtained with negligible field curvature.

Theoretically this is a near approach to an aberration-free system, and the fact that all surfaces except that of the primary are spherical makes for ease of manufacture. The correction of the telescope, however, is extremely sensitive to the centring of the correcting lenses, and in practice this would be the chief drawback of the design.

9.14 Chrétien telescope

Also sometimes referred to as the Ritchey-Chrétien telescope; invented during the 1920's, it is nothing more than an aplanatic Cassegrain giving wide fields at $f/6$ to $f/8$, from which both coma and spherical aberration are absent. The focus, as in the Cassegrain, is carried just behind the primary mirror, where it is visually accessible. Reduction of field curvature and astigmatism are gained at the expense of light grasp by increasing the size of the secondary mirror; as usually constructed, the compromise arrived at is $D_1/D_2 = 2$ approximately, which gives a fair degree of correction without intolerable light loss.

At $f/8$ the circle of confusion of the astigmatic image of a star lying 30' from the axis of a $31\frac{1}{2}$-in Chrétien is about 0·0021 ins in diameter. That this is nearly seven times as large as the spurious disc is immaterial so long as the instrument is not used visually, since diffusion in the photographic emulsion will prevent the resolution of two stellar images whose linear separation in the focal plane is less than 0·0020 ins. Further than 45′ from the optical axis, the image of the Chrétien is actually larger than that of the equivalent Newtonian would be; but whereas the latter is comatic (therefore asymmetrical as regards both shape and intensity), the former is symmetrical and uniformly illuminated.

The particular advantages of the Chrétien are its extreme compactness (tube length only about $J/2$) and the accessibility of the focal plane.

9.15 Couder reflector

Generally similar to the Schwartzschild, the Couder not only obeys the sine condition but is anastigmatic as well; freedom from astigmatism, however, is bought at the price of reintroduced field curvature, the radius of the focal surface being about $J/2$. Like the Schwartzschild, and unlike the Chrétien, the secondary is concave; it is, in fact, an ellipsoid. The foci of the two mirrors are brought together, so that $J=\dfrac{s}{2}$, where s is the separation of the two mirrors. It is thus a less compact instrument than the Chrétien, but since it works well at $f/3$ this is no great inconvenience. It does, however, share the more real weakness of the Schwartzschild of having a visually inaccessible focal plane.

9.16 Schmidt camera

The spherical mirror, whilst being free from the abaxial aberrations (of which the most important is coma), and therefore allowing the use of wide fields, nevertheless suffers from spherical aberration uniformly over the whole field. One method of correcting this is to deepen the curve into a paraboloid; but abaxial rays are then no longer axes of revolution and the coma thus introduced limits the field progressively as the relative aperture is increased. Schmidt's brilliant contribution to telescope design (B. 117), made in 1930, consisted in indicating how the spherical aberration of a concave spheroid could be corrected without the introduction of coma in exchange.

The result is an optical system giving good definition over phenomenally wide fields, and at relative apertures providing photographic speeds never before envisaged. For example, a $f/1$ paraboloid would have a well-defined field measurable in ″ arc, while the coma-free field of the $f/5$ 100-in reflector is only 7′·2, and that of the 200-in about 3′·6; yet a $f/1$ Schmidt can cover something like 20°, recording as much detail of extended objects in a few minutes as a first-class portrait lens would require several hours to show.

173

A concave spheroid possesses the following attributes:

(a) Spherical aberration, which can be reduced to below the Rayleigh limit of $\frac{\lambda}{4}$ by placing a diaphragm of aperture $\not> f/10$ on the optical axis, at the centre of curvature; for its restriction of the diameters of the pencils will mean that the central and peripheral beams are close enough together to be brought to virtually the same focus. This, however, is an impracticable solution, since at $f/10$ insufficient aperture is obtainable unless f is very large.

(b) Field curvature: the focal surface being a section of a sphere concentric with the mirror and of radius equal to $f = R/2$ (the arc FF' in Figure 58).

(c) Freedom from coma by satisfaction of the sine condition, and similarly anastigmatic and undistorted. Since the central ray of any pencil, no matter what its inclination to the axis of the mirror, has to pass through C, it is as much an axis of rotation of the system as AS; i.e. BT could equally well be considered 'the' axis, by tilting the diaphragm until perpendicular to BT. Optically, owing to the sphericity of the mirror and the smallness of the diaphragm, the two pencils centred on AS and BT are indistinguishable.

In the Schmidt camera (Figure 59): the advantages (c) are retained by employing a spherical mirror, M.

(b) is retained, whereby the photographic plate has to be warped by pressure to the correct curvature ($r = f$). The curvature of the focal surface can be practically eliminated, for relative apertures less than about $f/5$, by means of a plano-convex in contact with the photographic plate.

(a) is removed by means of a thin, aspherical, almost plane parallel correcting plate which replaces the diaphragm of the above account, i.e. is situated at the centre of curvature, C (the only point through which all rays pass)*; this introduces into the beam incident to the mirror spherical aberration equal to that of the mirror but of opposite sign. The design of the plate varies,† giving more or less chromatic aberration, but fundamentally it is equivalent to an axial convex lens surrounded by a ring of concave lenses; when its central and peripheral thicknesses are made equal, its power is zero. Being thin, it causes negligible light loss and negligible colour error down to about $f/1$; beyond this the elimination of chromatic aberration requires rather

* As the plate is moved away from the centre of curvature, the diameter of the corrected field decreases. When a very wide field is not required, this is a convenience which is made use of in some designs of Schmidt.

† Its curvature may be on the outer or inner surface, or distributed between them.

impracticably steep curves for the correcting plate, and it is preferable to make it in the form of an achromatic doublet, or to resort to the thick mirror or solid Schmidt modifications of the basic design. The correcting

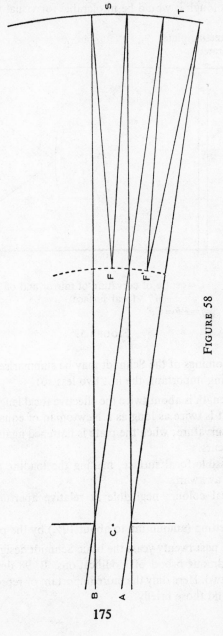

FIGURE 58

plate can provide theoretically perfect correction for only a single wave-length, μ varying with λ, but the residual chromatic inequality of spherical aberration is comparatively unimportant in a photographic instrument, though it would be intolerable for visual work.

Figure of correcting plate
(much exaggerated):

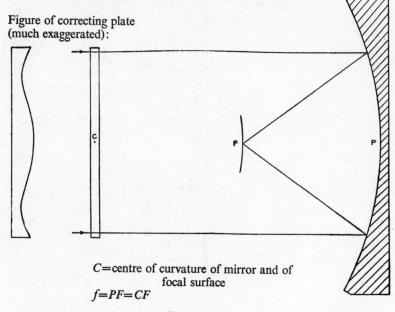

C=centre of curvature of mirror and of
focal surface
$f=PF=CF$

FIGURE 59

The shortcomings of the Schmidt may be summarised as follows, the first three being important, the last two less so:

(*a*) Tube length is about twice the effective focal length of the system, i.e. a Schmidt is twice as long as a Newtonian of equal f.

(*b*) Field curvature: when the plate is flattened again after exposure, distortion occurs.

(*c*) Inaccessible focal surface, making the loading and changing of plates rather awkward.

(*d*) Residual colour: negligible at relative apertures smaller than about $f/1$.

(*e*) Silhouetting (amounting to about 10%) by the plate holder.

During the past twenty years the basic Schmidt design has given birth to a multitudinous brood of modifications. B. 98 describes no fewer than twenty-two. Here only the more important or representative can be mentioned, and those briefly.

176

9.17 Thick-mirror Schmidt

Hendrix pointed out (e.g. B. 97) that the limit of about $f/1$ set to the ordinary Schmidt by the impracticability of correcting its chromatic aberration could be bypassed by utilising the extra refraction that would occur at the surface of a thick mirror, silvered on its rear surface (Figure 60).

It can be seen that the focal surface now lies at the surface of the

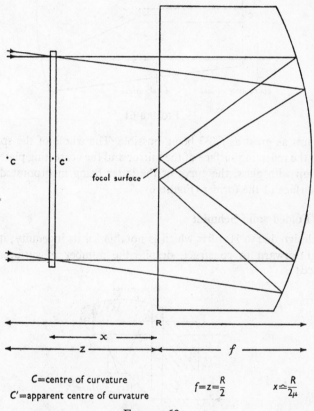

focal surface

$$C = \text{centre of curvature}$$
$$C' = \text{apparent centre of curvature}$$
$$f = z = \frac{R}{2} \qquad x \simeq \frac{R}{2\mu}$$

FIGURE 60

mirror, and the correcting plate at the apparent (not the geometrical) centre of curvature of the mirror.

The photographic speed of the system is increased by a factor of μ^2 (i.e. $2\frac{1}{2}$ to 3 times) by this means, thus in effect increasing the relative aperture: the speed of a $f/0 \cdot 6$ camera, for example, can be obtained with the correcting plate curves and the field curvature of a $f/1$ camera.

9.18 Solid Schmidt

The logical limit of the development from the basic to the thick-mirror Schmidt; it carries the same advantages as the latter, relative

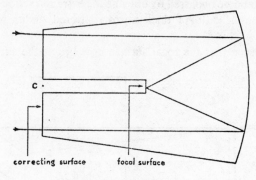

correcting surface   focal surface

FIGURE 61

apertures as great as $f/0·3$ being possible. The whole of the space between the reflecting surface of the mirror and the correcting plate is now filled up with glass, the curve of the latter being incorporated in the first surface of the former (Figure 61).

9.19 Folded solid Schmidt

A design due to Hendrix which is notable for its ingenuity; it is also straightforward to construct, despite the number of surfaces to be worked:

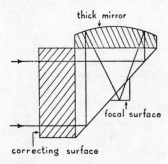

thick mirror

focal surface

correcting surface

FIGURE 62

9.20 Off-axis Schmidt

By scrapping rather more than half of the mirror and correcting plate (as in the off-axis reflecting telescope) it is possible to make the focus accessible (Figure 63).

178

Unless two or more cameras are being made at the same time, this design involves a great wastage of manufacturing time.

9.21 All-reflection Schmidt

By reversing the curve of the correcting plate and silvering it, a completely achromatic Schmidt is obtained (Figure 64). The relative

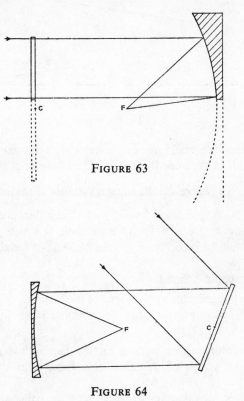

FIGURE 63

FIGURE 64

apertures obtainable with this design are limited by geometrical considerations.

9.22 Wright camera

Wright has shown (B. 131, 132) that two important advantages follow from the replacement of Schmidt's spherical mirror by an ellipsoid, the appropriate modifications being made to the correcting plate: the ratio tube-length/focal-length would be reduced from 2 to 1, and the field would be flattened (Figure 65). The focal plane now lies close to the correcting plate, immediately behind which the photographic plate is mounted.

The price paid for these advantages is the introduction of astigmatism,

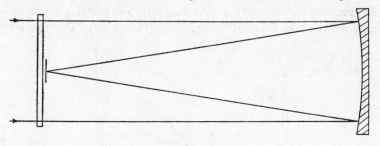

FIGURE 65

which limits the usable field; the colour error is also about twice that of a basic Schmidt of the same focal ratio. The diameter of the astigmatic star image $\propto \dfrac{D.d^2}{f}$ (where d is the angular distance of the image from the centre of the field), and amounts to about 1″ at a distance of 0°5 from the field centre at $f/4$. The practicable limit is about $f/3$. At $f/4$, however, it could be used visually in the Newtonian form.

9.23 Spectroscopic Schmidts

Among the innumerable applications of the Schmidt principle to the wants of the spectrographer may be mentioned the following (see further B. 94–5, 98, 122):

(*a*) The correcting figure worked on the face of the prism nearer the mirror, the centre of this face coinciding with the centre of curvature.

(*b*) The correcting figure incorporated with the collimating lens.

(*c*) Grating feeding a parallel pencil to the correcting plate of a normal Schmidt.

(*d*) Grating engraved on correcting mirror.

(*e*) Combined transmission grating and correcting plate, giving direct vision (Figure 66).

9.24 Maksutov telescope

In 1944 Maksutov described a family of aplanats in which the aspherical correcting plate of the Schmidt is replaced by a weakly

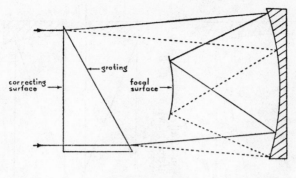

FIGURE 66

negative concave meniscus. The meniscus, of nearly constant thickness, is sensibly achromatic (1/1500, or less, the colour error of an equivalent achromatic OG), but contributes the positive spherical aberration required to neutralise the negative aberration of the mirror.

Advantages:

(*a*) Aplanatic (the suppression of coma is dependent upon the positioning of the meniscus).

(*b*) No aspherical surfaces (hence ease of manufacture).

(*c*) Negligible chromatic error.

(*d*) Shorter tube than the basic Schmidt, focal length for focal length.

(*e*) Can be used either visually (at about $f/8$) or photographically (to $f/3$ or $f/2$).

Its chief *disadvantage* is its curved focal surface (cf. the Wright camera).

Four variants are illustrated in Figure 67.

9.25 Schmidt-Maksutov, or Meniscus-Schmidt

A meniscus which is exactly concentric with the mirror introduces some primary colour, as well as residual spherical aberration. Its errors are, however, virtually uniform over the whole field. Linfoot and Hawkins have shown (B. 103) that by combining such a concentric meniscus-mirror system with an aspherical correcting plate at the centre of curvature, a 18° field at $f/1\cdot3$, at all points in which 97% of the light of each image is sent into a circle of confusion of smaller diameter than 12″, may be obtained. This is a considerable improvement on the performance of a conventional Schmidt of the same focal ratio. Not only

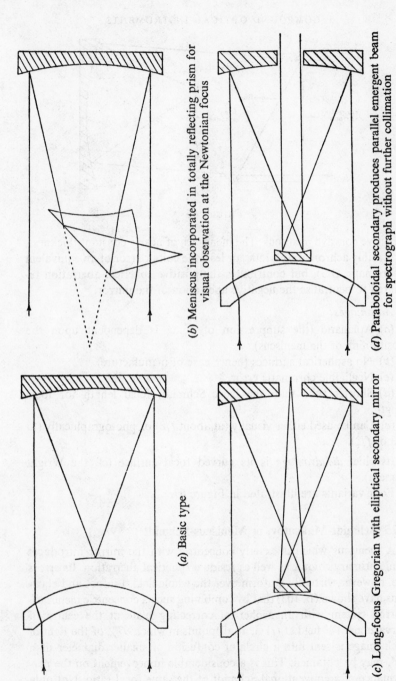

(a) Basic type

(b) Meniscus incorporated in totally reflecting prism for visual observation at the Newtonian focus

(c) Long-focus Gregorian with elliptical secondary mirror

(d) Paraboloidal secondary produces parallel emergent beam for spectrograph without further collimation

FIGURE 67

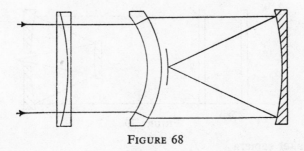

FIGURE 68

does the correcting plate (which is weaker, and therefore easier to manufacture, than that of the Schmidt) eliminate the residual spherical aberration, but if made as a doublet with a spherical interface it also corrects the primary colour of the concentric meniscus (Figure 68).

9.26 Schmidt-Cassegrain

Instruments of this type are in general more compact than the basic Schmidt, and they have the important advantage of an accessible

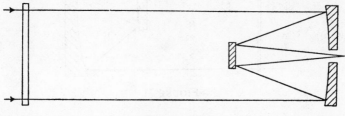

FIGURE 69

image; those which have the further advantage of a flat field are also more difficult to make (Figure 69).

With spherical mirrors concentric at C the focal surface is curved, and it can only be flattened at the cost of introducing coma. If, however, they are made slightly and equally aspherical, the radius of curvature of the secondary being increased till equal to that of the primary, a virtually flat-fielded aplanat is the result.

A further development of the Schmidt-Cassegrain, due to Linfoot, is a family of quadruple systems (two spherical mirrors with two aspherical correcting plates) which combine flat field and anastigmatism with apochromatism: this is made possible by arranging that the aspherical plates have compensating colour errors (Figure 70).

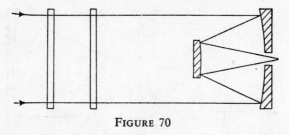

FIGURE 70

9.27 Baker camera

A family of two-mirror systems described by Baker (B. 78), with remarkable performances: aplanatic, anastigmatic, flat-fielded, and free from distortion (Figure 71).

With two spherical mirrors astigmatism is present but negligible; the tube is only three-quarters the length of a Schmidt of the same focal

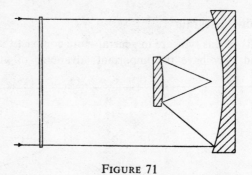

FIGURE 71

length; in one variant the tube length is reduced to $\frac{1}{3}f$. Figuring one mirror gives complete anastigmatism; figuring both gives anastigmatism and freedom from distortion. The colour error, however, is 3 to 4 times that of the Schmidt; on the other hand it is still very small, even at $f/2$.

184

SECTION 10

THE MOUNTING OF THE OPTICAL PARTS

10.1 Introduction

Object glasses, specula, and secondary reflectors should be mounted so that they are

(*i*) adequately supported to combat flexure;

(*ii*) held firmly enough for there to be no movement when the orientation of the telescope is changed in any possible way;

(*iii*) not so tightly gripped that the glass is subjected to mechanical strain;

(*iv*) adjustable in every respect required for collimation, centring, and squaring-on;

(*v*) as free as possible from the effects of differential cooling, dew deposit, and tube currents.

Two accessories of the objective cell may conveniently be mentioned at this point—the protective cover and, in the case of OGs, the dewcap —though there is little to say about them except that the objective must always be kept covered when not in use, but not be covered when dewed-up, and that the dewcaps usually supplied with small refractors are much too short to be any use.

10.2 Speculum cells

While being strong enough to give proper support to the speculum, the cell must not be so massive that it delays unduly the attainment of thermal equilibrium between the speculum and the night air. It must, further, provide uniform support for all parts of the speculum, limit the latter's possible movement to a few thousandths of an inch, yet not subject the glass to such constraint that thermal expansion and contraction set up strains in it. And this for all positions of the tube.

185

The detailed construction of cells for specula varies widely, but certain general principles can be stated. For mirrors of $D<8$ ins approximately, a three-point support is adequate (Figure 72). The supporting screws should lie on a circle, concentric with the mirror, of radius $D/3$. Each pair (lateral and supporting) of screws lies on a single radius of the cell, the three radii being separated by 120°. Round-ended brass screws should be used. Between the retaining screws and the rim of the mirror there should be a clearance of a few thousandths of an inch. There should also be a good clearance between the front lip of the mirror and the safety ring, if fitted.

For apertures of from 8 ins to 12 or 14 ins, the three-point support will cause some degree of image-distortion owing to the glass sagging between the points of support. The inner rear surface of the cell may be machined flat and covered with several thicknesses of blotting paper or flannel in the form of a ring with a central hole of diameter about $D/4$. Better still is some form of multi-point balanced support. Two such systems are illustrated in Figure 73, of which the second gives a better distribution of support. Nine points of support are here provided, each pressure-distributing plate being balanced at its centre of gravity on a ball bearing or, more simply, on a round-headed bolt locked to the cell floor. These points of balance should be located at $D/3$ from the centre of the mirror; the three support points on each plate may be small pads of washleather; the plates should be not less than $D/20$ thick.

For mirrors larger than about 15 ins the number of plates can be increased; individual plates of area about 1 square inch or so can then be covered with sheet lead to provide the bearing surfaces for the mirror.

Any inequality in the pressure exerted by the peripheral supports is liable to introduce astigmatism; insufficient central support, allowing the axial region of the mirror to sag, will cause spherical over-correction. Flexure with a three-point support will first be detected in the image by a slight brightening of the rings in three directions equidistant from one another; if gross, it can produce triangular star images.

The disadvantages of these orthodox types of speculum mounting in a solid cell will be referred to in section 10.4.

10.3 Mounting the speculum cell

The manner in which the cell is attached to the telescope tube depends very much upon the design and shape of the latter. Generally speaking, there must be three points of contact; at least two of these must be adjustable, the third having a little play to allow this adjustment to be made.

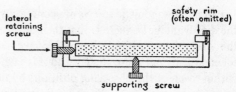

lateral retaining screw

safety rim (often omitted)

supporting screw

FIGURE 72

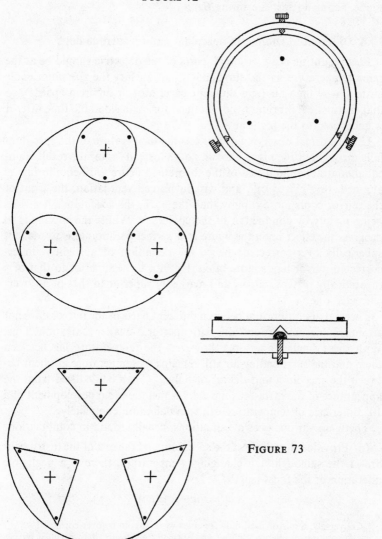

FIGURE 73

Pairs of antagonistic screws are commonly employed (Figure 74), but the disadvantage of these is that their adjustment cannot be made from the observing position. A more convenient arrangement is shown schematically in Figure 75. The rod A is carried up the telescope tube to within reach of the observing position; by turning it the cell flange can be pushed further from, or pulled nearer to, the upper end of the telescope, acting against the spring S.

10.4 Differential cooling of speculum and its surroundings

Ideally, during observation all parts of the objective should be at the same temperature as the surrounding air. In fact the speculum cools more slowly than air (and different parts of it at different speeds), so that thermal equilibrium is not attained for a considerable time after it is uncovered to the night air.

Under normal observing conditions the temperature of the speculum falls throughout the night, though not at a steady rate; the result is unequal contraction and loss of the theoretical figure until equilibrium is attained. In a closed cell, and without forced ventilation, the front of the mirror cools more rapidly than the back, the speculum's heat loss being mainly by conduction to the cooler air. Whilst the cooling is in progress the effect upon the figure of a perfect paraboloid is to make it spherically over-corrected; hence the definition of a slightly under-corrected mirror tends more often than not to be superior to that of a theoretically perfect mirror, and markedly superior to that of an over-corrected one.*

Thermal disequilibrium between different parts of the mirror does not attain troublesome proportions with apertures smaller than about 8 ins. With these smaller apertures, however, the temperature lag between them and the surrounding air still remains. A temperature gradient between the speculum and the air of only 2°C can be deduced from the appearance of the extra-focal image,† whilst the rapid development of a 10° difference of temperature can ruin definition completely.

There are various possible situations, usually found in combination:

(a) Provided the whole of the speculum and the rest of the instrument are at the same (though changing) temperature, there is a negligible alteration of the focal length:

$$\Delta f = F(\alpha - \alpha')\Delta t$$

---

* Conversely, a mirror used only for solar work should be over-corrected.

† Temperature gradients such as are normally encountered in practice have a negligible effect upon the focused image.

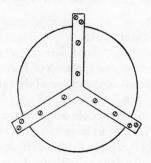

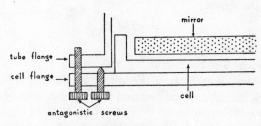

FIGURE 74

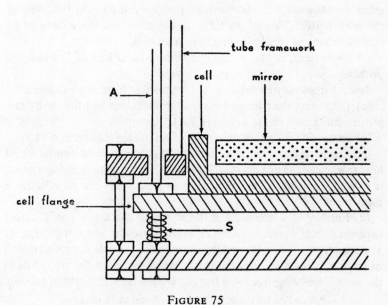

FIGURE 75

where F =focal length,

Δf =increase in focal length,

Δt =decrease in temperature (°C),

α =linear coefficient of expansion of mirror (glass: $7\cdot5\times10^{-6}$),

α' =linear coefficient of expansion of the tube (steel: $11\cdot1\times10^{-6}$; wood: negligible).

(b) Where there is a perceptible temperature gradient within the mirror (as there invariably is in large mirrors, and usually is in small ones). The difference between the front and back of the mirror is commonly several tenths of a °C. Assuming a linear gradient across the thickness of the mirror,

$$\Delta f = -2F^2\frac{\alpha}{e}\Delta t$$

where F, Δf, Δt are as before,

e =thickness of the mirror.

(c) When the two surfaces of the mirror are cooling rapidly and at unequal rates, an effect of astigmatism is produced. This is characteristically a trouble experienced by mirrors in exposed positions (e.g. coelostat). Forced ventilation of the rear surface is the cure.

(d) When the two surfaces are unequally warm across a diameter of the mirror, an effect of spherical aberration is produced: overcorrection when the temperature is falling. The magnitude of the effect depends upon the heat capacity of the cell and the thickness of the mirror. Reduction of the latter to the minimum value dictated by considerations of rigidity is therefore indicated.

(e) Short-term, erratic, and unpredictable variations of F, which are probably due, at any rate in part, to wind.

Several steps may be taken to reduce trouble from these causes:

(a) Increasing the contact between the speculum and the air by employing an 'open' cell (e.g. Figure 76*).

(b) Increasing it further by means of a fan (see also section 11.2).

(c) Employing a speculum material which is a good conductor of heat (e.g. metal, as against glass), so that the contraction, though great, is quickly completed. This is neither so practicable nor so effective a measure as:

(d) Employing a speculum material which has a low coefficient of expansion (e.g. Pyrex, 0·2 that of average crown; silica, 0·05 that of average crown). These materials further make possible the construction of large mirrors with sufficient thickness to overcome flexure, while at the same time being free from trouble arising from differential cooling.

* Based on Dr Haughton's 20-cm reflector at Teddington.

Refractors are comparatively little troubled by temperature differences between the OG and the night air: (*a*) they are of larger focal ratio

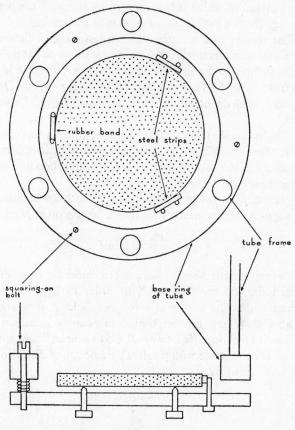

FIGURE 76

than the usual Newtonian; (*b*) the effects of contraction in a compound lens tend to cancel one another, which is obviously impossible where a single reflecting surface is involved; (*c*) an object glass being relatively much thinner than a speculum, thermal equilibrium is reached more rapidly. (See further, section 13.1.)

10.5 Flexure

Flexure in object glasses presents no problem with instruments of the size likely to be found in amateur hands. With mirrors, however, it is otherwise, and the manner in which they are supported in the telescope requires careful attention.

The amount of flexure in an objective is a function of four variables: (*i*) the material of which it is made; (*ii*) its size; (*iii*) the manner of its mounting; (*iv*) the inclination of its axis to the vertical.

(*i*) The rigidity of a material—i.e. its resistance to flexure under its own weight—is a function of d/E, where d is its density and E is Young's modulus (related to the elasticity of the material; its value in dynes/cm^2 varies from 4·9 to 5·9 for flint glass, 5·9 to 7·8 for crown). Optical glass, silica, and steel do not differ from one another significantly in this respect.

(*ii*) Neglecting the other factors, the flexibility of a mirror is a function of r^4/e^2, where $r = D/2$, and $e =$ thickness. Hence trouble from this cause varies as D^2 in the case of mirrors whose proportionate apertures and thicknesses are the same.

(*iii*) It has been found empirically that with an unbalanced three-point support, deformation of the image first becomes perceptible when

$$\frac{r^4}{e^2} \fallingdotseq 160 \text{ ins}^2$$

If the value exceeds about 160, a more elaborate form of overall or balanced support must be used. In order to keep the temperature gradients within the mirror at a reasonable level, it is undesirable to have $e > D/8$ for large mirrors, though this may be increased to as much as $D/5$ with small ones. Taking $e = D/6·5$ as a mean figure, the maximum value of D compatible with a tolerable amount of distortion from this cause is obtained by

$$\frac{r^4}{\left(\dfrac{r}{3·25}\right)^2} = 160$$

whence $\qquad\qquad r = 3·89$

or $\qquad\qquad D \fallingdotseq 8$ ins.

10.6 Object glass cells

These must be so constructed as to ensure that there is no possibility of lateral movement of the components, that the separation of the components is maintained at the calculated value, and that the planes of the components are maintained parallel, for all positions of the telescope. At the same time the cell must not subject the components to pressure strong enough to produce mechanical strain.

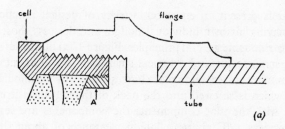

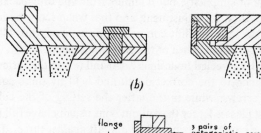

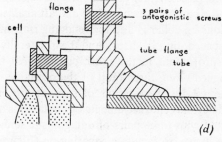

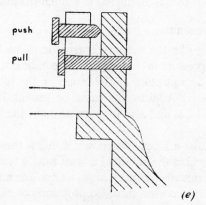

FIGURE 77

OG cells present an enormous variety of design, practically every maker having his own individual dodges and devices; those described here illustrate some general principles. Figure 77(a) illustrates schematically the main features of a typical small cell—up to about 4 ins, say. The components are kept in place longitudinally by the annular counter-cell A, which is screwed into the back of the cell until it is in gentle contact with the flint component; the components are separated by tinfoil wedges 120° apart; a lateral clearance of about 0·004 ins is allowed between the edges of the components and the cell. This design has the advantage of simplicity, but it suffers from the drawbacks that it contains no provision for adjustment, and that when screwing in A it is extremely easy to turn the flint component on its axis, thus altering the correct radial correspondence of the two components.

The latter fault is corrected in designs 77(b) and (c); in the former the collar A is machined to the correct depth and held by three equally-spaced retaining screws, while in the latter the setting of the collar is adjustable. Object glasses larger than about 4 ins usually have full three-point support: the components bear fore and aft against three equally-spaced projections on collar A and the retaining ring, are separated by three equally-spaced tinfoil strips, and bear laterally against three equally-spaced projections in the inner cellwall. All these points of support are aligned along three radii of the objective. In instruments larger than about 6 ins these points of support are often adjustable through the side of the cell. Instruments of about 4 ins and upward are also usually fitted with three pairs of antagonistic screws, holding the cell to the tube flange, to facilitate squaring-on (Figure 77(d) and (e)). It is important to remember that one screw of each pair must be slackened off before it is attempted to tighten the other.

10.7 Diagonal mounts

The diagonal, like the primary mirror, must be held securely but without undue pressure in its position in the optical train. The diagonal is normally held in position at the centre of the tube by a three- or four-legged spider. Adjustments must be provided for centring the diagonal in the tube, and also for varying its inclination to the optical axis.

Figure 78 shows a simple design which fulfils these conditions. The diagonal, a, is held at the mouth of a brass tube, b (cut back at an angle of 45°), by four turned-over projections on one side and by a brass plug, c, on the other. The central threaded projection of c is screwed into the cylinder, d, to which the spider is attached. The threaded outer ends of the spider, *e*, pass through slots in the telescope tube which run parallel with the tube's axis; they are secured by nuts on the outer side. These are used for centring the flat and also, by adjustment of *e* along the slot, for adjusting its inclination. The direction in which the reflected cone is

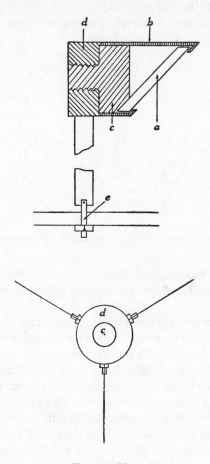

FIGURE 78

turned can also be adjusted by rotating *c* in *d*. This design is simple to construct at home; in professional instruments it is the invariable practice to make provision for all adjustments at the flat mounting, the spider being a fixture.

If a four-legged spider is fitted, it is better that the legs should not converge upon the central line of the flat mount: it is extremely difficult to deaden vibration of a flat suspended in this way, even when considerable tension is put on the supporting struts. A better arrangement is shown in Figure 79, in which the four legs do not radiate from the centre of the flat mounting, but are tangential to a circle concentric with, and slightly smaller than, the mounting. Much greater rigidity is obtained, and the necessity for heavy tensioning of the spider eliminated.

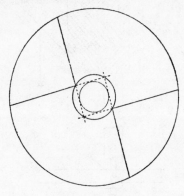

FIGURE 79

Many telescope makers, from Herschel onward, have preferred a single stout support for the diagonal, as giving greater rigidity and freedom from vibration.

If the eyepiece is not at times to assume awkward observing positions with certain types of mounting, there must be provision for rotation of the tube about its axis. With small instruments the whole tube may be rotated in the Dec axis cradle, but if D is more than 12 ins approximately, it will be found more satisfactory to restrict the rotation to a ring in the upper part of the tube on which the drawtube and spider are mounted. The moving parts demand accurate workmanship, since any divergence of the axis of rotation from the optical axis will result in the image moving in the field when the ring or tube is rotated.

SECTION 11

DEWING-UP AND TUBE CURRENTS

11.1 Dewing-up

The formation of dew upon any surface whose temperature is below the dewpoint of the surrounding air is explained in section 26.5. The condensation of dew particularly concerns the astronomer when it occurs on object glass, speculum, flat, camera lens, photographic plate, or ocular; such deposits initially reduce light grasp, and only begin to impair definition after a stage is reached at which the light loss has already become prohibitive.

Two conditions are to be distinguished, both of them liable to be encountered during observation:

(a) *Temperature of the air* above its dewpoint*:

Dew will not form on a surface warmer than the air.

Dew will not form on a surface at the same temperature as the air.

Dew may or may not form on a surface colder than the air, according as to whether it is below or above the dewpoint.

(b) *Temperature of the air below its dewpoint*:

Dew may or may not form on a surface warmer than the air, according as to whether it is below or above the dewpoint.

Dew will form on any surface at the same temperature as the air.

Dew will form on any surface cooler than the air.

In case (b) the deposit on exposed surfaces (e.g. object glass) will be heavier than that on surfaces which are relatively protected from the external mass of air (e.g. speculum). This is not so in case (a).

Assuming that during astronomical observation the temperature, if not stationary, is falling, a surface of low thermal conductivity will be less liable to dew-formation than one of high, since its temperature will lag further behind—i.e. be higher than—that of the air. Hence the aluminium film is less liable to dew than the silver film.

---

* The air is assumed to be unsaturated but with a finite water content, as non-conditioned air is, even in desert climates.

There are thus at least five factors tending to inhibit the formation of dew:

(a) removal of the moisture from the air in contact with the surface;

(b) the surface being slightly warmer than the air;

(c) the surface and its surroundings being of low thermal conductivity;

(d) shielding of the surface from the external air as much as possible;

(e) absence of dust particles, which act as condensation nuclei, from the surface.

The various methods that have from time to time been proposed for reducing the formation of dew on the optical surfaces of the telescope are applications of one or other of these principles.

(a) Replacing the natural moist air above the surface of a lens or mirror by a gentle stream of conditioned air. Franklin-Adams (B. 134) describes a somewhat complicated arrangement of this type: air from a pump or pressure bottle is passed through a dust filter, bubbled through pumice impregnated with concentrated sulphuric acid, passed through a tube of phosphorus pentoxide, a tube of soda lime (removing any traces of acid carried over), and a glass-wool plug, checked with litmus paper, and then passed via rubber tubes to jets situated round the objective.

The use of quicklime and calcium chloride has been suggested in two ways: (i) as a preventative of dew-formation, the hygroscopic substance being placed in a secondary cell below the speculum cell; this sounds messy and it is unlikely that it would be very effective; (ii) as a remover of already formed dew, the dewed part being placed in an airtight desiccator containing the calcium chloride; this is practicable in the case of oculars, less so with flats, and impracticable with objectives, besides being slower than the application of gentle warmth (see below).

(b) Keeping the surface at a slightly higher temperature than the air. Dew on a small flat can usually be removed by placing the hand on the supports or against the back of the diagonal for a few moments. An electric hair-dryer is extremely effective.

A flat pad of thoroughly dry, slightly warm flannel in the bottom of the objective's cover, which is then put in place (the telescope being swung downward to avoid contact between the pad and the lens or mirror) will remove dew in a few minutes. This should also be the procedure when closing down for the night. Wet optical surfaces should never be wiped, except, in the case of lenses, when deliberately cleaning them. Insulating the speculum from the surrounding air by enclosing it in a solid cell, desirable from this point of view, is ruled out by the necessity of reducing the temperature gradient between it and the air in the tube as much as possible.

Flats and prime focus plate holders can be maintained at, ideally, a fraction of a degree above the dewpoint by (*i*) a grid of resistance wire, mounted on vulcanite and separated by a small air space from an aluminium plate fixed to the back of the plate holder or flat; a rheostat in the circuit permits fine control of the temperature, which must at all times be the minimum that is effecive; (*ii*) a flash-lamp bulb, operated by a 4-volt battery, wrapped in tinfoil and separated from the back of the flat by several layers of tissue paper and cottonwool to ensure even distribution of heat. Similar bulbs mounted in small boxes round the outside of the dewcap of a refractor or camera objective will inhibit the formation of dew, but with a greater risk that the definition will be affected.

Oculars, when not in use, can be kept in inside pockets, where they are within reach of the warmth of the observer's body, though they pick up a good deal of fluffy dust; alternatively, but less conveniently, in the nearest warm room. Moisture condenses rapidly on a cold eye lens when the observer's warm face is brought near to it. If, during prolonged observation, an ocular becomes dewed, the simplest and quickest method of clearing it is an airtight jar containing warmed flannel or cottonwool. If excessive trouble from eye-lens condensation is experienced in cold weather when a rotating ocular adaptor is used, some form of gentle electric heating might be tried.*

(*c*) Employing materials of low thermal conductivity, with consequent slowness of cooling. Wooden tubes, for example, carry much greater freedom from dew-formation, both on the flat and the speculum, than metal. Jacketing the outside of the dewcap and upper end of a refractor tube, and the lower end of a reflector tube (the whole tube would be preferable from the point of view of dew elimination, but is impracticable for the reason discussed below), with layers of some non-conducting material such as flannel, or asbestos fibre between layers of flannel. Lining the OG dewcap with a non-conducting and preferably absorbent material, such as black blotting paper, is also a help.

(*d*) Minimum exposure of the surface to the external air, preventing excessive cooling by radiation to the open sky.

Object glasses: a long dewcap, which except in the case of the largest instruments can afford to be as long as $2\frac{1}{2}D$ to $3D$; those supplied by the makers are often too short (about $2D$), but form a convenient base for a homemade extension. On calm nights, however, lengthened

---

* Such a device, giving complete freedom from eye-lens condensation, is described in B. 135.

dewcaps have been known to suffer from 'tube current' trouble. Long caps are also impracticable with wide-angle cameras.

Flats: the further the diagonal is from the mouth of the tube, the less it will tend to dew up.

Specula: a closed tube is the ideal. This, however, is the opposite condition to that dictated by the necessity of reducing tube currents (section 11.2) to a minimum. The best compromise consists of a tube whose central section is of open or lattice construction, while the upper and lower ends, surrounding the flat and the speculum, are solid. The relatively protected position of the speculum in a partially closed tube, together with the latter's greater mass and therefore temperature lag, are the main reasons for its relative freedom from dew as compared with either OG or flat.

(e) Scrupulous cleanliness of all optical surfaces should be maintained at all times.

11.2 Tube currents

Impairment of definition by currents of unequally warm air in the telescope tube is a characteristic of the open-ended tubes of reflectors, being particularly troublesome when the aperture exceeds about 12 ins. The closed tubes of refractors permit greater homogeneity of the contained air, through which, moreover, the light has to pass only once. Small and moderate-sized refractors are virtually immune from tube currents, though trouble from this cause begins to make itself felt as the aperture is increased; the currents, though less noticeable than those of a reflector of the same size, tend to be more persistent.

It is unfortunate that the conditions militating against thermal air currents are among those that favour the formation of dew on the optical surfaces. The natural night-time temperature lag between the speculum, tube and mounting on the one hand, and the surrounding air on the other, while reducing the likelihood of dew condensation, brings with it the convection currents attendant upon any temperature gradient. These are favoured by the shape of a telescope, and by the fact that its axis is usually not far from the vertical. Since the upper part of the tube is, in an observing hut, more prone to cooling by radiation, the unstable situation arises wherein a mass of cold air is overlying a mass of warmer, and therefore less dense, air lower in the tube. A condition of circulation is thereby set up.

The elimination of tube currents takes precedence over the elimination of dew because their effect on the instrument's performance is more harmful, and because dew can be dissipated in various other ways without great difficulty.

Currents of relatively warm air rising from the speculum through colder layers ruin definition by virtue, not of their motion, but of the cause of that motion: their different temperatures, hence densities, hence refractive indices. Similarly, the velocity of the currents is immaterial to their effect upon the image. Any precautions tending to reduce the temperature differences between the telescope and the night air, and between the air in different regions of the tube, are likely to reduce tube currents. These and other helpful measures include:

(*a*) Keeping the instrument cool during the day. If it is mounted in the open, some sort of screen should be erected to shut off the direct rays of the Sun. An observing hut should be opened up well in advance of the time observing is to begin. The insulation of observing huts is dealt with elsewhere (section 17.2.4).

(*b*) Reducing the solid tube to an openwork structure. This may take a number of different forms, e.g. a solid tube perforated with as many holes as will not affect its rigidity and firmness; framework or lattice construction; tubular construction (see Figure 76), in which the longitudinal rods are locked in three rings, two of which hold the flat and speculum respectively, the third being bolted to the Dec axis.

Reduction of the solid structure of the tube not only reduces the mass of the material which is at a higher temperature than the air, but by allowing greater circulation accelerates the establishment of thermal equilibrium. While open tubes are perfectly satisfactory in an observing hut, in the open they are liable to meet with an unpleasant amount of trouble from dew.

(*c*) Lining the tube, if solid, and covering it externally with some material of low heat capacity—cardboard, three-ply, felt, sheet cork, etc. A compromise that may give satisfactory results is the piercing of such a lined tube with ventilation holes. A solid metal tube mounted in the open air is certainly the worst possible combination of circumstances.

(*d*) Breaking up the convection currents uniformly, and accelerating the cooling of the tube and mirror, by means of an electric fan blowing a steady stream of outside air up the tube from behind the speculum cell, or through the inspection cover just above the speculum. This is the only course open in the case of a solid tube, and proper ventilation can reduce very markedly those faults of performance which are too often assumed to be inherent in the reflector. Indeed, the performance of large refractors is similarly susceptible of improvement—the

Greenwich 26-in, for instance, is fitted with a fan drawing air down the tube from inlets just below the objective.

The disadvantages of a fan (which do not, however, outweigh the advantages) include: it cannot be attached to the tube on account of vibration, and its position has therefore to be constantly changed during observation; even so, it may cause vibration in a lightly constructed hut, especially noticeable when high magnifications are in use; it introduces dust; it introduces sharper variations in the effective focal length of the objective than are caused by the normal tube currents, and is hence of less value in photographic work than in visual.

On very still nights in an observing hut, a fan will often improve the performance even of a reflector with an open tube.

(e) Tube currents tend to cling to the walls; hence definition is less affected if the diameter of the tube is considerably larger than that of the mirror. For the same reason a square tube, which also presents fewer problems to the amateur telescope builder, is comparatively freer from visible currents. The improvement of definition that often results from stopping down the aperture probably depends to a great extent on the elimination of the affected peripheral rays.

(f) The temperature difference between the observer's body and the air is even greater than that between the instrument and the air. In the Newtonian, particularly, convection currents from the observer's body are liable to cross the path of the incident light. A screen of some light non-conducting material such as cardboard or fibreboard, attached to the upper end of the tube and projecting beyond its mouth, will eliminate most of this trouble; but it is only practicable in an observing hut, where there is no wind to cause unsteadiness. Out of doors, the best palliative is a reserve greatcoat to put on as soon as the first has taken up the body heat.

See also section 26.

11.3 Summary

Maximum freedom from dew is provided by a closed tube.

Maximum freedom from tube currents is provided by an open tube (less satisfactorily, by a closed tube and fan).

Maximum freedom from both is provided by an open tube mounted in an observing hut.

SECTION 12

NEWTONIAN AND CASSEGRAIN ADJUSTMENTS

12.1 Introduction

Much too great stress has on occasion been laid upon the difficulty of achieving and maintaining correct alignment of the optical components of the Newtonian. It is possible that, as a result, beginners have been scared off, and have equipped themselves with a 'safe' 3-in refractor instead of the economically equivalent 6-in or 9-in Newtonian, under the impression that the latter would always be getting out of adjustment.

The collimation of reflectors presents no difficulty if it is tackled systematically, and particularly if the devices mentioned below are incorporated in the telescope—a means of adjusting the set of the speculum from the observing position, and a removable upper section of the tube. Starting from scratch, a couple of hours should suffice for the lining-up of a Newtonian under adverse circumstances; given an assistant, or the devices already mentioned, this time can easily be cut in half. Once adjusted, a Newtonian should not need to be touched again for months, provided it is carefully and reasonably treated.

For a fuller treatment of the subject, see B. 100, 102.

12.2 Collimation of a Newtonian

The axes of the ocular and the paraboloid must intersect at the centre of the secondary reflecting surface, making equal angles with the normal to this surface at this point.

To achieve this, the following adjustments are available:

(a) of the mounting of the flat relative to the telescope tube;
(b) of the eye-end fitting itself (flat supports and ocular tube);
(c) of the inclination of the flat in two perpendicular planes;
(d) of the paraboloid cell relative to the axis of the telescope tube.

The procedure must be worked through systematically. Point the telescope at the day sky, so that the mirror is filled with fairly bright light. Remove the lenses from a HP ocular and screw it into the draw-tube, where it acts as a stop defining the exit pupil. Then:

(*i*) By means of adjustments (*a*) and (*b*) make the axis of the ocular pass through the centre of the flat. From the eye position at the ocular the outline of the flat is then concentric with the far (i.e. inner) end of the of the drawtube (see Figure 80).

(*ii*) By means of adjustment (*c*) make the reflection of the primary mirror in the flat concentric with the flat itself—and therefore with the inner end of the drawtube. It is essential that the whole cone of rays

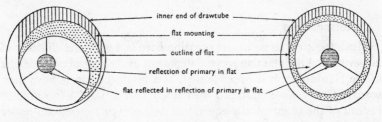

inner end of drawtube
flat mounting
outline of flat
reflection of primary in flat
flat reflected in reflection of primary in flat

| Before adjustment | FIGURE 80 | After adjustment |

whose base is the full aperture of the paraboloid be reflected from the flat. With the eye at the lens-less ocular, the edge of the speculum must be visible within the edge of the flat. It is of no importance if the flat is slightly larger than the minimum necessary to ensure this, since the only effect will be a negligible reduction of light grasp. But if the flat is slightly too small, so that a peripheral ring of the convergent cone is not intercepted by it, the light loss will be considerable. In the case of a $f/6$ Newtonian, for example, with the flat intercepting the cone at a distance from the prime focus equal to D (thus bringing the Newtonian focus well clear of the tube), a flat diameter one-tenth larger than the minimum necessary to intercept the whole cone will increase the light loss through silhouetting by about 0·5%; if, on the other hand, it is one-tenth smaller, the gain of about 0·75% through decreased silhouetting is offset by a loss of 30% due to a peripheral ring of the objective not contributing to the image at all.

(*iii*) By means of adjustment (*d*) make the reflection of the flat (a small black disc with the three legs of the spider radiating from it) in the primary mirror concentric with the primary mirror itself. To achieve this final adjustment with economy of time and temper one or other of the following should be provided: (1) an assistant to make the adjustments while their effect on the image is watched from the observing position, (2) means of making the adjustments from the observing position itself.* The visual judgment of the concentricity or otherwise of the primary mirror and the reflection of the flat is made much easier if a circular disc is pasted to the centre of the speculum, of such a size that the reflected flat shows a narrow dark rim round it. (Since this region of the speculum is silhouetted in any case, no damage is done to the usable silver surface.)

(*iv*) Fine adjustment is then carried out on a stellar image. If the displacement of the axis of the primary is slight, a stellar image at the centre of the field will probably exhibit coma only; if the displacement is gross, astigmatism may cloak the coma. Make adjustment (*d*) so as to displace the image at the centre of the field in the direction indicated by the comatic image:

Bring the image back to the centre of the field and repeat the process until the image is symmetrical. The interpretation of an astigmatic image is more difficult, but should not be necessary if the first approximation in daylight has been carried out. The orientation of the sagittal image indicates the diametrically opposed directions in one of which the adjustment must be made:

but whether towards *A* or towards *B* can only be discovered by detecting the eccentricity of the rings. This is much less easy to spot than the shape of the comatic image, but if the drawtube is racked out it should be quite plain in the extrafocal image:

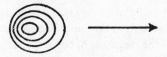

In case of doubt, discover by trial and error which adjustment reduces the astigmatism, and continue the tilting of the mirror in this direction until the image at the centre of the field shows no removable coma, or until the rings of the extrafocal image are as concentric, and as uniformly bright in all position angles, as they can be made.

It must be borne in mind that any alteration of the inclination of the

* See section 10.3.

primary mirror will affect the readings of the position circles. Hence the verniers must be reset after every readjustment of the speculum; so, if the readjustment has been considerable, must the finder.

12.3 Fine adjustment of the secondary (flat)

In the case of instruments of large relative aperture, adjustment (*i*) must be modified. The reason for this is that although a parallel pencil whose axis is incident to a 45° mirror at the centre of the latter will, after reflection, still be concentric with the mirror (Figure 81), a converging

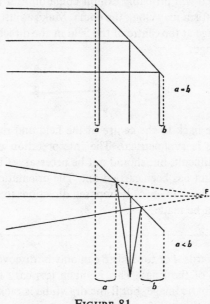

FIGURE 81

beam will no longer be so, since the axis of a cone does not pass centrally through an elliptical section of the cone. The eccentricity of the reflected beam is a function of the steepness of convergence in the incident beam, and therefore of relative aperture; for relative apertures smaller (i.e. focal ratios larger) than about $f/8$ it can be ignored.

The new procedure is as follows. A short tube which will fit snugly over the cell of the diagonal is fitted at one end with a pair of cross-wires which exactly define its centre. A hole cut in the side of the tube allows the cone, after reflection, to pass to the drawtube. A second pair of cross-wires are fitted across the mouth of the drawtube. In this way the true centre of the flat is defined, and it is on this point, rather than on the apparent centre, that the drawtube is aligned. In the remaining adjustments the reflection of the primary in the flat and the reflection of the flat in the primary are to be made concentric, not with the flat, but with the inner end of the drawtube.

12.4 Fine adjustment of the secondary (internally reflecting prism)

If the rays incident to the entrance and exit surfaces of the prism are not normal to these surfaces * (e.g. if the apex is not exactly 90°, or if the prism is tilted in a manner undetectable from the observing position), the procedure already described will not ensure against the introduction of false colour. A further check is required.

The inner edge of the drawtube is painted white, and a cloth mask arranged over the mouth of the telescope so that there is just enough light for this white ring to be seen. From the observing position the reflections of two white rings will now be observed in the prism, one from each of the faces bounding the apex. If these two circles are not concentric they must be made so, by means of adjustment (c). This will probably upset the other adjustments, and the two must be repeated alternately, the error being reduced each time, until both are satisfactory. This is liable to be a tedious and rather maddening job, and the device described by Peek in B. 100 simplifies and accelerates the adjustment very considerably. The upper section of the telescope tube, carrying the ocular tube and the prism mounting, is made detachable from the main part of the tube, to which it is fixed by bolts engaging in slots running lengthwise (i.e. parallel to the axis of the tube) and secured by wing nuts; they thus allow a margin of tilt adjustment of the upper section of the tube relative to the lower. The upper section is removed from the telescope and the squaring-on of the prism carried out in the manner just described. The section is then returned to the telescope and, by means of the wing nuts, locked in position so that the reflected image of the primary is concentric with the inner end of the drawtube. Finally, adjustment (iii) of section 12.2 completes the job.

See also sections 10.3, 10.7.

12.5 Collimation of a Cassegrain

In the case of a Cassegrain the equivalent adjustments are twofold: (i) the axis of the secondary must be made to coincide with the axis of the ocular tube (it being assumed both here and in the discussion of the

---

* Internally reflecting prisms cannot be used as secondary reflectors in a Newtonian whose relative aperture exceeds about $f/6$, for the reason that the convergence of the pencil at the focus is then too steep for this condition to be satisfied.

Newtonian adjustments that the axes of the ocular and the ocular tube are coincident); and (*ii*) the axis of the primary must pass through the centre of the secondary:

(*i*) Observe the day sky from the Cassegrain focus through a HP ocular from which the lenses have been removed. The whole of the primary should be reflected in the secondary, this reflection and the secondary being concentric. Any necessary adjustment is made to the secondary.

(*ii*) The image of the secondary in the primary must be concentric with the primary. Fine adjustment of the primary is made by observation of the comatic or astigmatic nature of a centrally placed star image, as described above.

SECTION 13

TESTING

13.1 Introduction

The most satisfactory method of testing an objective outside a laboratory is by means of the extrafocal images of a star—either natural or artificial (see section 35). This method is both more sensitive and less dependent on absence of atmospheric turbulence than the examination of the image in the focal plane.

If a natural star is used, it should be of moderate brightness (of the 2nd magnitude, say) and must be at or near the zenith. If far from the zenith, differential atmospheric refraction, resulting in dispersion, will introduce false colour: far from the zenith, a star image tends to exhibit a red fringe on its apparent upper side (i.e. the side nearest the horizon), and a greenish or bluish one on the lower. It is also important to keep the eye precisely central behind the ocular—i.e. the optical axis must pass centrally through the eye pupil—or false colour will be introduced. If the eye is displaced,* that side of the image towards which the displacement takes place will become tinged with red. Hence it is possible to suppress the atmospherically produced particoloured fringe surrounding images of objects which are far from the zenith by slightly displacing the eye downward from the centre of the eye lens—'downward' here meaning the direction such that the line from the centre of the eye pupil through the centre of the exit pupil will, if produced, pass through the zenith.

Turbulence must be very slight, or the image will not be fine enough for the more delicate and sensitive tests to be performed; if the seeing is not satisfactory, testing must be postponed until it is, or an artificial star used.

Magnification should not be so low that the rings are difficult to see and the exit pupil so large that aberrations in the eye and the ocular

---

* Too great a displacement of the pupil will result in effectively cutting down the aperture on one side.

become prominent, nor so high that secondary spectrum becomes troublesome. Where not otherwise stated, about $60D$ is satisfactory.

The telescope, and in particular the objective, must be allowed to reach thermal equilibrium before testing begins.* The length of time required varies with the initial temperature difference between the telescope and the night air, and also with aperture; in the case of a 6-in refractor, at least half an hour's exposure to the night air should be allowed; an hour would be safer. The outer surface of the crown component of a doublet cools fastest, then the back surface of the flint, and finally the internal surfaces. While these differential contractions are in progress, strains are set up in the objective which simulate the effects of spherical under-correction. Tube currents have been known to introduce effects mistakable for slight astigmatism. Similarly, when testing with an artificial star in sunlight it is essential that the telescope itself should be completely shaded; thermal effects within the tube, produced by neglecting this precaution, can cause great distortion of both focal and extrafocal images.

The order in which the tests are carried out is important. It is a waste of time, for example, trying to apply the very sensitive tests for the geometrical aberrations if the correct alignment of the objective has not first been assured. As always, where adjustments in several directions are involved, care must be taken to confirm an earlier adjustment if a later one may have disturbed it.

13.2 Telescope tube

Only such characteristics of the tube as affect the optical tests are considered here.

It must be confirmed that the axis of the drawtube of a refractor passes through the centre of the objective (in a Cassegrain it may be assumed that the ocular tube is normal to the mirror cell to which it is fixed). Fix cross-threads over the objective and the mouth of the drawtube; also fix a slip of plane unsilvered glass across the mouth of the drawtube in contact with the latter set of threads. If the eye is now placed on the optical axis about 10 ins behind the drawtube, its reflection will be seen dimly in the unsilvered slip. If the adjustment of the drawtube is satisfactory, it will be possible to find a position for the eye from which it sees the two sets of threads superimposed upon one another and bisecting its own reflection.

In the case of a Newtonian, an eye placed at the focus should see only

---

* See also section 10.4.

a narrow rim of the tube reflected round the image of speculum in the flat.

Finally, remove the ocular and see that none of the diaphragms in the tube is too large or, which comes to the same thing, too near the object glass. That is, the whole of the objective must be visible. To ensure that full illumination is obtained with the lowest powers, see that the whole objective is visible from the extreme edges of the eye-tube.

13.3 Oculars

The oculars used in the testing of an objective must be of established excellence.

With a star central in the field, twist the drawtube or, better, the ocular itself; since oculars are normally threaded this will alter the focus slightly, but this is not important, and the danger of jogging the whole instrument is smaller than in trying to twist the drawtube. Note whether the image of the star is moved in the field as a result of this rotation. Refocus, and see whether the appearance of the image has been modified in any way by the rotation of the ocular. If either has occurred (the alignment of the drawtube having already been established) the ocular should be rejected.

Shift the image to the edge of the field and expand the rings extrafocally; if they are not complete circles, or are brighter on one side than on the other, the ocular diaphragm is not passing the whole of the objective's field of full illumination. With the image still at the edge of the field, refocus the star. A Ramsden will then usually show chromatic inequality of magnification as a fringe of opposite colours on opposite sides of the image where the diameter of the field intersects the image. A Huyghenian showing this should be rejected.

13.4 Flats

Among the commoner causes of disappointing performance with reflectors are the use of unsuitable oculars and of a flat whose optical qualities do not match up to those of the speculum. The flat should therefore be checked before starting on the tests of the primary.

Three methods of testing the planeness of a flat are in common use:

(a) by Newton's rings in monochromatic light, using another flat of known quality;

(b) by knife-edge test of a spheroid of known quality, the flat intercepting the beam at 45°, whence any faults apparently detected in the sperical mirror must in fact be contributed by the flat;

211

(c) by Rayleigh's water test, which has the advantage that no second glass testing surface is required, though it is by no means easy to operate; it is in fact a variant of (a), in which the test flat of known planeness is replaced by a water surface.

The detailed operation of these tests is described in B. 101, 102.

13.5 Physical character of the object glass

Remove the ocular and point the telescope at the day sky; any physical defects in the object glass will then be strikingly visible to an eye placed at the focus. These can be seen in even greater detail in Foucault focograms (photographs of the appearance of the objective under the Foucault test—see section 13.17).

Bubbles, scratches, and minute particles in the glass which cause local variations of the refractive index and which appear visually as 'threads' are unimportant, since they are small compared with the entry pupil illuminating them; at most they produce an infinitesimal light loss and reduction of contrast through scattering. (The reverse is true of an ocular, where a bubble, for example, is necessarily large compared with the diameter of the pencil in which it is situated.) Large defects of this nature are unlikely to be encountered. If they are, the object glass should be rejected. Covering them with patches of adhesive tape or making irregularly shaped diaphragms to cut them out, as sometimes recommended, is useless.

Internal strain, with variations of refractive index, due to faulty annealing (the outer layers being allowed to set and cool while the interior of the lens still wants to contract further) can, if gross, cause double refraction—strikingly revealed by polarised light—or wings and flares attached to all bright images. No objective from a reputable maker need be suspect.

13.6 Mechanical distortion of the objective *

With a three-point support of the speculum, it may be impossible to eliminate a slight triangular distortion of the stellar image. In any case, the objective must be firmly held without being tightly pinched. An occasional and transient triangular deformation of stellar images has been described from time to time; its cause is unknown, but it is presumably of atmospheric origin; it was noticed in one case that, at an observing station in SE England, this effect only occurred when the wind was in the E or SE.

* See also section 10.

Excessive and/or unequal pressure from the supports of an object glass, in directions both parallel and perpendicular to the optical axis, is a cause of distorted (typically triangular) focal images and asymmetrical extrafocal rings; even astigmatism may occasionally be traced to unequal pressure on one or both components of an object glass. It must be sufficiently slack in its cell to ensure freedom from undue pressure when the cell is contracted to its minimum size, i.e. at the minimum temperature to which it may be exposed.

Sagging of an object glass under its own weight, at different orientations of the optical axis, will only be encountered in objectives of more than about 6 ins aperture. The flint element is always the more prone to this type of distortion, being both heavier and less robust (owing to its concave section) than the crown.

Symmetrical central sagging produces the effect of spherical aberration, and is allowed for in the calculation of the curves. Sagging between each point of support produces a triangular distortion of the image; if the crown is sagging, light passing through the objective in the vicinity of the points of support will be brought to a shorter focus than that from other parts of the objective; if the flint is sagging, light from the vicinity of the supports is brought to a longer focus. The resultant tendency for the two more or less to cancel one another is effective with medium apertures (say, 6 to 10 ins); with larger apertures the only solution is the provision of additional counterpoised points of support, ensuring that the weight of the lenses is distributed evenly over them at all times.

13.7 Centring and squaring-on of object glasses

Squaring-on neither has to be checked as frequently as with a reflector (once adjusted, with a newly acquired instrument, it should not require attention again for years), nor does it so seriously affect the quality of the image if imperfectly carried out. Means of adjustment (see section 10.6) are normally provided when the aperture exceeds 3 or 4 ins only; refractors of 6 ins aperture and larger should be fully adjustable—i.e. both components should be capable of adjustment parallel to, and perpendicular to, the optical axis. The adjustments are tiresome to make, since the screws are not within reach of the observing position; an obliging friend is a great help here.

The following maladjustments may be present singly or in combination:

(a) axes of the components parallel but not coincident: Figure 82(a);

(b) axes of the components inclined to one another, though the line joining their centres may pass through the centre of the ocular: Figure 82(b);

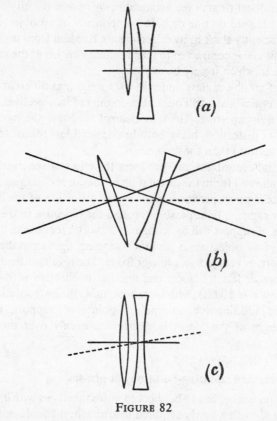

FIGURE 82

(c) axes of components coincident, but not passing through the centre of the ocular: Figure 82(c).

The final adjustments in the squaring-on and centring of an object glass require an acute and practised eye for the detection of the slight imperfections in the image.

(a) The flint component acts as a prism, the direction of the red dispersion being that in which its centre is displaced from the centre of the crown. The effect is therefore very similar to that produced by differential atmospheric refraction: the image of a bright star at the centre of the field is fringed with red on one side, while the other is bluish (though this is less easy to detect). In addition to the colour effect, the image will be noticeably elongated if the eccentricity of the components is great (exceeding about 0·1 mm at $f/15$).

If the manufacturer has made no provision for the lateral adjustment of the components, any error in this direction means that the objective must be returned to him.

(*b*) and (*c*) *Approximate checks:* If the telescope passes the following tests, the planes of the lenses are not grossly inclined to each other, nor is their axis grossly divergent from the line joining their centre and the centre of the field of the ocular:

(*i*) Put the telescope tube horizontal, and remove the ocular. Hold a candle flame between the eye and the front crown surface of the objective. The three images of the flame—reflected from surfaces 1, 2–3, and 4—should coincide when the line through the centre of the drawtube passes through the centre of the flame.

Alternatively, and rather more conveniently, a bright ring (such as the metal frame of a pair of spectacles on the nose of the tester) may be used in place of the flame. The reflected image of the ring and the circular ocular opening should be concentric.

(*ii*) Applicable with objectives whose fourth surface is concave towards the ocular. Cover the object glass; place a white card, illuminated by strong artificial light, about 6 to 10 ins behind the open end of the drawtube; this card is pierced with a ½-in hole, behind which the eye is placed. The card is moved about until the dark spot (reflection of the hole) on a white circle (reflection of the card, bounded by the edges of the drawtube), seen in the rear surface of the objective, is central within the circle of the drawtube. In this position the white circle and the objective will be concentric if the latter is correctly squared-on. If the circle is nearer one side of the objective, this side must be moved further from the ocular. The test is then repeated until a satisfactory adjustment is obtained.

(*iii*) Cover the object glass. Over the open end of the drawtube place a cardboard mask pierced with two circular holes situated on a radius (Figure 83). The eye is placed behind *A* and shaded from a source of light directed into *B*. When *AB* is vertical, three closely adjacent images of *B* will be seen in the objective, and the line

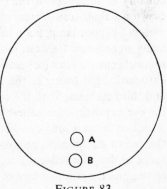

FIGURE 83

215

joining these should be vertical. If satisfactory, repeat with *AB* horizontal, when the line joining the images should likewise be horizontal.

(*b*) *and* (*c*) *Fine adjustment:* As usual, the highest degree of precision in the detection and correction of instrumental shortcomings is provided by the examination of a stellar image:

(*i*) Object glass with inherent internal coma: (*c*) is indicated by a comatic image at the centre of the field. Determine in which direction the rings are brightest, and reduce the distance of this side of the objective from the ocular. This will reduce, but may not remove, the coma because (*b*) also causes coma—as well as astigmatism, which may only become visible after the joint squaring-on of the objective has been

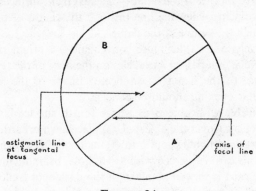

FIGURE 84

improved. Of the two regions of the objective situated at right angles to the tangental focal line (*A* and *B* in Figure 84), one is too near the ocular, and one too far. The object glass is removed from its cell and the equality of thickness of the inter-lens air-wedges tested by means of Newton's rings, using monochromatic light such as is provided by a mercury-vapour lamp, neon tube, or sodium flame, or more cheaply by using an ordinary 100-watt lamp and observing the objective through a monochromatic filter (see section 27.2). The rings are invisible in polychromatic light owing to the relatively great separation of the second and third surfaces. With the source of the illumination held as close to the eye as possible (asbestos sheet between the temple and the flame, if a sodium flame is used), and both source and eye opposite the centre of the crown component, approach the objective until the eye is at the focus of the joint reflecting surface provided by the rear surface of the crown and the front surface of the flint; the objective will then fill with light, although the individual reflections of the source will still be visible.

If the Newton's rings are not concentric with the object glass, adjustment is required. Press gently on the edge of the crown component above each of the separating wedges in turn, to discover which decreases the eccentricity of the objective and the system of rings. Then insert a slip of tinfoil between the crown lens and each of the other two spacers, repeating the performance until the concentricity of rings and objective is satisfactory. If c_2 and c_3 are so different as to result in great crowding of the rings towards the edge, the estimation of concentricity or otherwise will be difficult to make; in this case, mark the centre of the crown lens with a minute ink-spot, and judge the concentricity by relation of this to the innermost rings.

The total thickness of the spacers should not be greater than that necessary just to separate the centres of the components if $c_2 > c_3$ (e.g. Littrow)—in which case the spacers will be of roughly the thickness of a calling card—or just to separate their edges if $c_2 = c_3$ (e.g. Clairaut) or $c_2 < c_3$—in which case the spacers will be of tinfoil about 0·1 mm thick. The star test is sensitive to variations in the thickness of the wedges of only 0·01 mm at $f/15$.

When returning the objective to its cell it is advisable to check the tensioning of the counter-cell behind the flint component by the same method, since there is enough 'give' in the spacers completely to wreck the adjustment if undue and unequal pressure is exerted. The necessary pressure to hold the objective firm will inevitably distort the rings somewhat, but a degree of distortion that would involve the complete destruction of the image in a reflector (with its single optical surface) is immaterial in the case of an object glass.

On reassembly, confirm that the joint squaring-on of the components has not been interfered with (it is advisable to mark the setting of the cell relative to the tube before removing it), while improving their relative squaring-on. If reobservation of a star indicates that the astigmatism has only been improved, and not removed, the procedure must be repeated until satisfaction is obtained.

(*ii*) Object glass with non-existent or inherent external coma: (*b*) is indicated by comatic images at the centre of the field. The line joining the points in the field through which the axes of the crown and flint components pass is parallel to the axis of the image, the former lying towards the apex of the image, and the region where the components are too close to each other being indicated by the broad end of the image (Figure 85). The necessary adjustments of the thickness of the wedges are made—by the Newton's rings method described above—and a further test made on a star.

(c) Coma having been removed, any trace of astigmatism in the image at the centre of the field indicates inexact squaring-on of the objective

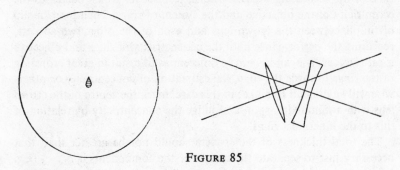

FIGURE 85

as a whole. Slacken the cell-holding screws and pull in the side of the objective marked *A* very slightly (Figure 84); if the condition is aggravated, push *A* further out from the ocular and pull in *B*. By trial and error, the astigmatism may be reduced to zero.

13.8 Alignment of the optical parts of reflectors

See section 12.

13.9 Spherical aberration

The necessity of allowing the objective to attain thermal equilibrium before testing has already been stressed.

A method of detecting residual spherical aberration by the alternate employment of a diaphragm and a central stop, involving a change of focus, is described in the literature.* It is not satisfactory, however, owing to the narrowness of the emergent pencil when all but the central area of the objective is stopped out, which allows an appreciable movement of the drawtube without affecting the sharpness of the image to an extent that is perceptible to the eye. The test is therefore insensitive, and since a simple and extremely sensitive test is available it has nothing to recommend it.

This test involves the comparison of the image of a star inside and outside the focal plane. The latter being the region of uncorrected colour in an 'achromatic', and the extrafocal image being tinged with purplish flare, the use of a green or yellow filter facilitates the observation and increases the accuracy with which the test can be made. The drawtube

* It was the method used by Herschel for testing his specula.

218

is racked quickly in and out from the focus, so that the appearance of the image on either side of the focus can be easily compared. Then:

(a) *Under-corrected:*

Inside focus: light concentrated in the periphery of the image; weak central disc; rings increasing in brightness from the centre outward, being extremely prominent at the outer edges of the diffraction pattern.*

Outside focus: light concentrated at the centre of the image; central disc nearly as small and bright as at the focus; brightness of the rings diminishing from the centre outward, the periphery being faint and ill-defined.

(b) *Over-corrected:*

Inside focus: as for outside focus (above).

Outside focus: as for inside focus (above).

(c) *Perfectly corrected:*

No appreciable difference between the strength of the rings on either side of the focus, the only difference between the two images being (in a refractor) the secondary colour—a greenish or greenish-yellow fringe outside, and maybe a reddish fringe inside, the focus. The concentration of secondary colour at the centre of the image outside focus may produce the impression that the objective is under-corrected; hence the value of a yellow filter.

13.10 Zonal aberration

The test is as for spherical aberration; but even more than the latter it requires an acute and practised eye. Zonal errors are revealed by the anomalous brightness or faintness of one or more rings in the extrafocal image; that is to say, the change in brightness of the rings from the spurious disc to the edge of the image is not uniform. Any anomaly of this nature will obviously be reversed inside and outside focus.

The position in the inside-focus image of an anomalously bright zone corresponds with the position of the zone in the objective that is focusing short; and conversely, that in the outside-focus image, to a zone in the objective that is focusing long.

The test is facilitated by employing a bright star—mag 1, if available —and by a greater displacement of the drawtube than when testing for spherical aberration, a dozen or more rings being expanded. The errors are also more easily detectable on the 'bright' side of the focus—i.e.

---

* A spurious effect simulating spherical aberration, due to a temperature gradient in the air immediately surrounding the flat of a Newtonian, is described in B. 144.

that on which they show as brighter-than-normal rings, rather than as darker-than-normal.

13.11 Coma

This is unlikely to be encountered in visual instruments, owing to their comparatively restricted fields. The appearance of the comatic image is described in section 4.5. If coma is found to be intrusive, it is more likely that the squaring-on is faulty than that the coma is inherent in the objective.

13.12 Astigmatism

Astigmatism that cannot be removed by adjustment of the squaring-on of the objective (and it is only in the so-called coma-free objectives that faulty squaring-on simulates astigmatism in the extrafocal image on the axis) must be inherent in one or more members of the optical train. If the ocular used for testing has previously been given a clean bill of health it must lie in (a) one or more surfaces of the OG (or be caused by varying refractive index across one or both components), (b) the speculum and/or secondary mirror, (c) the eye.

(c) can be eliminated by changing the orientation of the telescopic field with respect to the line joining the eyes; it is preferable to do this with an image-rotator (see section 8.11), since rotation of the head may induce or modify astigmatism in the eye. If this changes the orientation of the astigmatic planes relative to a fixed direction in the objective, the astigmatism lies at least partly in the eye. Try the other eye, which may possibly be free from astigmatism. Changing to a higher magnification will almost certainly reduce or altogether suppress the astigmatism if it is due to the eye (owing to the smaller area of the cornea and crystalline lens being illuminated by the exit pupil), in which case the observer must discover for himself the minimum magnification that his eye will allow him to use, unless the trouble is corrected by spectacles. Astigmatism that remains under high powers is invariably inherent in the instrument rather than in the observer.

If the cause is (a), and only one component is astigmatic, no improvement of the condition can be made; but if both lenses contribute to the astigmatism, rotation of one of them relative to the other will find an orientation in which the two astigmatisms are opposed, and the trouble minimised, if not entirely suppressed.

Similarly with (b), if the speculum alone is astigmatic nothing can be done about it; if it is contributed to by both the speculum and the secondary mirror (Newtonian flat or Cassegrain hyperboloid), a relative orientation of the two can be found by trial which gives minimum astigmatism; though in the case of the Newtonian with its elliptical secondary it is easier to scrap it and get another. If it is due to either mirror alone, the direction of the focal line will rotate with rotation of the mirror.

If the astigmatism is only slight, examination of the focal image, or either the intrafocal or extrafocal image alone, will not easily reveal the error. In the former it will cause a slight blurring of the normal diffraction pattern, and in the latter an elliptical expansion of the rings. As with spherical aberration, the fault is most easily spotted by racking the ocular backwards and forwards across the focus: the expansion of the image in two mutually perpendicular directions is then easily detected.

Any departure whatever from absolute circularity in the extrafocal images, if not due to faulty alignment of the optical parts or to the flat or secondary mirror, justifies the rejection of the objective.

13.13 Chromatic aberration

There remains, in the case of a refractor, the examination of the secondary spectrum, which should be the only residual aberration of any importance. With a magnification of $60D$ to $70D$ examine the focal image and the images inside and outside focus of a white A-type star, about mag 2:

(*a*) *Well-corrected visual objective*:

At focus: yellowish disc; halo, if any, purplish or indigo (not red or green).

Inside focus (2–3 rings expanded): yellowish disc with faint and narrow reddish-purple border.

Outside focus (2–3 rings expanded): tiny red disc forms at centre.

Outside focus (3–4 rings expanded): mauve centre; no red fringe; greenish or greenish-yellow border.

Outside focus (5–6 rings expanded): bluish flare overlying all the central part of the image.

(*b*) *Under-corrected*:

Inside focus: white or greenish-white disc; red fringe (scarlet rather than purplish-red) more prominent.

(*c*) *Over-corrected*:

Inside focus: deep yellow rings; greenish-blue background.

With a blue-white star the reddish-purple fringe inside the focus may be more nearly blue, while with a red star it may be so marked as to compare with that of an under-corrected image of a solar-type star.

Further, the lower the magnification (hence the larger the exit pupil) the more the aberration of the eye and ocular will introduce colour under the guise of under-correction in the objective. The choice of test object and magnification is therefore of great importance in the examination of the secondary spectrum. The colour corrections of different objectives nevertheless vary between certain accepted limits, while still rating as good, since to this limited extent the nature of the correction is a matter of taste.

13.14 Testing a photovisual objective

(*a*) *Eccentricity of the component lenses:* The image at the centre of the plate, viewed with a HP ocular mounted axially, should be a brilliantly greenish-yellow point, surrounded by a uniform purplish-red fringe. If this fringe is reddish on one side and bluish on the other, the components of the objective are not concentric. The amateur is advised to return the objective to the makers for adjustment, rather than to attempt it himself. Correct centring is of great importance in a three-component objective.

(*b*) *Squaring-on of the objective as a whole:* Examine the axial image for astigmatism and coma with a HP ocular mounted at the centre of the plate, using a green filter.

(*c*) *Squaring-on of the plate holder:* Insert a plane mirror into the plate holder; mark its centre and the centre of the objective with a very small disc of white paper, dampened to make them stick. On looking down the tube, with the eye in front of the objective, the reflection of the pupil of the eye should be in the centre of the mirror.

(*d*) *Coma:* Mount a HP ocular near the edge of the field, inclined inward at a sufficient angle for its axis to pass through the centre of the objective; if the emergent pencil does not, in this way, pass through the ocular axially, the ocular itself will contribute coma. Using a green filter, look for traces of coma. Astigmatism in its pure state, with no trace of unilateral comatic brightening of the rings, should be seen.

13.15 Testing anastigmats and short-focus camera lenses generally

Compound photographic objectives have no means of adjusting the relative positions of the lenses in the cells, this being a highly skilled task performed with great stringency before the objective is released from the manufacturer's shop.

The squaring-on of the objective as a whole, however, must be carefully checked—especially in the case of a home-made camera. The best method, being both simple and accurate, is as follows. A plate carrier whose back has been removed is loaded with an exposed plate, the centre of which is marked by scratching through the emulsion along its two diagonals. The camera is then clamped to a table with its back towards the light. If the objective is observed through a bright circular object such as a curtain ring—the eye being placed directly in front of the objective, and the ring between it and the objective—the following will be seen: the two white lines scratched on the plate, whose point of intersection marks the latter's centre, and a series of reflections of the ring at the several surfaces of the objective. If the components of the objective are in correct mutual adjustment all the reflections of the ring will be concentric (if they are not, the lens must be returned to the makers); if, in addition, the lens as a whole is correctly squared-on, these concentric reflections will be centred on the intersection of the diagonals. If they are not, it is a matter of a few minutes' experimenting with the cell-adjusting screws to make them so.

An alternative method—which is less satisfactory, since its accuracy is dependent on that of a spirit-level, and also because it is rather tiresome to operate—is as follows. Remove the objective; point the camera approximately at the zenith; level the plate holder by means of a spirit-level laid at mutually perpendicular orientations on a sheet of glass inserted in the holder. Clamp the camera securely in this position, remove the spirit-level, replace the objective, and test the horizontality of the plane of the objective by laying the level across it in two mutually perpendicular directions. Any adjustments needed to make the plane of the lens horizontal can be effected by inserting tinfoil shims between the cell flange and the camera box to which it is attached.

13.16 Testing resolving performance

See section 2.5.

13.17 Laboratory methods of testing

For simplicity and sensitivity there is nothing to beat the examination of the extrafocal images of point sources for discovering the characteristics of an objective. It is not, however, convenient for the optician who wishes to determine the nature and magnitude of the errors in his objective with a view to correcting them by subsequent work. To meet his requirements a number of laboratory methods have been developed. But since these are primarily of interest to the instrument-maker, and are comparatively unsuitable for the investigation of the performance of a newly acquired telescope, the reader must be content merely with references to other sources.

The Foucault test, for which a knife-edge and an artificial star are set up at the centre of curvature, is probably the most generally used by amateurs. Good general accounts are to be found in B. 90, 101, 151; B. 153 describes a convenient form of apparatus for carrying out the test; B. 145, 146 describe an autocollimating application of Foucault's principle.

The Hartmann test is less convenient to operate than the Foucault, involving photographic work on either side of the focus, and the measurement of the plates. It allows the focal length of any desired zone of the objective to be determined, and is perhaps the most commonly used test where large objectives are concerned. B. 90, 102 contain good general accounts; B. 147 should also be referred to.

The Zernike test is an ingenious use of diffraction to show the high and low regions of the objective's surface in different colours. See B. 102, 154, 138.

Autocollimation methods, in which the error is doubled by arranging for the beam to pass through, or be reflected from, the objective twice, suffer from the disadvantage that a flat as large as the objective is required. Burch has described a method of overcoming this difficulty with quarter-size compensating mirrors in B. 139; see also B. 140.

The Ronchi test, like the Foucault, requires no complicated equipment. It is an interference method, employing a coarse grating (40 to 200 lines per cm). For general accounts of the method, see B. 102, 136. Like the Foucault, also, the Ronchi test is qualitative; for a quantitative development of it, see B. 102, 141. A useful variant of the Ronchi test is described in B. 149.

A conveniently operated null test, a development of the Foucault test, is described by Dall in B. 142.

SECTION 14

MOUNTINGS

14.1 Introduction

The mounting of the optical parts of a telescope has two functions to perform: (a) the maintenance of these parts in their correct, fixed relative positions; (b) the provision of means of moving them as a whole in any required direction.

The optical parts themselves, mounted in some form of cell, are maintained in their correct relative positions by mounting them in a single framework (solid or open tube, circular, square, or polygonal in section). In some types of mounting, and particularly those for small instruments, the tube is supported by an altazimuth or equatorial head, permitting its movement about two mutually perpendicular axes; this head is then supported by some form of tripod or column, connecting it with the ground. In some larger mountings the distinction between the head and the support cannot be made, while in other types of mounting in which the movement of the eyepiece is restricted or suppressed (as in the Sun telescope), the tube may be eliminated and the optical parts mounted independently of one another.

14.2 Mountings: general qualifications

(a) The primary characteristics of any worthwhile mounting are a rock-like stability and robustness of construction, combined with smoothness, uniformity, and ease of movement about both axes.

(b) Overhanging of centres of gravity, involving either the use of counterpoises or excessive strain upon one end of an axis, tends away from stability; other things being equal, a mounting not possessing this characteristic is preferable to one that does.

(c) The first point being assured, characteristics of secondary importance may be considered. First among these is the possibility of long uninterrupted runs in RA—preferably from horizon to horizon.

(*d*) Accessibility of all parts of the sky, notably the region of the pole and the zenith.

(*e*) Compactness.

(*f*) Comfortable observing positions in all positions of the telescope.

(*g*) Applicability to as many types of telescope, and arrangement of the optical parts, as possible.

(*h*) Simplicity and cheapness, and/or adaptability to home construction.

14.3 Altazimuth

If it is to be possible to point a telescope in any direction it must be movable in two planes, and these are most conveniently made perpendicular to one another. In the altazimuth mounting they are respectively the horizontal (azimuth being measured westward from S) and the vertical (altitude being measured from the horizontal plane passing through the observer).

The commonest varieties of altazimuth head incorporate two of the following structural features: (*a*) altitude axis reduced to a single pivot situated below the tube at its point of balance; (*b*) altitude axis defined by a pair of trunnions attached to opposite sides of the tube, between which lies the tube's point of balance; (*c*) columnar azimuth axis; (*d*) azimuth axis in the form of a substantial circular disc running in a ring bearing. (*b*) and (*d*) are illustrated in Figure 87, (*a*) and (*c*) in Figure 86.

In the model illustrated in Figure 86 the centre of gravity, *C*, is not at the point of support, and considerable tensioning of the bolt *B* is there-

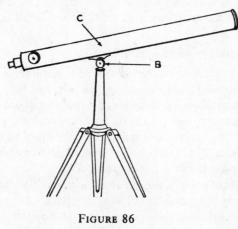

FIGURE 86

fore required to keep the tube stable in altitude. This militates against smooth and easy motion. The elongated column supporting the azimuth bearing would also introduce unsteadiness with any but the smallest instruments.

In the model shown in Figure 87, on the other hand, the tube swings between two points of support—a much more satisfactory arrangement.

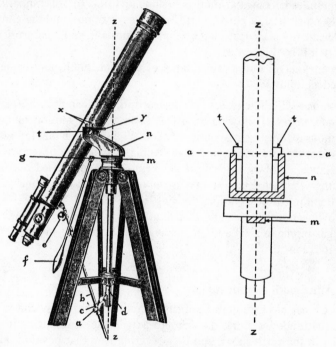

FIGURE 87
(By courtesy of Broadhurst, Clarkson & Co. Ltd.)

These trunnions, *t, t*, lie in a horizontal plane, their median line being the altitude axis, *aa*. The trunnions are supported by a pair of arms, *n, n*, which are bent back at an angle to the azimuth axis, *zz*, in order to allow the telescope to be directed at the zenith. The arms are mounted on the horizontal platform, *m*, which carries the bearings for the short and relatively massive azimuth axis. This bearing is of the plain annular type; freedom of movement about the altitude axis is controlled by means of the screws *x, x* which bear down upon the metal leaf, *y*.

The greater tube diameter of the Newtonian, compared with the sort of refractor usually found in amateur hands, necessitates some modification in the design of the head, but the same principles of solidity and lack of vibration, combined with ease of movement, apply (Figure 88).

Steadying rods and slow motions (section 14.24) improve the performance of any altazimuth head by increasing stability and smoothness of motion, while facilitating the following of the object in its diurnal motion. Indeed, if the altazimuth is to stand up to any sort of comparison with the equatorial, it must be fitted with slow motions, and these must be conveniently placed; the latter point is important, and it is surprising how often it is disregarded.

The advantages and disadvantages of the altazimuth may be summarised as follows:

Advantages: Simplicity, ease of construction, and therefore cheapness.

Disadvantages: Following an object in its diurnal motion involves continuous readjustment about both axes. While it is true that following soon becomes automatic—and that Denning's planetary observations, for instance, or Markwick's work on Venus, were all carried out with azimuths—the fact remains that a driven equatorial leaves *all* one's time and attention for observation, drawing, etc.

Two ingenious homemade altazimuths—one using wood and the other gas-piping (a rigid and readily available material)—are described in B. 164, 184.

14.4 Altazimuth modifications

Most types of altazimuth head can be more or less easily converted into equatorials by tilting the azimuth axis in the meridian until it is parallel to the earth's axis, and then securing it in that position. Such conversions, though cheap and easily contrived, are at best only make-shifts, and in view of the ease with which a solid and well-designed equatorial head can be made (if a professionally manufactured one cannot be afforded) they have little to commend them.

If the head is of the pillar type, it can be locked in a groove made in the surface of a hardwood prism whose base angle, ϕ, is equal to the latitude of the observing site, and whose side SN is horizontal and oriented south and north (Figure 89). The head is then mounted on a tripod or pier.

If the head is of the fork and base-plate type, the base can be counter-sunk and screwed to the face of a truncated hardwood prism whose apex angle, ϕ, is equal to the latitude of the observing site and whose side SN is oriented south and north (Figure 90). A counterpoise in RA is required, and this, unless it is forked, will interfere with the observation of low Decs.

The modification known as the Earl of Crawford's equatorial is

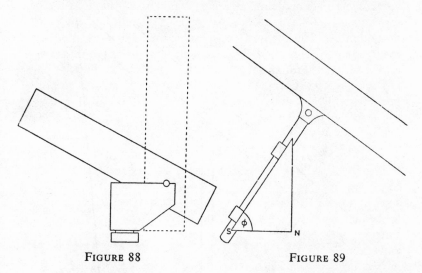

FIGURE 88   FIGURE 89

illustrated in Figure 91. The rod *XY* lies in the meridian; the point of attachment of the wire, *B*, is such that the angle *CBX* equals the latitude of the observing site; the length of the wire, *AB*, is adjusted according

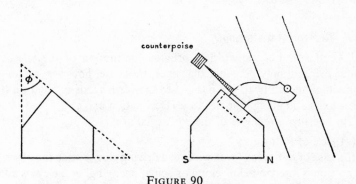

FIGURE 90

to the Declination of the object; the wire is kept taut by a counterpoise at *W*. A force applied horizontally to the eye end of the telescope will then make it follow the diurnal motion of a star. This method is only applicable to objects south of the zenith.

229

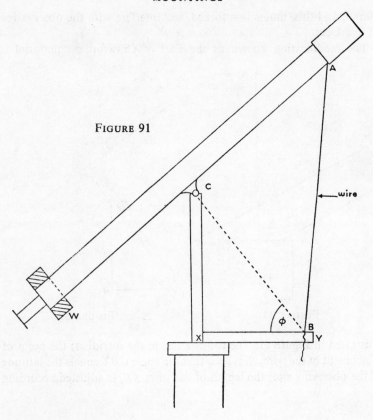

FIGURE 91

14.5 Equatorial mountings

Like altazimuths, equatorials provide for motion of the telescope about two mutually perpendicular axes; these are, however, no longer horizontal and vertical, but parallel to the earth's axis of rotation (the polar axis) and at right angles to it (Declination axis).

The specific *advantages* of the equatorial are:

(*a*) ease of following;

(*b*) ease of re-finding (even without a driving clock) when the observation is interrupted by cloud or ocular changing, or the eye is taken from the telescope for whatever reason;

(*c*) possibility of a mechanical drive, applied to one axis only;

(*d*) possibility of finding and locating approximate positions by means of graduated circles.

It is even more a *sine qua non* of photographic work than of visual

230

observation and though not strictly essential to the latter, its advantages over the altazimuth are so overwhelming that for all serious work two thoughts will not be given to the choice.

14.6 The German mounting

Invented by Fraunhofer. The telescope is carried at one end of the Dec axis, which is approximately bisected by the polar axis; thus the centre of gravity is offset with respect to the polar axis, and the weight of the telescope has to be balanced by a counterpoise at the other end of the Dec axis. This is also a very convenient place to mount subsidiary apparatus such as cameras, since no increase in the total counterpoise weight is then involved.

A typical German head, suitable for small instruments, is illustrated in Figure 92. The polar axis is carried on bearings inside the housing, a; the Dec axis is similarly mounted in b. The cradle, c, carrying the telescope, must be of sufficient length to give adequate support to the tube, and to ensure that it is always normal to the Dec axis. Declination and hour circles (see section 14.23) are at d and e respectively. The counterpoise f balances the weight of the telescope and cradle; accurate counterpoising is important, especially if the polar axis is clock-driven. Slow motion (see section 14.24) in RA is applied by the rod g, through a universal joint and worm, to the driving circle h. Platform i is for a driving clock (see section 15). Slow motion in Dec is applied by the rod j. Both axes are supplied with clamps. Adjustment of the elevation of the polar axis for any latitude from 0° to 70° can be made by slackening off the bolt k and reading the angle between the polar axis and the horizontal from scale l; most heads also provide for 5° to 10° of adjustment in azimuth. Levels (m, n) are also provided to facilitate the setting up of the head (see section 16).

The whole sky is visible from the two positions of the telescope (E and W of the pier) but not from each. With normally designed mountings, a refractor placed W of the polar axis cannot pass the meridian when directed at altitudes greater than a few degrees N of the zenith; reflectors can generally pass the meridian at higher altitudes than a refractor. At least half an hour's run past the meridian at all but very high Decs is desirable, but cannot always be provided; the operative factor, for altitudes N of the zenith, is the clearance between the eye end of the telescope and the mounting; for other altitudes it is often the obstruction caused by the presence of the driving gear.

Reversing the telescope involves a rotation of the polar axis through 12^h and of the Dec axis through 180°. The tube and its parts—e.g. field diaphragm or photographic plate—are therefore inverted relative to the field. The interruption to observation is an inconvenience in visual work, but is even more serious for the photographer, since the exposure

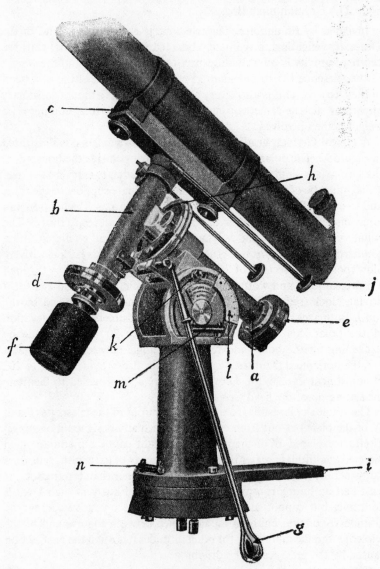

FIGURE 92
(By courtesy of W. Watson and Sons Ltd.)

cannot be continued on the other side of the mounting owing to the 180°
rotation of the plate.

Objects E of the meridian should be observed with the tube W of the
mounting. Observation can then be continued till the object is on or a
little past the meridian; the telescope is then moved back eastwards 6^h,
turned over in Dec, and brought once more to the meridian by a further
6^h eastward rotation of the polar axis; it is then reset in Declination,
and the observation continued.

Among the *advantages* of the German mounting are:

(*a*) easily portable with small instruments (cf. English mounting);

(*b*) only a single supporting pier or tripod is required (cf. English
mounting);

FIGURE 93

The author's clock-driven German equatorial,
mounting a 4½-in refractor and camera

(c) compact (cf. English mounting);

(d) visibility of the whole hemisphere of the sky.

Its *disadvantages* are:

(a) necessity of reversing when passing the meridian;

(b) overhung centres of gravity, involving a counterpoise on the Dec axis and overloading of the upper end of the polar axis.

Among the numerous descriptions of home-made German heads which are to be found in the literature, the following may be specially mentioned: B. 160, a wooden head for a small refractor, very similar to that described in B. 439; B. 181, a metal head suitable for apertures up to about 5 ins; B. 156–7, lightweight heads suitable only for a camera and small guide telescope.

14.7 The English mounting

Sometimes also called the Yoke mounting (Figure 94). The points of support of the tube are on either side of it, and the Dec axis is borne in a cradle which is supported at both ends; this cradle, or yoke, constitutes

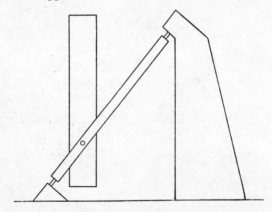

FIGURE 94

the polar axis. Thus no counterpoise is required, and both ends of the polar axis take the weight of the telescope, making for greater stability and freedom from flexure in the axes. It is particularly suitable, therefore, for large instruments (e.g. the 100-in Hooker telescope at Mt Wilson).

While reversal at the meridian is eliminated, the region round the pole is inaccessible*—a drawback which has led to the development of the

* If Airy's 12¾-in at Greenwich is regarded as an English equatorial, then this statement requires modification, since the pole is accessible with it. Airy's mounting may, however, perhaps more correctly be regarded as a halfway stage to the Horseshoe mounting.

Horseshoe mounting (see section 14.11). Also it is not a design that is well adapted to use with refractors or Cassegrains, owing to the great height involved.

14.8 The modified English mounting

Also known as the English, or Cross-axis, mounting (Figure 95). The yoke of the English mounting has here been reduced to a single beam, at whose centre lies the Dec axis bearing; as in the German mounting,

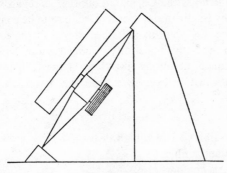

FIGURE 95

the Dec axis must be counterpoised. The 72-in reflector of the Dominion Astrophysical Observatory (Victoria, B.C.) is a well-known example; one of the earliest was Admiral Smyth's equatorial at Bedford (*c.* 1830).

Advantages:
  (*a*) two-point support of the polar axis reduces the heavy load on its upper end (cf. German);
  (*b*) polar region accessible (cf. English);
  (*c*) equally suitable for refractor and Newtonian (cf. English).

Disadvantage: Dec axis counterpoise (cf. English).

14.9 The Astrographic* mounting

Designed to combine the advantages of the English and the German mountings by means of the overhanging polar axis:
  (*a*) no reversal at the meridian, which involves not merely the interruption but the termination of exposures;
  (*b*) compactness.

  * The term is used here in its wider sense. The mountings of the photographic telescopes used for the International Astrographic Chart (Carte du Ciel) were not generally of this form.

Condition (*a*) is satisfied if the upper part, *A*, of the pier (Figure 96) is longer than the half-length of the guide telescope. The astrographic is not, therefore, a design which is generally applicable to telescopes of normal focal length.

14.10 The Fork mounting

Whereas the Modified English mounting consists in effect of one side of a Yoke mounting, the Fork is in effect the lower half of an English mounting; it resembles the German mounting as regards its polar axis bearings, and the English as regards its Declination bearings (Figure 97). To avoid flexure, the polar axis, its bearings, and the fork have to be massive. Given the requisite solidity of construction, this is one of the most useful forms of mountings for Newtonians. The 48-in Schmidt at Mt Palomar, the Mt Wilson 60-in, and the new Lick 120-in are all mounted in this way.

Advantages:

(*a*) pole accessible to Newtonian or camera;

(*b*) no meridian trouble;

(*c*) no Dec axis counterpoise.

Disadvantages:

(*a*) weight of axis and fork bear outside the points of support of the polar axis; flexure of the fork difficult to overcome with large instruments;

(*b*) unsuitability for refractors and Cassegrains, there being a wide region, centred on the pole, which is inaccessible.

An interesting Fork mounting, homemade out of reinforced concrete, is described in B. 168.

14.11 The Horseshoe mounting

Also known as the Open Yoke type. The mounting designed by Russell Porter for the Mt Palomar 200-in is the prototype. It aims at combining the advantages of the German and English types, while avoiding the weakness of the Fork (Figure 98).

Advantages:

(*a*) no meridian reversal;

(*b*) polar region accessible;

(*c*) all masses lie between their points of support;

(*d*) the three-point support of the yoke provides great rigidity in all positions of the telescope.

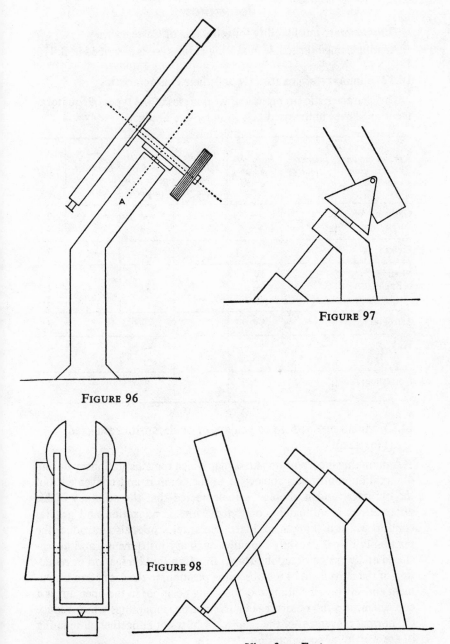

FIGURE 96

FIGURE 97

FIGURE 98

View from South (tele-
scope tube omitted to
simplify the diagram)

View from East

Disadvantage: unsuitability for refractors or Cassegrains.

An interesting homemade mounting of this type is described in B. 170.

14.12 Summary of equatorials and their characteristics

The characteristic strengths and weaknesses of the types of equatorial mentioned are summarised below under five heads:

A=Newtonian B=Cassegrain or refr.	Accessibility of pole	No reversal at meridian	Tube inside Dec axis supports	Dec axis inside polar axis supports	Compactness	Total
German: A	*	—	—	—	*	2
B	*	—	—	—	*	2
English: A	—	*	*	*	—	3
Modified English: A	*	—	—	*	—	2
B	*	—	—	*	—	2
Astrographic	*	*	—	—	*	3
Fork: A	*	*	*	—	*	4
B	—	*	*	—	*	3
Horseshoe: A	*	*	*	*	—	4

14.13 Mountings with fixed oculars or oculars with restricted movement

A mounting incorporating an ocular which remains stationary during observation contributes somewhat to the comfort and convenience of the observer; and it should be remembered that the accuracy of his observations is intimately connected with his relaxation and general comfort whilst making them. Fixed-ocular telescopes also introduce the possibility of permanently mounted auxiliary instruments, and an enclosed and warmed observing-room. But even the elimination or reduction of the movement of the observing position, in the case of an instrument entirely (i.e. including the observing position) in the open air, is a convenience to the observer. For this reason, mountings which reduce the normal movement of the eyepiece, whilst not altogether eliminating it, are worthy of consideration.

The possibilities of design are summarised in the following Table. Mountings of type (*a*) (known as polar telescopes, for obvious reasons) provide a completely stationary observing position; so do telescopes fed by a siderostat or coelostat. Types (*b*) and (*c*) reduce the normal range of ocular displacement.

The two disadvantages to which these mountings as a class are subject (they are not individually subject to both, or even either, of them) are:

(*a*) difficulty in determining the orientation of the field;

(*b*) complexity and size, with resultant expense.

Axis of ocular	Observing position	Axis of ocular directed toward	Locus of ocular for all positions of the telescope
(*a*) in polar axis	fixed (ocular rotates)	N/S celestial pole	point
(*b*) parallel to polar axis	restricted	N/S celestial pole	circle centre: polar axis radius: distance of ocular axis from polar axis
(*c*) in Dec axis	restricted	polar axis	circle centre: polar axis radius: distance of ocular from polar axis
(*d*) *telescope fed by siderostat*	fixed (field rotates)	usually horizontal and in meridian	point
(*e*) *telescope fed by coelostat*	fixed (no field rotation)	usually horizontal, any azimuth, or in meridian	

14.14 The Grubb polar mounting

Structurally the simplest of the polar telescopes. Economically, however, it must be beyond the reach of most amateurs. Although some types of polar telescope, not adapted to the provision of an enclosed observing-room, are quite easily encompassed by the amateur pocket and home-making ingenuity, this is not true of the mountings which, like the Grubb, can conveniently isolate the observer from the intrument. It is unfortunate that the small observer-enclosed polar refractor with which Messrs Ottway were experimenting after the war has never been put into commercial production.

At the lower end of the Grubb telescope (Figure 99) is an equatorially mounted plane mirror, to whose English-type mounting the telescope is attached: *PP'* is the polar axis, *DD'* the Dec axis. Relative to a vertical line through the eyepiece, the field rotates; but if the ocular rotates together with the rest of the instrument once in 24 hours there will be no relative motion between it and the field. The same result would be achieved by disconnecting the main telescope from the flat mounting,

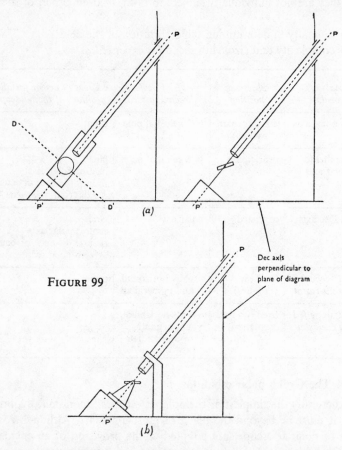

(a)

FIGURE 99

Dec axis perpendicular to plane of diagram

(b)

fixing it on a permanent mounting, and simply rotating the eyepiece at the same angular velocity as the mirror by means of a small driving clock (Figure 99(b)).

A typical existing example of this design of mounting is the 12-in polar telescope at Harvard, designed in its details by Gerrish.

The specific weaknesses of the design are:

(a) the position of the objective near the ground makes it very susceptible to convection currents, with deterioration of seeing;

240

(*b*) polar region inaccessible;

(*c*) low Decs are unobservable unless the flat is large compared with the objective.

14.15 Cambridge (Sheepshanks) coudé

The main tube (Figure 100) forms the polar axis, *PP'*. A side tube, carrying the objective, is mounted so as to be able to move in N and S directions, i.e. it is mounted on a Dec axis. At the junction of the two

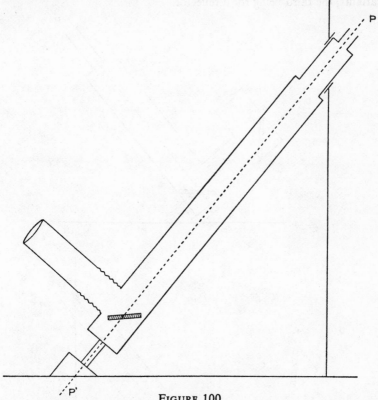

FIGURE 100
Dec axis perpendicular to plane of diagram
through centre of flat

axes is situated a rotatable flat which is connected to the Dec axis by a somewhat intricate linkage which ensures that as the latter is turned, the normal to the mirror always bisects the angle between the axes of the two sections of the tube.

As against the Grubb mounting, its objective is raised higher from the ground. On the other hand, the polar regions are still inaccessible, loss of aperture will occur at low S Decs unless the flat is considerably larger than the objective, and the image suffers a one-plane reversal (owing to the single reflection).

14.16 Paris (Loewy) coudé

The *coudé* (or 'elbowed') principle of telescope design was initiated by Loewy at the Paris Observatory in 1891. Figure 101 shows three variants, the third being for a reflector.

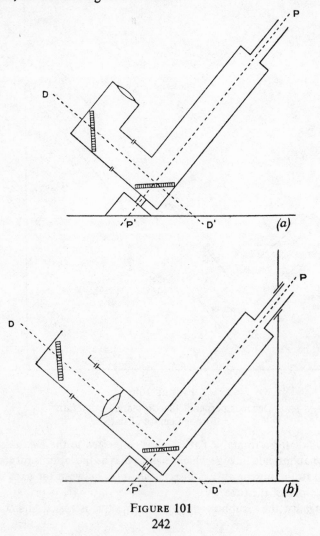

FIGURE 101

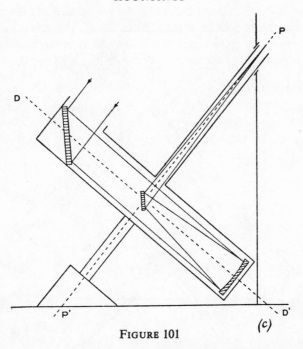

FIGURE 101

(c)

Its characteristic feature is the two large flats, permanently mounted at right angles to one another. The second reflection corrects the mirror-reversal of the first, and the introduction of a second flat makes all Decs accessible, though with an unavoidable small loss of light grasp. It shares the general disadvantages of polar telescopes: it is cumbersome, expensive, and outside the scope of most amateurs' constructional ability. Its specific disadvantage is that the orientation of its field varies with the Dec of the centre of the field. For a given Dec, however, the field remains motionless relative to the ocular. Reversal is generally necessary soon after the meridian passage of an object.

14.17 The Ranyard mounting

Two variants of this type of mounting are shown in Figure 102. Others are described in B.182. The cone of rays from the primary mirror is intercepted first by a small convex mirror, mounted axially near the focus, and then by a still smaller flat which reflects it up or down the hollow polar axis. The flat is geared to the Dec axis by a linkage which ensures that it rotates at half the angular velocity of the axis. An incidental advantage of this mounting is the large equivalent focal length provided.

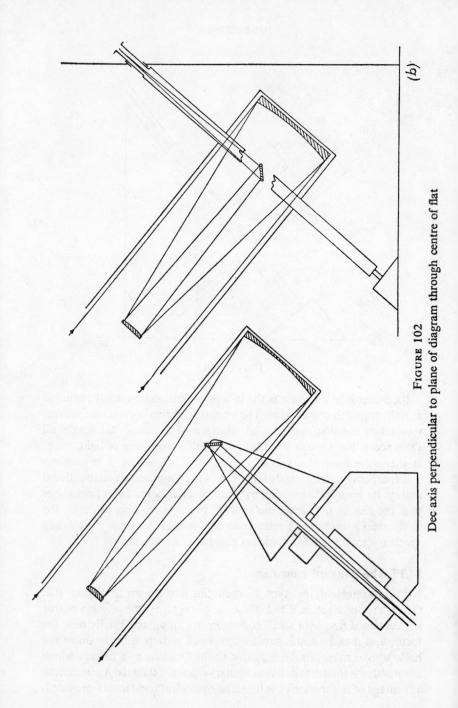

FIGURE 102

Dec axis perpendicular to plane of diagram through centre of flat

14.18 Pasadena and Springfield mountings

The optical arrangement of these two polar mountings is the same (Figures 103, 104). In the latter—which was designed by Russell Porter,

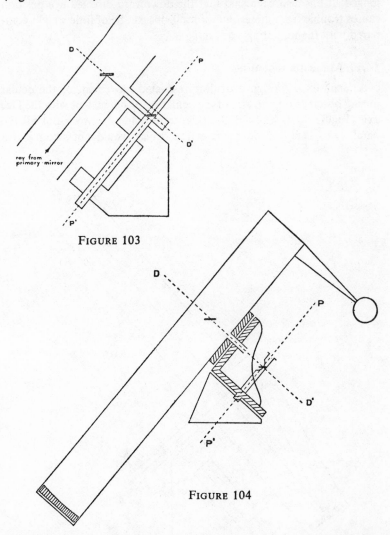

FIGURE 103

FIGURE 104

designer of the Mt Palomar 200-in—the Dec axis is reduced to a minimum, in the interests of stability. Both it and the polar axis are reduced to mere studs; these are attached to circular plates, again for stability. For further details of construction, see B. 101–2.

The odd number of reflections introduces mirror-image reversal, and the light loss at three reflecting surfaces is far from negligible. A smoother drive would probably be obtained if the polar axis were prolonged. It has also been said that the downward-directed eye position causes trouble from motes, but of itself this can be of little weight compared with the mounting's advantages.

14.19 Manent's mounting

A 'semi-fixed' design providing restricted movement of the ocular during observation, the axis of the ocular being coincident with the Dec axis (Figure 105). Thus, if the tube were made to sweep out all the parallels of Dec and hour circles the ocular would not sweep out a

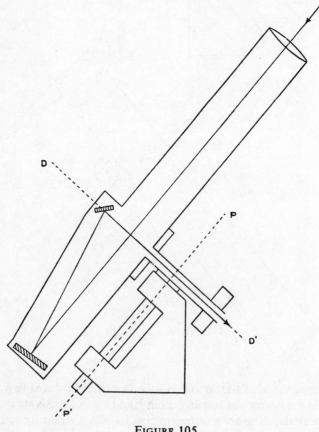

FIGURE 105

sphere, as with ordinary mountings, nor a point, as with polar tele-scopes, but a circle. Additional advantages are its great compactness (allowing a long-focus optical system to be housed in a small tube) and the protection offered to the reflecting surfaces by their enclosed position—though this is not now such a telling advantage as it was in the days before aluminising.

Chief weaknesses of the design are: only one end of the Dec axis being available for the ocular, this assumes awkward positions when the tube is far from the meridian, and a diagonal has to be used; the loss of light at the two additional reflecting surfaces.

14.20 Pickering's mounting

In this design (Figure 106) the ocular may be mounted in either of the hollow Dec trunnions—the flat being rotatable about the optical axis of the primary mirror—which gives the choice of a less uncomfortable

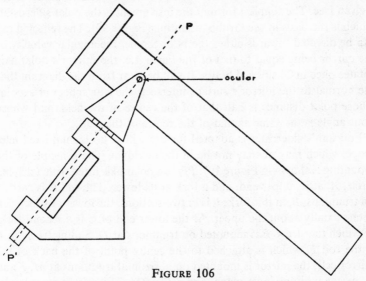

FIGURE 106

Dec axis perpendicular to plane of diagram through centre of flat and ocular

observing position when the tube lies far out of the meridian. The dis-placement of the ocular from the polar axis being so small, the observing position remains at virtually the same height above the ground at all times (cf. Newtonian). A further advantage of this mounting is the comparative ease with which it can be homemade. Being restricted to

Declinations S of the zenith it is of value to the lunar or planetary, rather than the sidereal, observer.

Apart from the characteristic drawbacks of the fork type of mounting, Pickering's mounting involves the use of a heavy counterpoise, or alternatively a considerably longer tube than that of the equivalent Newtonian. A fuller description will be found in B. 175 or 176.

14.21 Siderostats

These are a development for astronomical purposes of the more primitive heliostat, which is chiefly used by surveyors. The heliostat—a two-mirror arrangement, one mirror being equatorially mounted—reflects the image towards the northern horizon (usually between NE and NW). Its disadvantage is that the field rotates in a non-uniform manner (the rate of rotation varying with hour angle), which makes its orientation impossible, or very difficult, to establish.

The field of a siderostat also rotates, but this rotation is uniform for a given Dec. The simplest form of the instrument is the polar siderostat, which is the basis of the Grubb mounting (Figure 99). The reflected ray can be directed towards either the N or S pole, the angular velocity of the mirror being equal to that of the Earth, i.e. the mirror's polar axis rotates once in 24 sidereal hours. It will be clear from the diagram that the normal to the mirror's surface intersects the star sphere at a point whose polar distance is half that of the centre of the field, and whose hour angle is the same as that of the centre of the field.

Foucault's siderostat is adapted for use with a horizontal fixed telescope, which may or may not lie in the meridian. The principle of the mounting is shown in Figure 107. PP' is a polar axis fitted with a driving circle, A, at its upper end, and a fork at its lower. The rod C is carried on trunnions, B, in this fork; it is in two sections, the lower of which can rotate axially about the upper. At the lower end of C is a second fork, in which the sleeve E is mounted on trunnions at D. Sliding freely in E is the rod F, which is attached to the centre point of the back of the mirror cell; the mirror is mounted on horizontal trunnions at M. F and C always lie in a single plane, and $BD=BM$. The mirror can also be swung in a horizontal plane about G, which lies vertically below B.

If the telescope is mounted horizontally in the meridian, the mirror mounting is revolved about G until M is due south of B, i.e. BM lies in the same vertical plane as the polar axis. Since triangle DBM always remains isosceles, if BD is maintained parallel to XY, the mirror will orient itself in such a manner that the reflected ray remains parallel to BM. The polar axis is then driven to maintain the plane $PP'D$ parallel to the hour circle of the object; angle $PP'D = \delta + 90°$. If, now, $\phi =$ latitude of the observing station, and $\theta =$ the polar distance of the centre of the field:

> field is fixed when $\theta = \phi$,
> field rotates clockwise when $\theta < \phi$,
> field rotates anticlockwise when $\theta > \phi$.

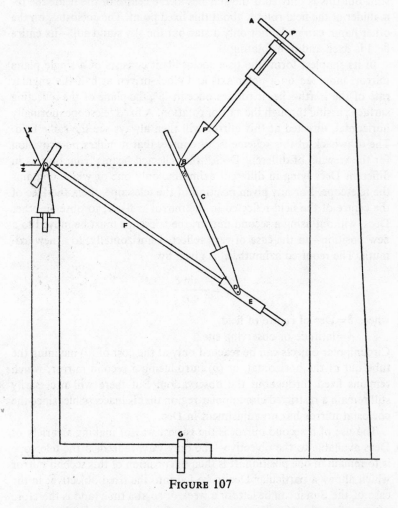

FIGURE 107

If it is required to set up the telescope in some azimuth outside the meridian, BM is simply moved into the required direction by revolving the mirror mounting about G. The pressure of C against F forces the mirror at the same time to turn about its two axes, the sleeve E both turning and sliding on F.

14.22 Coelostats

The siderostat, as its name implies, can make the image of a star stand still. But this is only so if the star lies at the centre of the field; the remainder of the field rotates about this fixed point. The coelostat, on the other hand, can make not only a star but the sky stand still—its entire field is fixed and non-rotating.

In its simplest form (the true coelostat) it consists of a single plane mirror, mounted on a polar axis and clock-driven at half the angular rate of the Earth—i.e. it rotates once in 48^h, the plane of the reflecting surface passing through the axis of rotation. A fixed telescope, normally horizontal, directed at this mirror will then always see the same field. The drawback of this scheme is, of course, that it makes no provision for the viewing of different Decs,* the reflected beams from objects in different Decs lying in different azimuths, only one of which contains the telescope. For any given position of the telescope, then, the Dec of the centre of the field reflected in the mirror is fixed; to observe other Decs, without using a second mirror, the telescope must be moved to a new position—in the case of rays reflected horizontally, to a new azimuth. The required azimuth, a, is given by

$$a = -\frac{\sin \delta}{\cos \phi}$$

where $\delta=$ Dec of centre of field,
   $\phi=$ latitude of observing site.
Circumpolar objects can be reached only at the cost of (a) inclining the tube out of the horizontal, or (b) introducing a second mirror, which remains fixed throughout the observation. But there will necessarily still remain a restricted circumpolar region that is inaccessible, since the coelostat mirror has no adjustment in Dec.

The use of a second mirror is the easiest way of making a variety of Decs available to the objective—the only way, indeed, if the telescope is to remain in one position. It is then the position of this second mirror which allows a particular Dec to be fed into the fixed objective; in the case of the Sun it can be left for a week or so at a time (and is therefore often called the 'fixed' mirror—as in fact it always is through an

---

* This is no disadvantage in the observation of solar eclipses, where the single-mirror coelostat is perfectly effective.

observation) owing to the Sun's slow motion in Dec. In changing from one Dec to another, both the position and inclination of the fixed mirror must be altered.

The Sun presents the most favourable conditions for observation by means of a coelostat with a fixed secondary flat; not only is its rate of change of Dec slow, but it never leaves approximately equatorial regions. The coelostat can be arranged to deliver either a horizontal (e.g. the Snow telescope) or a vertical beam to the objective (e.g. the Mt Wilson Tower telescopes).

14.23 Setting circles

If the usefulness of a telescope is increased 100% by mounting it equatorially, this is primarily because of the increased ease of following an object in its diurnal motion, increased accuracy in photographic work, and the possibility of using certain accessories which are impracticable with an altazimuth. The addition of circles to the equatorial increases the instrument's usefulness by perhaps a further 10%. Circles are a convenience, but by no means a necessity. Given a good star map, all the preliminary work of identifying the neighbourhood of a faint object can be done during the day, and the amount of observing time wasted in locating it need be very small. Only for finding objects during the daytime do circles have an unquestionable advantage.

In any case, very finely divided circles are not only unnecessary but may be positive nuisance, owing to the time spent poring over the verniers and the probable destruction of the eye's dark adaptation by the use of a fairly bright light to make the graduations visible. Position circles are, after all, for finding and not for accurate position measures—for which the correct instrument is the micrometer—and this function they perform perfectly satisfactorily if they permit the identification, not of a point object, but of a LP field.

The order of precision in the graduation of commercially manufactured circles is:

Diameter of circle	Dec circle	Hour circle
4 ins	5′	20^s
5 ,,	5′	15^s
6 ,,	1′	5^s
8 ,,	20″	5^s
12 ,,	20″	2^s

Finely divided circles justify the fitting of verniers, but it must be remembered that this is only true if the adjustment of the equatorial axes is of a comparable precision. Circles of sufficient precision to enable an object to be picked up with the finder or in a LP field can easily be home-made; they do not require reading verniers, a simple index mark being sufficient.

Two obviously faulty procedures should be avoided in the graduation of circles: (a) if the graduations are to be made direct on the rim of the circle with the aid of some form of protractor, the latter must be larger, not smaller, than the circle; otherwise errors in the graduation of the protractor will be increased, not reduced, in their projection on the circle; (b) if the graduations are made on a long strip of some material which is to be wrapped round and fixed to the rim of the circle, the two end points must be accurately fixed at a distance apart equal to the circumference of the circle, and the space between them filled in with the requisite number of equal graduations; to start at one end and work along to the other will inevitably introduce cumulative error which will be discovered when 360° is reached.

Method (b) is easier to carry out than (a). The strip to be graduated may be of thin brass, marked with a steel engraving tool; celluloid, painted black on one side, on which the graduations are cut with a sharp knife; paper, on which the graduations are marked with a fine-nibbed pen, which is fixed to the circle rim and protected from the weather with glue size.

Graduations can easily be made by any of these methods at 0·05-in intervals; about 0·1-in intervals, however, are generally to be preferred. Assuming that the Dec circle is to be graduated in degrees at 0·1-in intervals, and the hour circle in 0·1-in divisions of 6^m (preferable, for interpolation by eye, to 5^m intervals), the diameters of the two circles will be respectively 11·46 and 7·64 ins. Rather than decide on the exact linear value of one graduation and then calculate the diameter of the circle, however, it is better to decide on a convenient circle diameter which will allow the graduations to be about 0·1 ins, and then to subdivide a distance equal to its circumference. Figure 108 illustrates, to scale, the appearance of the graduations of a Dec and hour circle, both 12 ins in diameter: they are easy to read at a glance (the graduations are 0·105 and 0·157 ins respectively), and with interpolation by eye can be read to 15' and 2^m.

The method of marking the circles is as follows. Wrap the strip round the rim of the circle with an overlap of about an inch, and mark off the circumference on it. Pin the strip to a drawing-board and lay off a perpendicular to it through one of the two marks indicating the circle's circumference. On the edge of a separate sheet of paper divide any convenient length, greater than the circumference of the setting circle, into the required number of graduations (e.g. in the case of the 12-in hour circle already mentioned, the graduations on the subsidiary scale could be 0·2 ins apart). Now pin the scale to the drawing-board so that one

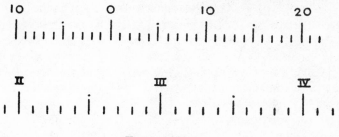

FIGURE 108

end of it coincides with one end-mark of the strip, and the other coincides with the perpendicular through the other end-mark. With a set-square, drop a perpendicular to the strip from every graduation of the scale. These are the required divisions of the circle.

Circles may be graduated in any of the ways illustrated in Figure 109. But bearing in mind that the Dec circle must read 0° and the hour circle 0^h when the telescope is lying in the meridian with the celestial equator passing through the centre of the field, and that, furthermore, reversal of the instrument involves a 180° rotation of the Dec axis and a 12^h rotation of the polar axis, it will be seen that method (a) (for both Dec and hour circles) is the most convenient to use.

With either the polar axis circle fixed to the axis and its index to the mounting, or vice versa (the circle then reading hour angle), the procedure for finding an object whose coordinates are known is as follows:

(i) Swing the telescope in Dec until it is pointing at the celestial equator (0°). Depress or elevate the telescope through an angle equal to δ, and clamp. The actual reading of the circle will depend on the way in which it is graduated: in case (a) (Figure 109) it will be δ; in case (b) it will only be δ in one position of the telescope; in case (c) it will only be δ in one position of the telescope and on one side (N or S) of the equator.

(ii) The RA of the meridian, Local Sidereal Time, is given by the observatory clock (T); the RA of the object is assumed known (α). Then:

If α > T, object is E of the meridian: swing telescope to W side of the pier.

If $T>\alpha$, object is W of the meridian: swing telescope to E side of the pier.

Start the driving clock (if any). Set the telescope in the meridian (hour angle 0ʰ) and swing it eastwards through the angle $\alpha-T$ or westwards through the angle $T-\alpha$, as the case may be. Clamp. The object should

Declination circle

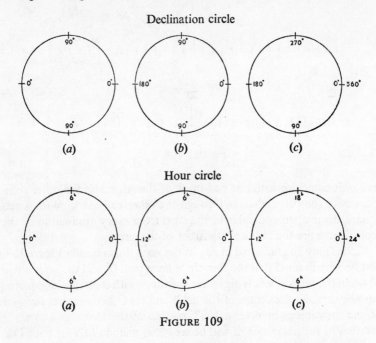

Hour circle

FIGURE 109

then be in the field of the finder. Bring it to the intersection of the cross-wires with the slow motions. It will then be in the telescopic field.

In practice it is usual to add, say, 1ᵐ to T, to cover the time taken in setting the telescope. Then the object will be in the field of the finder at time $T+1^m$.

To discover the approximate coordinates of an object at the centre of the field, the procedure is simply the reverse of the foregoing: δ is given by (or quickly calculated from) the Dec circle reading; $\alpha=T\pm H$ (according to whether the object is E or W of the meridian), where T is given by the observatory clock and H by the hour circle.

With a fixed hour circle a separate calculation of $\alpha-T$ or $T-\alpha$ thus has to be made each time an object is found by means of its coordinates. To avoid this inconvenience a slipping polar axis circle with two verniers, or two reading indices, is usually employed. Off the single scale T is read by one index and RA by the other, eliminating the separate calculation of H each time. One index is fixed permanently to some immovable part of the mounting—generally, for convenience, central on the S side; the circle rotates freely on the polar axis but can be clamped in any position relative to this fixed index; a second circle, ungraduated but carrying an index, rotates with the polar axis; this index coincides with the fixed index when the telescope lies in the meridian. The procedure, having set the telescope to the correct Dec, is then as follows: set the telescope in the meridian (same circle reading with the two indices): slip the circle till the fixed index reads α; clamp the circle; rotate the polar axis till the second index reads T. The object should then lie in the field of the finder.

14.24 Steadying rods and slow motions

Steadying rods are fitted to most altazimuths. They attach the eye end of the telescope tube to some point on the mounting—the pillar, the tripod, or the trunnion fork. Individual makers have their own particular designs, but the general principles are illustrated in Figure 87. When the locking screw a is slackened, the rod b can slide freely in the sleeve c, which is in turn connected with the sleeve d; c can rotate in a vertical plane and d in a horizontal plane, allowing the tube to be moved in altitude and azimuth.

With instruments over about 3 ins aperture, slow motions are so useful as to be virtually essential. Altazimuths require slow motions in both altitude and azimuth; equatorials require slow motion in RA to correct variations or inaccuracies in the drive, but a slow motion in Dec is less important providing the mounting itself is accurately adjusted. Whatever the type of mounting, the slow motions must be efficient: in particular, they must be slow enough, and there must be no looseness or backlash in their operation.

An altazimuth steadying rod is easily converted to a device for applying slow motion adjustments by the addition at e (convenient to the observer's hand) of a rack-and-pinion action, as in Figure 87. Alternatively, the end of the rod may be threaded and connected to the telescope by a long sleeve, threaded internally and milled externally. Rotation of this sleeve will then vary its position along the rod, applying slow motion in altitude (Figure 110). A slow motion in altitude for a reflector mounted on an altazimuth is illustrated in Figure 111; it, or something similar, can very easily be made at home. The set-screw a clamps the sleeve b permanently to the collar c; this in turn is fixed to the hinge d. The telescope tube is so balanced that its upper end tends to drop, forcing the milled head *e*—which is threaded to the rod *f*—against the top of the sleeve *b*. Rotation of *e* will provide slow motion in altitude.

Slow motion in azimuth (see Figure 87) is usually provided by means

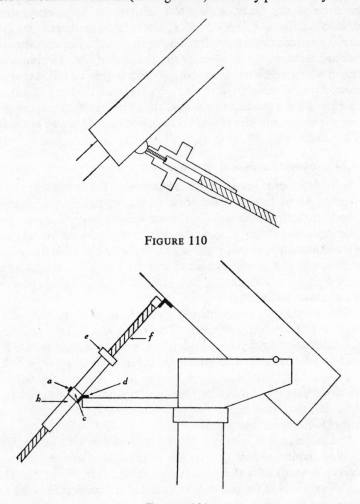

FIGURE 110

FIGURE 111

of an endless screw, actuated by a hand-rod *f*, to which it is attached by a universal joint (often called a Hooke's joint), and meshing with a toothed collar incorporated in the azimuth axis. The quick motion in azimuth is independent of the slow motion screw.

With equatorial heads, slow motion adjustments are commonly

applied by means of a worm and sector, or worm and whole wheel. The worm is controlled from the observing position by a flexible connection, a rod and universal joint, or some similar device. In the case of a power-driven equatorial, provision must be made for the slow motion rotation of the polar axis in either direction without interfering with the drive. This is usually achieved either by modifying the rate of drive (e.g. by means of a rheostat in the case of an electric drive, or by frequency control—see section 15) or by means of manual adjustment working through some form of differential. This may consist of 'sun-and-planet' pinions, similar to that in the back axle of a car. A common form, operated by cords, is illustrated in Figure 112. The pinion C is driven by the motor. The shaft F transmits the drive to the telescope via gearing and worm,

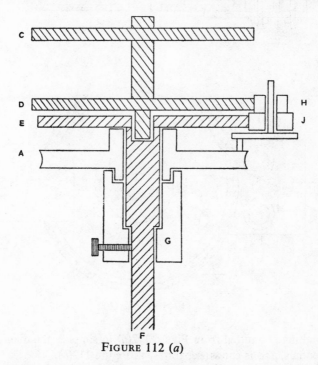

FIGURE 112 (a)

not shown in the diagram. Between C and F lies the differential whereby the drive can be accelerated or retarded without altering the speed of C and interfering with the motor. D is mounted on the same shaft as C. Its rotation is transferred to E, which is mounted on F, by means of the two small pinions H and J. These are mounted on A which, supported on the flange G, can rotate freely about F. Rotation of A will make H and J

travel round the pinions E and D, either in the direction of the rotation of C or against it. The transmission of the drive from D to E can therefore be accelerated or retarded by suitable rotation of A. This rotation is effected by means of a continuous cord, operated by the observer; its

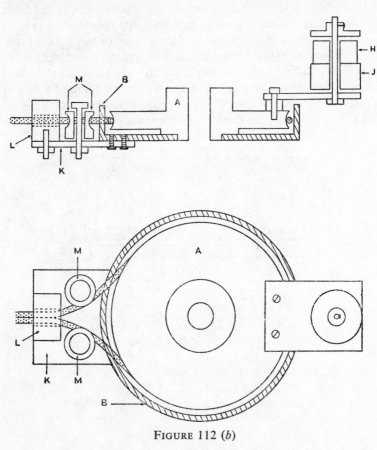

FIGURE 112 (*b*)

mounting is omitted from Figure 112(*a*) to simplify the diagram to its essentials, and is shown separately in Figure 112(*b*).

A is mounted in a shallow tray B, to which is fixed the platform K. On K are mounted a guide for the cords, L, and two freely rotating pulleys, M. The cord passes completely round A, between the pulleys, and out through the guide to the observer. If the two ends of the cord are joined together, it should be pulled through before observation starts, so that the join is in the neighbourhood of the observer's hands.

A slow motion in Dec—or in RA in the case of a hand-driven equatorial—can easily be made with a worm wheel of not less than about 200 teeth, grub-screwed to the axis, and actuated by a worm to the end of which a rod is attached through a universal joint, which may be nothing more complicated than a short length of tough spiral spring.

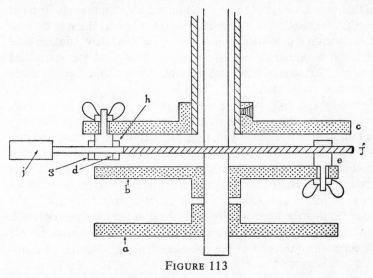

FIGURE 113

Another easily made Dec slow motion or manual drive is illustrated in Figure 113. As a drive it is less satisfactory than the foregoing, however, since it cannot provide a drive for unlimited periods without resetting. Two flanges, a and b, are screwed on to the threaded end of the axis; a third, c, is fixed to the threaded end of the axis housing by a grub-screw; a is simply a locking device which, when screwed up against b, prevents the latter from turning on the axis. Holes are drilled near the edges of b and c; through these are passed the brass pillars d and e, and held by wing nuts slackly enough for the pillars to rotate. Both pillars are drilled, the hole in e being threaded. The long threaded rod, f, is held firmly in position in d by the flanges g and h, which are soldered to it. The end of the driving screw, j, is connected to the observing position by the usual flexible cable or rod and universal joint. This driving handle can be brought to either side of the mounting as required, by slackening k, turning b and c as far as required, locking c to the axis housing again, and screwing up a against b. Rotation of j will then drive f in either direction through e, causing b to rotate either clockwise or anticlockwise relative to c.

259

14.25 The support for the head

An altazimuth head, a German equatorial head, or a fork mounting, if small, has to be supported at a convenient height above the ground.

Solidity and insusceptibility to vibration—whether from the wind, traffic, or the movements of the instrument itself—are all-important. A portable tripod, designed to be stood on a table, window-sill, or other makeshift foundation, is worse than useless. The best thing to do with such a mounting is to weld or bolt the toes of the 'claw' to three steel plates and mount these on 2-in morticed hardwood stays arranged radially at 120° to each other and mounted firmly on the top of a single pier sunk in the ground.

Generally speaking, a single vertical pier is preferable to a tripod, since it renders the zenith accessible by elevating the telescope from any azimuth. A stout wooden tripod, firmly bedded in the ground, and preferably fitted with stretchers (see Figure 87), serves quite well with altazimuth or German heads carrying instruments of up to about 4 ins aperture. In the case of an equatorial it is preferable to fit the points of the tripod into sockets in the top of concrete columns sunk in the ground, so that the horizontality of the base of the head can be found again immediately, after the instrument has for any reason been moved; truly accurate adjustment of the polar axis cannot, however, be maintained in this way.

Vertical piers can be constructed of masonry, concrete cast in a mould (it should be reinforced), or a steel pipe or pillar embedded in concrete. An iron pillar is the most affected by traffic vibration; a wooden tripod, the least. Vibration of this sort consists of surface waves, and if instead of sinking a concrete pier 4 ft in the ground it is sunk to a somewhat greater depth in the bottom of an 8-ft hole (comfortably larger than the pier, and revetted), complete freedom from vibration can be secured, even in the vicinity of a railway or a highway carrying heavy traffic. Since this involves an additional 8 or 9 ft of concrete, it is a solution that will only be applied if the problem is acute.

Vibration caused by underground trains can only be circumvented by moving the observing site to another district, or by restricting observation to those few hours when the trains are not running—neither a very palatable solution.

If observation is carried out in a hut (see section 17) the pier must pass through a hole in the floor large enough to avoid all contact between the mounting and the floor-boards.

SECTION 15

TELESCOPE DRIVES

15.1 Introduction

The drive is a mechanism that imparts to the polar axis of an equatorial a rotation whose angular velocity is such that the axis completes one turn in 24 sidereal hours; provision for varying the speed to the solar or lunar rate may also be incorporated.

Essentially it consists of (a) a source of power; (b) transmission connecting (a) with the polar axis; (c) some form of governor to control the drive at the required rate and to smooth out short-period variations in the output of (a).

Varying degrees of precision are demanded, according to the work that is going to be done with the telescope: (a) for visual work, it should at least be accurate enough to keep an object near the centre of a HP field for, say, half an hour; (b) photographic work with short-focus cameras demands greater precision, if guiding is not to become intolerably tedious; (c) photographic work with long-focus cameras demands the highest precision possible.

For (b) and (c) a drive may be considered essential—though surprising and perhaps rather pointless feats have been achieved with hand guiding; for (a) it is not essential, but it is so easily effected and so convenient (especially when making drawings) that, once experienced, it will not subsequently be forgone.

15.2 Sources of power and methods of control

For approximately accurate driving, such as is an enormous boon to the visual worker, some most extraordinary mechanisms have from time to time been brought into successful service: the works of grandfather clocks, Meccano, air-bladders with adjustable exhaust cocks, falling weights regulated by the flow of sand or mercury from a cylinder with an adjustable vent, and the like. With such drives, not suitable for large instruments, some form of belt transmission is adequate.

Omitting the more fantastic, methods in common use include:

(a) clockwork, mechanical governors, and falling weights in a variety of combinations;

(b) electric motors, controlled either by mechanical governor or pendulum;

(c) synchronous motors, relying on a frequency-regulated grid supply or employing either some form of frequency regulation or a frequency-controlled private supply.

15.2.1 Clockwork and falling weights: Ordinary clockwork (e.g. the works of an alarm clock) is neither accurate nor robust enough to be suitable as a source of power for telescopes. It can, however, be used with very light photographic apparatus with, at most, a small guide telescope (e.g. B. 156–7). More importantly, clockwork can be used, not to drive the telescope, but to control the rate at which an independent source of power becomes available to turn the polar axis. At its simplest such an arrangement need consist of no more than a wire attached to the rim of a driving circle, thence passing over a pulley in the plane of the circle and the vertical plane through the polar axis, to a spindle mounted on the winding shaft of an alarm clock; it is given several turns round this spindle in the direction opposite to that in which the clock is wound up; a weight is attached to its free end. A slightly more elaborate arrangement, easily constructed at home, is described in B. 177.

Arrangements of this sort, while having the advantage that the weight and not the clock is driving the telescope (hence adequate power can be obtained merely by increasing the weight), are nevertheless suitable only for small instruments (4-in or less); they are a blessing to the visual observer, but are not reliable enough for photographic work.

Clockwork (controlled by pendulum or escapement), whether used as a source of power or as a governor, introduces jerkiness into the drive, though this may be of negligible magnitude. An escapement beating 4 times per second, for example, advances the telescope in jerks whose angular value is $0.25 \times 15'' = 3''75$. Star images on the plate of an 18-in focus camera would therefore be short lines of length

$$\frac{18}{206,265} \cdot 3.75 \cos \delta$$

or 0.0003 in at the equator. This is quite negligible, and would be swamped by other errors in a drive of this sort, quite apart from atmospheric turbulence and emulsion diffusion.

On the other hand a falling weight regulated, not by clockwork, but by an isochronous governor, will provide a smooth drive—as will other types of drive, such as the pendulum-controlled electric drives described below. The governor of a gramophone motor may be made use of (though it is unlikely to be perfectly isochronous) by removing the spring from the motor and winding the weight cable round a drum or spindle mounted on its shaft. If a suitable transmission is built up at home, the governor must be driven off a shaft revolving at several r.p.s.

15.2.2 Electric motors: Good results can be obtained from small electric motors of the type used in radiograms and electric fans. Gramophone governors are not normally isochronous, nor particularly reliable in other respects, but they are quite good enough for visual work and short-focus photography at least.

Among pendulum-controlled electric drives the Gerrish drive, in its numerous forms, is perhaps the most commonly employed. The original Gerrish drive was needlessly complicated, and faults correspondingly difficult to locate. The Harvard modification of the Gerrish principle is much simpler, while retaining its specific advantages. This principle is simple, and extremely neat. The motor circuit contains an electromagnet-operated switch which is closed once in each swing by the pendulum; later in the swing it is reopened by a cam geared to the motor. This gearing is such that at the correct speed of drive the cam revolves once in the period of swing of the pendulum, so that the switch is opened and closed once per swing. The driving impulse occurs, during each swing, between the midswing position of the pendulum and the breaking of the switch by the cam. Control of the rate of drive is obtained by virtue of the fact that if the motor is running slow this interval is longer than normal, and the motor is therefore running under power for a longer fraction of the period of swing than if it is running fast. Hence the speed of the drive imparted to the telescope by the motor is constantly and automatically regulated to the speed of swing of the pendulum. This speed to which the system regulates itself can be adjusted very simply be means of weights on the pendulum platform. Neither voltage nor frequency control is required.

The circuit is illustrated in Figure 114. The sequence of events as the pendulum swings from A to B is as follows:

Pendulum at A: contact D, pivoted at C, is closed; hence the electromagnet E is energised; it therefore closes the switch G, mounted on the arm pivoted at H, at the same time opening F; hence the mains circuit is broken and the motor, M, is idling (the inertia of its flywheel keeps it running, though at reducing speed).

Pendulum passes midswing: permanent magnet attached to the bob opens D, which then falls back into the closed position again; this breaking of the electromagnet circuit allows the arm pivoted at H to drop, thus closing switch F; the gap at G is wide enough for the contact not to be closed by the re-operation of the electromagnet, which occurs immediately the pendulum has passed midswing. The mains circuit now being closed, the motor starts driving, accelerating the cam J to which it is geared; this in turn raises the arm carrying the switch, breaking F and closing G. The electromagnet holds the pivoted arm in this position until the pendulum returns to the midswing position from B, when the cycle is repeated.

By a slight modification of the circuit the permanent magnet on the pendulum bob can be replaced by a mercury-bath contact at D; the circuit is then closed, not opened, at midswing.

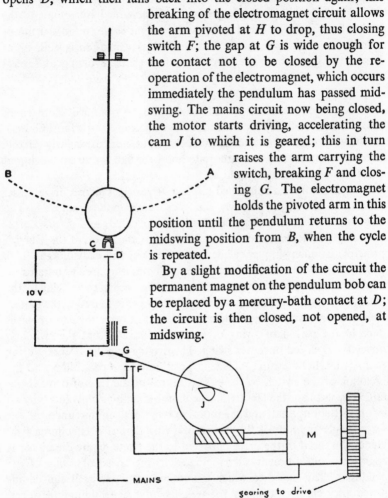

FIGURE 114

The action of the cam will be retarded or accelerated if the contacts F and G are rotated round the centre of the cam in the direction of, or against, its own rotation. This introduces the possibility of effecting slow-motion adjustments in RA without having to rely on a differential gearing. The use of the contact-breaker from a motor-cycle magneto for this purpose is described in B. 169.

Three useful papers on various applications of the Gerrish principle are B. 165–7.

Another type of pendulum-controlled electric drive, employing a commutator, is illustrated in Figure 115. As with the Gerrish drive,

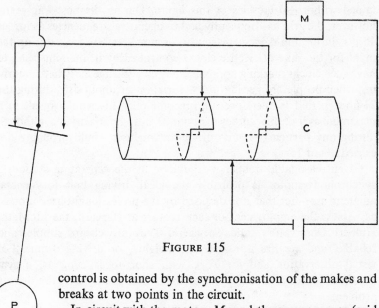

FIGURE 115

control is obtained by the synchronisation of the makes and breaks at two points in the circuit.

In circuit with the motor, M, and the power source (grid or battery) are a pendulum, P, and a commutator, C. C is either geared separately to M so as to rotate at x r.p.s., or is operated by one of the gears of the transmission rotating at this speed; the correct time of swing of the pendulum is then $\frac{x}{2}$ secs. With the motor working at a running speed slightly in excess of that required, the commutator will switch the current from one branch of the circuit to the other slightly before the latter is closed by the pendulum, with the result that the motor will idle for a fraction of the period $\frac{x}{2}$ secs.

15.2.3 Synchronous motors: These are simply electric motors whose running speed is, within certain limits, dependent solely upon the frequency of the current supplied to them. The speed of a synchronous motor is, in particular, unaffected by temperature variations over any normal range, or by small variations in its load; when a certain level of overloading is reached the motor will suddenly stall, and for this reason a slip clutch should be incorporated in the transmission.

The effectiveness of a synchronous motor for telescope drives therefore depends upon the precision of the frequency-regulation of the main electricity supply. Theoretically the frequency of the grid is constant (50 cycles per second), and before the war it was in fact maintained within $\pm 0 \cdot 1$ cycles of this figure. During the post-war years, however, the grid has frequently had to operate at frequencies as low as 48 c/s during the day, increasing to as much as 50·75 c/s during the night for the sake of electric clocks controlled by it, the aim being to have such clocks marking approximately correct time early in the morning when people are getting up. A variation of about $5\frac{1}{2}\%$ during the 24-hour period is therefore not uncommon. This would result in a maximum average inaccuracy of about 50' in an hour's driving, although during any particular hour's driving at night it would probably not exceed about 15'.

Given reasonable frequency-regulation at the generating station, a synchronous motor is probably the ideal driving unit for general amateur use—not that it is despised by the professional observatories, as witness the employment of such motors at Harvard, the McMath-Hulbert Observatory, and elsewhere. They are cheap, simple, and reliable, and require no governor, weights, or driving drums. For visual observation and for work with short-focus cameras, a synchronous motor is probably one of the easiest solutions of the driving problem.

Unfortunately, however, the grid, as we have seen, is not to be relied upon at the present time. Frequency control by the observer himself is therefore worth considering, though in normal times this is an unnecessary refinement for most amateur work. Cox, in B. 159, describes a simple method of control through a thermionic tube oscillator.

Synchronous motors are manufactured primarily for use in electric clocks, and in this country can be obtained from the makers of Smith's English Clocks or Synclocks, or from the Synchronome Company (see section 40). The motors of Smith's clocks are made to operate at from 2 to 250 volts, and at frequencies of 25, 40, and 60 as well as 50 c/s; final speeds of 1 r.p.h. or 1, 2, 4, or 16 r.p.m. can be provided; torques at 1 r.p.m. vary from 4 to 7 in/lbs. Synclocks are standardised at 50 c/s and 200–250 volts, but others can be supplied to order. The Synchronome motors also operate at 200–250 volts and 50 c/s; the 1 r.p.m. torque of the heavy-type movement motor is 2·5 in/lbs. These are all fractional-horsepower motors, and their consumption is normally something between 2 and 12 watts.

The performance of these small motors is surprising. In one case, two 2-in/lb motors drove a 10-in Cassegrain carrying a 4-in refractor and photographic equipment weighing, in all, more than 9 cwt.

15.3 Control of the drive

Methods of control already mentioned include: clockwork control of a weight drive; pendulum control of an electric drive; frequency-regulation in the case of a synchronous motor; and mechanical governors.

Many other methods have been used successfully—e.g. hydraulic control by the escape of oil from a cylinder through an adjustable valve—but these four remain the perennial stand-byes. Something further must be said about mechanical governors. Their function is to absorb any excess of power that may be developed by the motor over and above that required at the driving circle. To carry out this function efficiently, a governor must be isochronous—i.e. so constructed that an infinitesimal increase in its speed of rotation corresponds with an ability to absorb a finite amount of power. Whether or not a governor is isochronous can be tested by disengaging the driving circle and counting the r.p.m. of any convenient gear in the transmission when different driving forces (e.g. different weights, in the case of a falling-weight drive) are applied. If it is isochronous this speed will, within certain limits, be independent of the driving force. An efficient mechanical governor should permit an increase of angular velocity not exceeding a fraction of 1% when the power of the drive is doubled.

Various modifications of two types of governor are in common use: Foucault's air-resistance governor, and the friction governor. The principle of the former is explained by Figure 116. Two light metal vanes, A, are mounted on a crossbar attached to a vertical axis XY which is driven off the transmission. Being pivoted at B, the blades are free to swing out from XY, but when at rest the two springs C keep them vertical by contact with the support D, also attached to XY. The bar E, pivoted at each end to one of the vane stems, ensures that their inclination to XY be equal, no matter what its value may be. F, F are weights fixed to the vanes, and G, G are smaller weights whose position on the vane stems is adjustable by screw. When the rotation of XY is slow, the vanes will be vertical, F, F lying vertically below the pivotal points B, B. As the speed of rotation is increased, a limiting value (depending upon the distance BG for a given value of the weight F) will be reached at which the vanes leave D; with even an infinitesimal increase of the angular velocity past this threshold value they will fly out to their fully open position. The resistance offered by the air to any point on the vanes, being a function of the square of the angular velocity of that point,

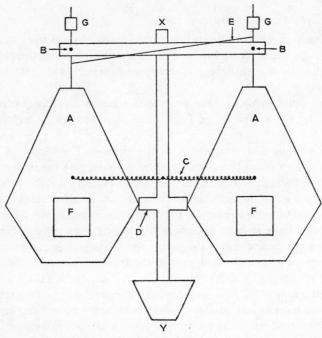

FIGURE 116

also rises sharply and exerts the necessary braking effect upon their rotation. Regulation of the governor involves adjustment of the position of G, G till the critical angular velocity at which the vanes open is that required for the drive of the telescope. The governor as a whole is enclosed in a box protecting it from draughts and dew.

The principle of the friction governor is shown diagrammatically in Figure 117. The axle of the governor, XY, is operated by the transmission of the drive. To a cross-arm, B, an angle arm, C, is pivoted at D. Attached to the upper end of C is a plastic, wood, cork, or paper shoe, E, which bears against the rim of plate F, mounted independently of the governor itself. Pivoted at G is an arm H, to which a weight A is attached. When stationary, and at low angular velocities, A rests against J, and the pressure of E against F is nil. As the speed of rotation is increased, a threshold is reached (its value depending on the length of *H*) at which *A* flies out to some position *A'*; the further it is displaced from the rest

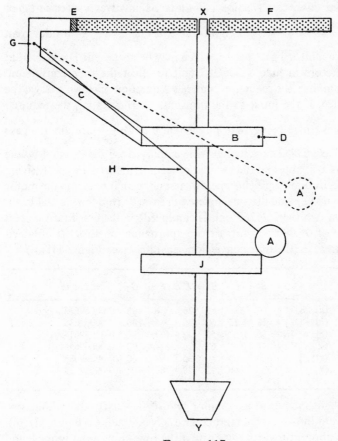

FIGURE 117
For the sake of clarity the right-hand arm *C* and its
attached weight are omitted

position *GA*, the greater will the pressure of *E* against *F* become, thus retarding the rotation of *XY* until it drops below the threshold value once more.

15.4 Transmission gearing

Where the power is supplied by some source regulated to mean time, a conversion to sidereal time is necessary between it and the driving circle, in addition to the reduction from 1 r.p.m. (or whatever the speed of the motor spindle is) to the polar axis's 1 r.p.d.

The tropical year consisting of 366·24222 sidereal days and 365·24222 mean solar days, a conversion of MT to ST involves a speeding up of $\frac{366·24222}{365·24222} = 1·0027379$. Taking this as equivalent to $\frac{366}{365}$ introduces an error of roughly 0·1 secs per hr, which may be neglected. The gear train must therefore include a 366/365 unit to effect the time conversion. Apart from that, the reduction is merely a question of slowing down the initial drive. If the input to the transmission is x r.p.m., the required reduction is (there being 1440 minutes in a day) $\frac{x}{1440}$, or, in the case $x = 1$, 0·00069635. The actual gearing employed will be dictated by the spur gears available and the number of teeth in the worm wheel (driving circle). Six examples are given below, assuming an input from the motor of 1 r.p.m. (MT), and driving circles with from 96 (too few) to 504 teeth. In the last two the 366/365 unit is employed; in the first four the total reduction of 0·00069635 (or a close approximation to it) is obtained without treating the time conversion separately (see Figure 118(*a*)):

	A	B	C	D	E	F	G	Reduction
(*a*)	24	60	28	66	26	66	96	0·00069636
(*b*)	14	18	15	66	26	66	100	0·00069636
(*c*)	26	56	32	56	32	127	96	0·00069634
(*d*)	12	14	25	63	26	127	100	0·00069634
(*e*)	15	365	366	30	26	52	360	0·00069635
(*f*)	15	365	366	15	28	80	504	0·00069635

By a slight modification, splining two of the shafts and adding one gear, drive at the solar or lunar rate can easily be obtained (Figure 118(*b*)). Case (*f*), above, provides solar rate if the time-conversion gears B and C are cut out: D is disengaged from C and slipped along the splined shaft to engage with A. Lunar rate can be obtained by disengaging E from F and sliding H into its place; in the case of (*f*), where F has 80 teeth, an approximation is obtained by giving H 27 teeth; this is the most that can be hoped for, owing to the wide range of variation of the lunar rate with the hour angle of the Moon.

The transmission should be kept as simple as possible; its accuracy is of paramount importance, since irregularity in one gear will introduce a periodic variation in the speed of the drive. The drive is transferred to the polar axis via a worm wheel (the driving circle) and a tangential screw (worm). The worm should be single thread—i.e. in one rotation it turns a worm wheel of x teeth through $1/x$ of a revolution.

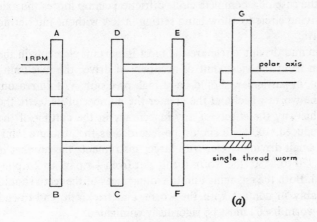

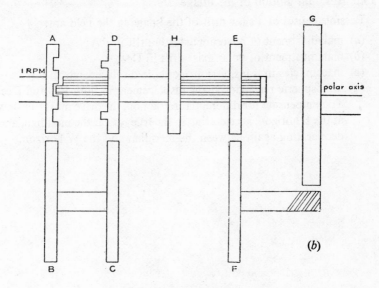

FIGURE 118

If the driving circle is reduced to a segment (as is often the case in English and fork mountings) whose angle is θ, the maximum length of

271

uninterrupted run that is possible is $\dfrac{\theta}{15}$ hours. In such a case the worm has to be disengaged in order to reset the segment to the beginning of its run; in the case of a complete circle a friction clamp looses the axis from the driving circle to allow hand setting in RA without interfering with the drive.

The worm and driving circle are the most important elements in the transmission from the viewpoint of accuracy of drive; the machining of the worm, in particular, should be of high precision. Within reason, the larger the worm wheel and the greater the number of its teeth, the better; in this way the effects of any inaccuracy in the cutting of the worm are reduced. 100 teeth should be regarded as the absolute minimum for a small driving circle; with larger instruments the number is more often 720 to 2880 (the worm turning at from $\frac{1}{2}$ r.p.m. to 2 r.p.m. respectively). Both the engaging and the idling faces of the teeth should be comfortably in contact with the worm, and backlash (end-to-end play in the worm itself) must be rigorously eliminated.

15.5 Residual motion of the image

Possible causes of a slow drift of the image in the field are:

(*a*) maladjustment of governor: steady drift in RA;

(*b*) maladjustment of polar axis: drift in Dec;

(*c*) inaccuracies in transmission: periodic displacements in RA;

(*d*) atmospheric refraction: slow displacement in both RA and Dec. To compensate for differential refraction, the drive must run slow at the E horizon, accelerating to the true rate at the meridian, and decelerating again between the meridian and the W horizon.

SECTION 16

SETTING AN EQUATORIAL HEAD

16.1 Introduction

The successive stages in the setting up of an equatorial are:

(1) approximate orientation of the polar axis in altitude and azimuth;
(2) zero setting of the Dec circle index or vernier;
(3) final adjustment of the polar axis in altitude;
(4) final adjustment of the polar axis in azimuth;
(5) zero setting of the hour circle index or vernier;
(6) checking collimation: the perpendicularity of the optical and Dec axes;
(7) checking that the polar and Dec axes are mutually perpendicular.

These adjustments between them ensure that the polar axis is parallel to the Earth's axis of rotation, and that the Dec axis is perpendicular both to the polar axis and to the plane containing the optical axis. The order in which the operations are carried out is important: if the mounting is not fitted with circles, stages (2) and (5) are of course omitted; if, on the other hand, either of the circles is used in determining a particular error, its index error must first be determined; with one exception the methods described here for carrying out stages (3) and (4) require at most the Dec circle, while the hour circle is employed in stages (6) and (7).

Most small mountings make no provision for the adjustment of the optical axis relative to the Dec axis, or of the Dec axis relative to the polar axis, these being attended to by the makers. In the case of new instruments, at least, it is safe to assume that they are correct, and stages (6) and (7) may be omitted.

If any particular stage involves a large shift in the position of the head, it is likely that prior adjustments will have been affected, and these must therefore be repeated. A large change in the azimuth of the head in stage (4), for example, will also affect the adjustment made in stage (3); stages (3) and (4) must therefore be repeated as fine adjustments after the first coarse adjustment of (4).

A variety of methods are available to choose from at each stage; these vary among themselves in accuracy and in convenience and speed of operation. It is assumed in the following instructions that a HP and a LP ocular are available (if possible both, and certainly the former, provided with cross-webs), and that from the observing site there is a clear view of the pole and of the meridian point of the equator.

16.2 Approximate orientation of polar axis

(*i*) *Azimuth:*

(*a*) The equatorial head aligned on a meridian mark on the ground with the aid of a prismatic compass, the current magnetic variation being taken from the Ordnance Survey sheet of the district in which the observing site is situated.

(*b*) Meridian marked by the shadow of a plummet at true noon (i.e. equation of time applied to LMT).

(*c*) Meridian defined by two plummets aligned upon Polaris at upper or lower culmination.

(*d*) With hour circle reading 0^h 0^m 0^s—or, in the absence of circles, with the Dec axis made horizontal with a striding level—bring a star of known RA, α, to the centre of a LP field when LST$=\alpha$, by rotating the whole head in azimuth.

(*ii*) *Altitude:*

(*a*) The latitude of the observing site, correct to the smallest fraction of a degree to which the latitude scale on the mounting can be read, is obtained from a 1-in O.S. map. The base of the head having been levelled, this angle is then set on the latitude scale.

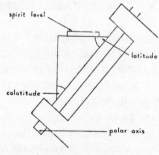

FIGURE 119

274

(*b*) If the mounting is not fitted with a latitude scale, employ a home-made clinometer as in Figure 119.

Note: Before making any adjustments of the head, the base supporting it must be shown to be at least approximately horizontal by levels placed on it in, and at right angles to, the meridian.

16.3 Index error of Declination circle

(*a*) Select any fairly bright star near to the meridian (whose altitude, therefore, will vary by a negligible amount during a few minutes). With the telescope E of the pier, bring the star to the intersection of the webs of the HP ocular; clamp, and read the Dec circle (δ'). Reverse the telescope, return the star to the cross-wires, and re-read the Dec circle (δ''). Assuming that $\delta' \neq \delta''$, adjust the vernier or index to read $\dfrac{\delta' + \delta''}{2}$ when the star is at the intersection of the webs, or is bisected by the horizontal web if the webs have been oriented NS and EW. Repeat, until the readings E and W of the pier agree, or treat any residual error as a correction to be applied, with a change of sign, to all Dec circle readings. It is advisable, as a check, to repeat the procedure with a star of widely different Dec from the first. A distant and well-defined terrestrial object could equally well be used.

(*b*) The base supporting the head having been levelled both NS and EW, mark the head's position in azimuth. Then turn it through 180°, so that the upper end of the polar axis is lying due S of the lower. Bring Polaris, at its upper or lower culmination, to the intersection of the webs; clamp, and read the Dec vernier. Reverse the telescope and repeat. As in (*a*), if the two readings do not agree, set the vernier to their mean value when Polaris is at the centre of the field, and repeat.

(*c*) A method that can be used by day, but is not otherwise convenient. Remove the OG cell and ocular; attach a plummet to the intersection of a wooden crossbar laid symmetrically over the top of the telescope. Adjust the tube so that the plummet falls centrally through the drawtube opening; the tube is then directed at the zenith. Set the Dec vernier to read the lattitude of the observing site; reverse the telescope; readjust its position relative to the plummet and re-read the vernier. If the two readings are different, set the vernier to their mean. Repeat until the circle reading is the same on either side of the pier, the plummet falling centrally through the ocular flange. Note the index error, if any, and apply it as a correction to all subsequent Dec circle readings. Alternatively, two brackets may be fixed to the tube by collars, as in Figure 120, *CD* being parallel to the optical axis; a plummet suspended from point *A* on one bracket will pass centrally through the hole *B* in the other, provided the telescope is vertical and that $CA = DB$.

Alternatively, again, the telescope may be pointed to the zenith by placing a spirit-level across the cell of the objective and centring the bubble in both E–W and N–S directions. Given an accurately constructed level the results are likely to be as accurate as those obtainable from the use of a plummet, which in practice is not easy to manipulate with precision.

16.4 Adjustment of the polar axis in altitude

(*a*) *Without circles:*

The North Celestial Pole lies within about 3^m of RA of the line joining Polaris and η UMa. At $\pm 6^h$ ST of the culmination of η, therefore, the altitudes of Polaris and the NCP are very nearly equal.

Set the Dec axis in the vertical plane through the polar axis, using a plummet, and clamp the polar axis. Adjust the elevation of the polar axis until Polaris can be brought to the intersection of the webs by rotation of the Dec axis only.

The inaccuracy in the derived altitude which is involved by assuming that the two stars and the NCP lie on the same hour circle is of the order of $50''$: elevation too small if η is W of the pole, too great if it is E.

(*b*) *Without circles:*

A quick method of correcting both the elevation and the azimuth of the polar axis in a single operation.

FIGURE 120

Bring Polaris to the centre of the LP field and clamp the Dec axis. Rotate the polar axis through a wide angle, and note whether the image has moved relative to the webs. Reduce any such movement to a minimum by adjustment of the telescope in Dec; carry out the fine adjustment with the HP ocular. The tube and the polar axis are now virtually parallel.

Clamp both axes and adjust the head bodily (elevation and azimuth of the polar axis) so as to bring the NCP to the centre of the field. The pole lies at a distance of 58' from Polaris in the direction of η UMa.

276

The adjustment can be made by eye estimation (the field diameter being known), with the aid of a micrometer, or by eye in conjunction with a map of the polar star field (see section 37).

(*c*) *Without circles:*

Orient the webs of the LP ocular NS, EW by trailing an equatorial star near the meridian (star over web, or web over star with Dec axis clamped).

Select a pair of stars in the NE or NW sky, differing in RA by several hours and in Dec by at most a few ' arc. Set the *f* member of the pair at the intersection of the webs and clamp the Dec axis. Rotate the polar axis to bring the *p* to the NS web. Knowing the difference in Dec of the two stars, estimate whether it is too far N or S in the field. Then:

p star too far N	both stars E	elevation of polar axis too great.
„ „ „ S	of meridian	„ „ „ small.
p star too far N	both stars W	elevation of polar axis too small.
„ „ „ S	of meridian	„ „ „ great.

Adjust the elevation of the axis by half the amount required to bring the *p* star to its estimated correct position on the NS web. Correct, if necessary, the orientation of the webs, and repeat the procedure first with the LP and finally with the HP ocular; in the latter case it may be necessary to use another pair of stars, of smaller $\Delta\delta$.

The practical inconvenience of this method lies in the necessity of selecting pairs of stars of closely similar Dec, which involves reference to one of the more comprehensive star catalogues.

(*d*) *Without circles:*

Observe a near-equatorial star when on the meridian and when at equal distances E and W. At the first (E) observation bring the star to the intersection of the webs, oriented NS and EW, and clamp the Dec axis.

The second and third observations consist in noting whether the star has shifted N or S from the EW web and, if so, estimating by how much. Then:

If field position of star at meridian is midway between its positions when E and W of the meridian: elevation of polar axis is correct.

If position at meridian is N of the mean of its positions when E and W of the meridian: elevation of polar axis is too great.

If position at meridian is S of the mean of its positions when E and W of the meridian: elevation of the polar axis is too small.

Having reduced the error, and if necessary reoriented the EW web, repeat the observations at increased intervals before and after culmination.

(e) Without circles:

The most generally satisfactory method, and one capable of revealing very small errors in the elevation of the polar axis, is the following:

Select two stars about 6^h E and W of the meridian respectively, and in about 45° N Dec. Orient the HP ocular with its webs NS and EW by trailing an equatorial star on the meridian.

Bring the E star to the intersection of the webs, clamp Dec and follow the star with the driving clock or RA slow motion; note whether it tends to drift N or S from the EW web. If the elevation of the axis is nearly correct this drift will be slow, and some time will pass before it becomes perceptible.

Repeat the observation with the W star. Then:

If E star drifts N
„ W „ „ S } elevation of polar axis is too great.

If E star drifts S
„ W „ „ N } elevation of polar axis is too small.

Correction of the elevation of the axis must be carried out by trial and error.

(f) With circles:

Select a star of known Dec, δ, and, to minimise the effects of refraction and of the still only approximate azimuth adjustment, small zenith distance.

Set δ on the circle, the index error having been eliminated or allowed for, and clamp the Dec axis.

Adjust the elevation of the polar axis so that the star, when on or very close to the meridian, can be brought to the intersection of the webs of of the HP ocular by sweeping in RA only.

Reverse the telescope and repeat; finally check with another star.

(g) With circles:

With the set-up described in section 16.3, para *(c)*, set the Dec axis horizontal by means of a striding level, and clamp the polar axis.

Rotate the telescope about the Dec axis to make the plummet fall centrally through the hole in the lower bracket.

The Dec circle reading should now be the latitude of the observing site. If it is not, set this reading on the Dec circle and bring the plummet back to the centre of the lower bracket hole by adjusting the elevation of the polar axis.

Alternatively, a spirit-level may be used in the manner described in section 16.3, para. (c).

16.5 Adjustment of the polar axis in azimuth

(a) *Without circles:*

The same assumption is made as in section 16.4, para (a). The Dec axis is set horizontal with a striding level, and the polar axis clamped.

At the upper or lower culmination of η UMa, the azimuth of the head is adjusted so that Polaris can be brought to the intersection of the webs by the Dec slow motion.

If the observation is made at upper culmination, the N end of the axis will lie about 50″ too far W; if at the lower, 50″ too far E.

(b) *Without circles:*

See section 16.4, para (b).

(c) *Without circles:*

Orient the webs NS, EW on an equatorial star at the meridian. Select two near-equatorial stars of known α and δ, situated at equal distances (about 30^m) E and W of the meridian, and differing in Dec by not more than a few ′ arc.

Bring the E star to the intersection of the webs and clamp the Dec axis. Swing the telescope westward in RA to bring the second star into the field. Knowing $\Delta\delta$, estimate whether it lies too far N or S in the field. Then:

> If W star is too far S: polar axis is lying NE–SW.
> „ „ „ „ N: „ „ „ NW–SE.

Reduce the error by half, by means of azimuth adjustment of the polar axis. Repeat with a second pair of stars, differing in Dec from the first. Finally, repeat with pairs of stars of increasing $\Delta\alpha$ until no error is detectable over arcs of 6^h to 8^h. This method is open to the same practical objection as that mentioned under section 16.4, para. (c).

A single star, observed when E and again when W of the meridian, could of course be used instead of a pair of stars, but this procedure takes longer. In the early stages of correcting the azimuth it is sufficient to note the N or S drift of a single star (preferably near the zenith) when crossing the meridian.

(d) *Without circles:*

Employs the same principle as section 16.4, para. (c); but the first setting of the telescope in Dec is made on a star at the meridian, and the comparisons are made with stars at equal distances E and W of the meridian. Then:

If E star lies too far N
„ W „ „ „ S } polar axis is lying NE–SW.

If E star lies too far S
„ W „ „ „ N } polar axis is lying NW–SE.

(e) *Without circles:*

Orient the webs NS, EW on an equatorial star at the meridian. Select two stars of known α, δ, and Δα of a few minutes, situated respectively in S Dec and high N Dec.

When they are in the neighbourhood of the meridian, clamp the telescope in RA slightly ahead of the *p* star, and time its transit at the NS web.

Rotate the telescope about the Dec axis to pick up the *f* star, and time its transit. Then:

If transit of *f* star occurs too early: polar axis is lying NE–SW.
„ „ „ „ „ late : „ „ „ NW–SE.

(f) *With circles:*

Select a star of known α, δ, situated roughly midway between the meridian and the E or W horizon, and between the zenith and the horizon.

Set the Dec circle to δ, and clamp. Adjust the azimuth of the polar axis so that the star can be brought to the intersection of the webs by motion in RA alone.

Repeat with a second star, situated in approximately the same position across the meridian, making the final adjustments with the HP ocular.

(g) *With circles:*

A method of great precision (given finely and accurately divided circles), which requires the prior elimination, or determination, of the hour circle index error (see section 16.6). It involves the observation of a circumpolar star of known α, δ. A few minutes before upper or lower culmination note the time of its transit at the NS web of the HP ocular (T_1), and the hour circle reading (t_1). Reverse the telescope and note T_2, t_2.

Then
$$a = \frac{1}{15}\left(\alpha - \frac{T_1 + T_2}{2} - \Delta T + \frac{t_1' + t_2'}{2}\right)\frac{\cos \delta}{\sin\phi \pm \delta}$$

where a = required azimuth correction in ″ arc,

t_1', t_2' = the values of t_1, t_2 corrected for index error,

ΔT = clock correction,

ϕ = latitude of the observing site.

The sign of δ in the denominator of the last term is $+$ for lower culminations, $-$ for upper. If the derived value of a is $+$, the polar axis is oriented NW–SE; if $-$, NE–SW.

Knowing the precise value of the required adjustment, and not merely its direction, involves a great saving of time, since the micrometer web can be laid off a'' from some reference point on the horizon, and the whole head rotated in azimuth until it is returned to the web, thus eliminating an often rather lengthy process of trial and error.

16.6 Index error of the hour circle

(a) Set the Dec axis horizontal by means of a striding level (bubble central in both positions of the level), and clamp the polar axis. Set the hour circle index or vernier to zero; if two verniers are fitted, one reading local ST and the other RA, their zero readings must coincide.

(b) Calculate the exact time of culmination of a star whose RA is known. Just before culmination, bring it to the intersection of the webs in a HP ocular, clamp both axes, and follow it with slow motion in RA. At the calculated instant of meridian transit stop following the star; set the hour circle vernier to zero, and the RA vernier, if present, to the RA of the star.

(c) Set-up as described in section 16.3, para (a). Note the hour circle reading, t. Reverse the telescope and reread the hour circle, t'. If $t \neq t'$, either 5, 6, or 7 (section 16.1) may require adjustment. Assuming that the two latter are in fact correct, the index error is

$$\frac{(t+t')}{2} - 24^\text{h}\ 0^\text{m}\ 0^\text{s}$$

which can either be removed by setting the circle to the reading $\dfrac{t+t'}{2}$ and adjusting the vernier to read $24^\text{h}\ 0^\text{m}\ 0^\text{s}$, or can be applied with a change of sign as a constant correction to all hour circle readings.

16.7 Collimation

With circles:

Select an equatorial star near the meridian; clamp the telescope ahead of it, and time its transit at the NS web (T_1). Read the hour circle (t_1).

Reverse the telescope; again clamp it ahead of the star, time its transit at the NS web (T_2) and read the hour circle (t_2).

Then if the collimation error is zero,

$$T_2 - T_1 = t_2 - t_1$$

If T_2-T_1 is greater/less than t_2-t_1, the angle between the OG end of the tube and the Dec axis is more than/less than a right angle. The collimation error

$$\frac{(T_2-T_1)-(t_2-t_1)}{2}$$

may be applied as a correction, or may be removed if the requisite adjustment is provided. With small instruments, errors of collimation can be corrected by inserting tinfoil or thin metal shims inside one end of the cradle, with frequent trials on a star. The procedure is bad for the temper, but practicable.

If the difference between (T_2-T_1) and (t_2-t_1) is expressed in seconds of time, and $b=$ length of the cradle in inches, the distance that the end of the cradle must be displaced from its existing position is, in inches.

$$15b\left[\frac{(T_2-T_1)-(t_2-t_1)}{206,265}\right]$$

16.8 Declination and polar axes perpendicular

(a) With circles:

Set the Dec axis horizontal with a striding level, and read the hour circle (t_1). Reverse the telescope, level the axis, then reread the hour circle (t_2).

Then it should be that

$$t_1=t_2 \qquad (12^{\mathrm{h}}\text{ graduation})$$
or $$t_1=t_2\pm12^{\mathrm{h}}\ (24^{\mathrm{h}}\text{ graduation})$$

Most small mountings make no provision for the correction of this error, and a faulty mounting should be returned to the makers or to any optical firm specialising in repair work.

(b) With circles:

This error may be distinguished from collimation error by observations of the type described in section 16.7. Both errors will have the effect of introducing inequality between (T_2-T_1) and (t_2-t_1). But if the cause of the discrepancy is lack of perpendicularity between the instrumental axes, the error will be a function of δ; more precisely, of $\tan\delta$. Therefore repeat the observations with a star not less than 45° from the equator. The magnitude of the error is

$$\frac{(T_2-T_1)-(t_2-t_1)}{2\tan\delta}$$

282

16.9 Limits of accuracy required

The more accurate of the methods described above are capable of reducing the angle between the polar axis and the parallel to the Earth's axis passing through the mounting, to less than 1'. Providing this angle does not exceed 2' or 3' the setting will be perfectly good enough for ordinary visual observation under all circumstances.

Even for photographic work a divergence between the polar and terrestrial axes not exceeding a few ' arc is permissible. The effect of such divergence is to make the photographic plate rotate slowly about the image of the guide star, so that the images of all other stars in the field will not be points but arcs; the lengths of these arcs are a function of their distance from the guide star, and of the guide star from the pole. From the curve in Figure 121 may be read the maximum permissible

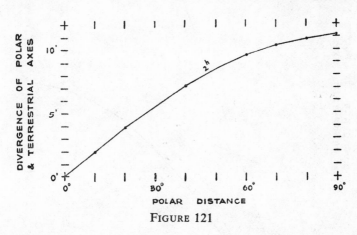

FIGURE 121

error in the setting of the polar axis at various polar distances, assuming that a length of trail of 0·0025 ins for an image situated 1·5 ins from the guide star in the focal plane, during an exposure of 2^h, is tolerable. Figure 122 reveals the approximate values of the permissible divergence of the polar axis, using the same criterion, for a variety of exposure times.

For micrometer work involving the measurement of position angles, as for photography, the tolerance for the orientation of the polar axis decreases rapidly with decreasing polar distance. Such work demands the greatest possible precision in the adjustment of the polar axis, which must be directed at the true pole—i.e. the effect of refraction must be taken into account (see section 16.10). Precision of this order is only

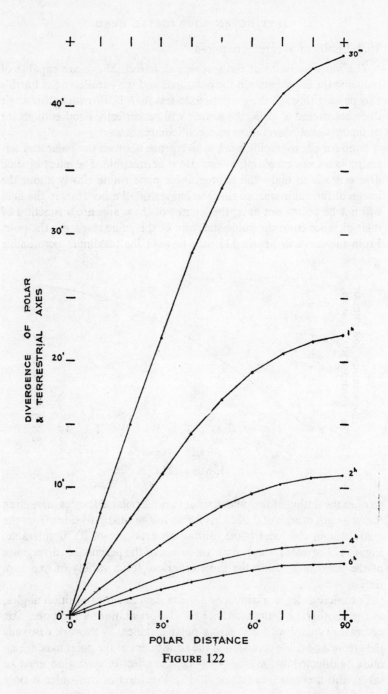

FIGURE 122

obtainable if the instrument is fitted with finely divided circles, or by photographic methods based on star trails (see B. 163, 216).

16.10 Effect of refraction

The effect of atmospheric refraction is to increase the observed zenith distance of a star as it approaches its culmination, and to decrease it as it passes from culmination towards its setting. Hence a precisely adjusted polar axis would necessitate southward adjustments in Dec while following a star E of the meridian, and northward adjustments when the star is W of the meridian. At the same time the clock rate is falsified, since refraction keeps the star apparently above the horizon for a slightly longer period than it in fact is.

For photographic and ordinary visual observation it is convenient to effect a partial compensation of this by increasing the elevation of the polar axis slightly, so as to direct it at a fictitious pole as affected by refraction. The effects of refraction are virtually nil over an area of sky centred on the meridian in (for our latitudes) Dec about 20° N, whose radius is about 30°. Hence, if as a final adjustment the elevation of the polar axis is increased enough to keep, say, Arcturus, Aldebaran, or γ Leo at the intersection of the webs without use of the Dec slow motion—or, alternatively, to use one of these stars in the first place—then the effect of refraction will be largely eliminated over a wide area of the southern sky.

See also B. 183, 216.

16.11 Portable equatorials

Many users of small equatorials prefer to remove the head and telescope complete and keep it indoors when not in use, rather than go to the trouble and expense of making a shed which can be run off, folded back, or otherwise removed during observation.

The head can without difficulty be reset with sufficient accuracy (to within about 10', say) for visual work—i.e. mechanical drive or slow-motion operation in RA will maintain an object near the centre of the field for considerable periods without any readjustment in Dec being required. The head is first set accurately, by any of the methods already described. It should be fitted with two adjustable levels, one in the NS plane containing the polar axis and the other at right angles to it. The bubble in each level is then centralised by adjustment of the level itself. When, subsequently, it is desired to set up the head again, it is placed on the mounting and the two levels are made horizontal. It remains to set the axis in the correct azimuth. Set the hour circle to 0^h or 12^h—or set the Dec axis horizontal with the help of a striding level—and clamp. Turn the tube about the Dec axis until it is approximately horizontal. Sight down the tube with a prismatic compass* whose graticule is bisecting the concentric flat and primary mirror. Rotate the head in azimuth until the compass (±the current magnetic variation for the place) indicates that the tube is lying NS.

* If the proximity of the iron in the mounting makes the compass erratic, it is preferable to adjust the azimuth by setting on the Sun, or, after dark, on a star of known hour angle.

SECTION 17

OBSERVING HUTS AND SITES

17.1 Observing sites

If one were fortunate enough to have unrestricted choice of a telescope site, the following conditions could be satisfied in full:

(a) remoteness from road and rail traffic (see section 14.25);
(b) remoteness from any urban area, especially in the direction of the prevailing wind, and no inhabited buildings in the immediate vicinity;
(c) absence of all artificial lights not controllable from the site;
(d) clear view of the whole stellar hemisphere—at any rate of the whole meridian from the NCP to the S horizon, the greater part of the southern sky, and if possible the W and E horizons;
(e) protection from the wind;
(f) surroundings planted with low vegetation, or at least grassed.

The choice in practice usually being more or less restricted, precedence should be given to the satisfaction of (c) and (d) so far as possible. Condition (e) loses importance when the telescope is housed in a domed observing hut, and when the telescope is largely exposed during observation can often be partially satisfied by means of a movable canvas screen, mounted on a wood framework and secured by guy ropes.

Condition (c) is important in all types of work, but particularly so in photography and the observation of faint objects. If possible, any extraneous light should be screened from the observing site. Working lights must be dim, and controllable from the telescope. For a general light about the telescope, a low-wattage bulb may be used; it should be mounted at a sufficient distance to provide a distributed illumination at the telescope equal to about two Full Moons. For the illumination of charts, notebooks, etc, an ordinary torch can be used, the bulb being covered with enough thicknesses of red tissue paper to reduce the intensity to the required level. Alternatively, and more conveniently, a a shaded bulb-holder may be mounted on the drawing-board, portable desk, or whatever the observer uses when at the instrument.

For observations requiring maximum dark adaptation it is a good plan to wear an eye-shade which can be flapped up against the forehead when observing, and down over the observing eye before the working light is switched on for consulting charts, making notes, etc. If the Sun and also faint objects, such as comets or variables, are habitually observed, it is advisable to use the right eye exclusively for the one type of work, and the left eye for the other.

17.2 Observing huts

Small altazimuths can be moved indoors between observing spells, without trouble or waste of time; small, and not very precisely adjusted, equatorials can be set up in a matter of five minutes or so at the outset of each night's observations. Larger instruments cannot conveniently be removed from their mountings, and these must be protected from the weather when not in use. The minimum requirement is a small hut—or more accurately, perhaps, a large box—of just sufficient size to cover the telescope and its mounting, which can be removed when the instrument is wanted.

While considering the construction of a cover of this sort, it is worth remembering that protection is not only required by the instrument, and not only when it is not in use. In short:

(a) the instrument must be protected from the weather when it is not in use;

(b) the instrument must be shielded from the wind when in use;

(c) any form of protection tending to reduce dewing-up during observation is desirable;

(d) a comfortable observer tends also to be a good one, and even a simple and quite unpretentious observing hut vastly improves observing conditions, by giving protection from the wind, some protection from dew, and a permanent home for papers, notebooks, charts, etc.

In view of these considerations the construction of an observing hut—rather than a mere storage box—is strongly recommended wherever conditions allow.

A small observing hut is easily made at home by anyone whose carpentry is of the hammer-and-saw level of proficiency. The difficulties are likely to be economic and spatial, rather than constructional. Timber today is still very expensive, and seems likely to remain so, while some types of hut (though not all) require two or three times the ground area covered by the instrument itself. Recommended timber construction for small huts consists of: 4-in × 3-in bed-plates resting on a brick or concrete foundation, with damp-course between (it may be advisable to have the foundations put down by a professional bricklayer); 3×2 floor joists; air-space between floor and ground; 3×2 uprights, with tongued-and-grooved or weatherboard walls. Construction of the roof depends upon the method of opening to the sky that is employed: tarred roofing felt on ¼-in boards supported by 3×2 joists is adequate for fixed or sliding roofs, while domes may be of 3-ply bent on to a 1-in thick framework or painted fabric (Ruberoid or doped canvas). Windows are not necessary, though one is certainly a convenience; the door must of course open outward; the pier supporting the telescope must be carried through the floor from a masonry foundation, and there must be a clearance of ½ in or so between it and the floor so that no vibration is transmitted from the floor to the instrument, there being no physical contact between the telescope and the hut structure.

17.2.1 Size: The minimum requisite size of an observing hut is determined by the focal length of the instrument, its optical construction (refracting, Newtonian, Cassegrain-Newtonian, etc), the type of mounting, and, in some cases where the instrument is extremely compact, the height of the observer. The walls are commonly made just low enough for the optical axis to clear them when the telescope is horizontal; but it is better to make them somewhat higher than this in the interests of wind protection, which is a more valuable advantage than the ability to observe objects within a degree or two of the horizon.

The Table below sets out the approximate dimensions (in feet and

Instrument	Focal ratio	Tube length	Height of mounting	Diameter of floor	Height of dome
		ft ins	ft ins	ft ins	ft ins
12-in refractor	$f/15$	15 0	10 6	21 0	21 0
12-in Newtonian	$f/5$	5 0	3 6	7 0	7 0
12-in Cassegrain	$f/5: f/10$	3 6	4 9	9 6	9 6
	$f/25$	4 0	5 0	10 0	10 0
12-in Cassegrain-Newtonian	$f/5: f/10$	3 6	4 3	5 6	7 0
	$f/25$	4 0	4 0	6 0	7 0
12-in Schmidt (with guide telescope)	$f/1$	2 0	4 0	8 0	8 0
12-in off-axis reflector (section 9.9)	$f/5$	2 6	4 3	8 6	8 6

inches) of a domed hut to house different types of 12-in instrument mounted on an altazimuth head; these figures are to be regarded as comparative rather than exact. If the mounting were a German equatorial, the dimensions would have to be increased slightly to allow for the alternative positions of the instrument. A clearance of 3 ft between the ocular and the walls (tube horizontal) is allowed in the case of instruments whose ocular lies on the optical axis of the objective, and of 1 ft in the case of instruments whose observing position is at the side of the tube; in the latter case the height of the mounting is adjusted to give an ocular position 6 ft above the floor when the telescope is directed at the zenith. If the quoted minimum dimensions were in fact employed, it would be necessary to incorporate a small annexe—even if this were no more than a cupboard-like projection from one wall— to accommodate a desk, bookshelf, accessories, etc. It should also be noted that these dimensions are the minimum dictated by the instrument, and in some cases would have to be increased to accommodate the observer comfortably.

17.2.2 Types of hut: A few basic types which have been proved successful are briefly described. Modifications to suit local conditions and personal requirements are possible in innumerable ways.

(a) Rectangular hut with roof, either pent or ridge, which slides off on horizontal rails supported by uprights mounted in concrete foundations. The roof may either move as a single unit or may be divided centrally, the two halves being removable in opposite directions. In the former case, rails should be provided, if possible, on both sides of the hut, so that the roof can be slid off either to eastward or westward; if rails can only be provided on one side of the hut this should be the N side. In the latter case the overlap at the roof joint needs careful weatherproofing, especially if it is central and therefore directly above the instrument when the hut is closed. The 4×3 rails should be surfaced with galvanised sheeting (26-gauge can be cut quite easily with tinsnips); side-plates and end-stops prevent the roof from jumping the rails; roller-skate wheels are a convenient means of mounting the roof on the rails. The roof may be boarded and felted, but a great gain in lightness is obtained by stretching fabric directly on to a wood framework. Pins or U-clamps should be provided to lock the roof securely to the wall top-plates when it is in the closed position.

Advantages of this arrangement are the wide opening provided, resulting in infrequent adjustment of the roof during prolonged observations or photographic exposures, plenty of sky room for auxiliary instruments such as wide-angle cameras, and rapid establishment of thermal equilibrium between the interior of the hut and the night air. Its disadvantages are the comparatively large ground area required, and reduced protection from wind and dew.

(*b*) A rather more cumbersome arrangement is to mount the entire hut on rollers, and remove it bodily, a door at one end being large enough to clear the telescope and mounting. While in position over the telescope it is securely anchored by pins passing through the bed-plates and rails, and fitting into sockets in the foundations.

This type cannot truly be described as an observing hut at all, since it is not there during observation. It is, in fact, a mere storage box, and gives no protection to telescope or observer when the instrument is in use.

(*c*) Domed hut, the dome consisting of two quarter-spheres, both revolvable one inside the other on concentric rails. This arrangement has the advantage—shared by all domed, as against sliding-roof, huts—that the ground area required is no greater than that of the hut itself. It has the further advantage over the usual dome-and-shutter type that a greatly increased sky area is exposed.

Its main disadvantage is that frequent adjustments of the dome are required when observation is being carried out in the vicinity of the zenith, which is itself inaccessible.

(*d*) Dome-and-shutter type. The drawbacks of costliness and difficulty of construction have often been rather over-emphasised. Furthermore, by the right choice of materials domes can be made both light and absolutely weatherproof. They provide maximum protection from the wind and greater comfort for the observer, including reduced trouble from dew-formation. Their chief disadvantages are relatively frequent interruptions to observation—too wide a slit reduces the strength of the dome below the danger level—and increased trouble from thermal currents, particularly with small domes.

Figures 123 and 124 illustrate two designs of dome structure, the latter being preferable since its shutter opening passes the zenith. The members of the dome framework may most easily be built up of sections cut from 1 × 12 planks, and stressed at the joins. A 12-ft dome running on roller-skate wheels can quite easily be turned by hand. It is preferable to mount the rollers on the top-plate of the wall rather than on the base of the dome; with this arrangement, any tendency of the dome base-plate to warp or sag between its points of support can be corrected by occasionally changing its orientation when not in use. With sufficiently wide and numerous rollers it is not necessary to provide a metal rail for them to bear against—the wooden base-plate of the dome supplying a hard enough running surface. Further rollers mounted vertically to bear against the side of the dome base-plate act as thrust bearings. The junction of the dome and the wall must be protected from the weather by a galvanised strip nailed round the bottom of the dome and overlapping the wall by 6 ins or so.

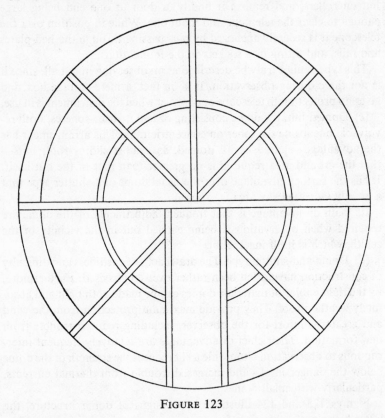

FIGURE 123

Of the many systems for opening and closing the shutter which are easy both to construct and to operate, two are illustrated in Figures 125 and 126. The lower section of the shutter, BC (Figure 125), merely hinges outward; it is retained in the closed position by means of a cord from B to a cleat on the inner side of the observatory wall; by means of this cord, also, it can be lowered gently into the open position. The counterpoise, D, is in the form of a roller, drilled centrally; the cord E is attached to a wire harness passing through D, so that as the shutter is raised until F (the other point of attachment of the cord) is directly

292

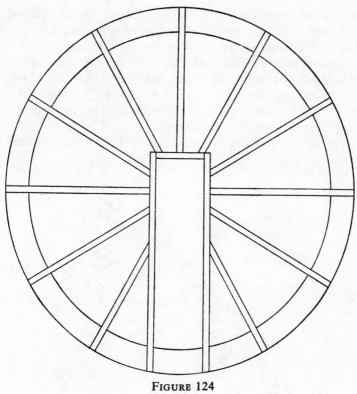

FIGURE 124

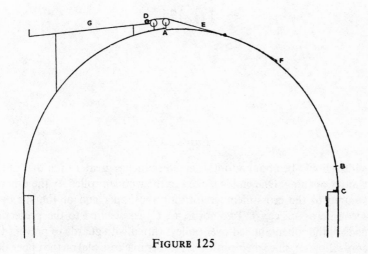

FIGURE 125

below A, the counterpoise will have rolled down the full extent of its guide rails, G. As the opening of the shutter is continued, D will be pulled back into the position shown in the Figure. The shutter AB runs on castors or skate wheels in guide rails set on the inner side of the dome and continued down the sides of the slit. The shutter can be moved either by a pole with a hook at its end which engages a screw-eye near A or B, or by cords attached in these positions and running over pulleys mounted on the inner side of the dome.

Figure 126 illustrates a roller-blind type of shutter. If the diameter of the dome is d, a strip of canvas of length $\frac{\pi d}{4}+1$ ft will be required; its

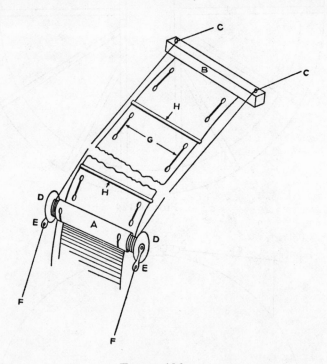

FIGURE 126

width should be about a foot or eighteen inches greater than that of the shutter opening. One end is tacked to the wooden roller A; the other is secured to the cross-member forming the upper end of the slit by a second cross-piece, B. Two cords, C, C, are secured to the projecting ends of the roller, and led over pulleys (fitted with guards to prevent the cords slipping, since their position is rather inaccessible) on the upper side of *B*, and thence to a cleat on the hut wall. When these cords are pulled together the canvas will roll up on *A*, which will travel towards the upper end of the slit; the cords are prevented from slipping off *A* by the discs *D, D*, screwed to the ends of the roller. The metal links *E, E* are attached to *A* so that they can swing freely, i.e. do not turn when *A* revolves. The cords, *F, F* attached to these links are for lowering the shutter and holding it in the closed position; they are attached to a cleat on the wall below the slit. A third pair of cords, *G*, is threaded through eyelets in the canvas, their ends being attached to the dome at the top of the slit and their free ends secured to a cleat at the bottom of the slit. These must be cast off before the shutter is opened; when secured, they hold the shutter tightly against the edges of the slit even in a high wind. *H, H* are battens tacked to the shutter at approximately 1-ft intervals to keep the canvas stretched across the slit without sagging.

(*e*) A cylindrical dome is considerably easier to construct than a hemispherical one, while retaining the particular advantages of domed

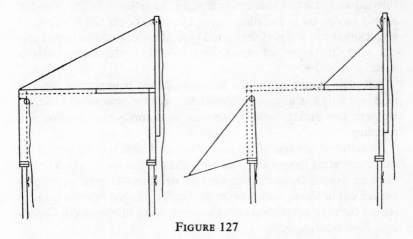

FIGURE 127

as against sliding-roof huts. Both the horizontal and vertical sections of the shutter can be hinged, which again makes for ease of construction (Figure 127).

(*f*) The whole hut, rather than just the dome, may rotate; indeed, with this type of construction, there is no dome as such. The observing slit in the roof lies on the same diameter of the hut as the door, and is continuous with it—i.e. there is no horizontal jamb (or at any rate a removable one, giving strength to the hut when the telescope is not in use) between slit and door. Overall height of the hut can in this way be reduced. Rails can be avoided by setting vertically mounted roller bearings at intervals round a circular concrete foundation; these bear against the inner side of the wall base-plate. Small truck wheels mounted underneath the base-plate carry the weight of the hut and run directly on the concrete foundation.

17.2.3 Temperature control: The temperature conditions to aim for are: minimum excess, due to solar heating, of internal over external temperature during the day; speedy reduction of any such excess to zero after sunset.

Experiments have shown (B. 185–7) that as regards the first of these conditions, fresh white paint is markedly superior to aluminium paint —the only other competitor in the field. In urban atmospheres, however, this advantage is probably lost rather quickly, though no figures are available, and aluminium has the advantage of being more durable when applied to a rough and porous surface, such as millboard.

To satisfy the second condition—minimum temperature lag between internal and external temperatures when the latter is falling—the hut should be opened to its fullest extent, including doors and windows, at about sundown; if it is of the domed type, the slit should be turned into the wind. In respect of this second condition, aluminium paint is superior to white.

Adequate ventilation of the hut when closed is important, since it tends not only to reduce the difference between the external and internal temperatures during the day, but also to decrease condensation and sweating.

The internal painting of the roof or dome, though it can influence the hut temperature to a small extent—by affecting the rate at which heat is radiated inward from the sun-warmed structure—is most commonly carried out in black, since a white or metallic-finished expanse tends to reduce the dark adaptation of the eye even when illuminated indirectly by a faint working light.

SECTION 18

MICROMETERS

18.1 Principle of micrometer measures

Unlike the meridian instrument, which measures the absolute co-ordinates of celestial bodies, the micrometer determines the position of one object relative to another. For the observation to be of value, this second object (the comparison star or reference object) must at least be bright enough to be located with certainty on subsequent occasions; if possible, it should be brighter than the magnitude threshold of whatever catalogue is available. Its coordinates being known, and the micrometer having measured the difference between its RA and Dec ($\Delta\alpha$, $\Delta\delta$) and those of the unknown object, the RA and Dec of the latter can be derived.

Even when the RA and Dec of the unknown object are derived in this way, it is desirable that the observed quantities, $\Delta\alpha$ and $\Delta\delta$, as well as the designation and coordinates of the reference object (as measured from a chart or quoted from a catalogue), and the epoch of the chart or catalogue, should also be recorded. Otherwise the derived position must be reduced to the apparent place at the epoch of observation (see section 30); this is somewhat tedious, and may not be necessary unless the measures were made with great accuracy (as with a filar micrometer) or the epoch of the reference object's coordinates is separated by many years from that of the observation. The process, briefly, is as follows:

(a) The position of the unknown object derived from the measured $\Delta\alpha$ and $\Delta\delta$ is its mean place for the epoch of the catalogue or chart from which the coordinates of the reference object were obtained. First, the catalogue position of the reference object must be reduced to the mean place on 1 January of the year of observation.

(b) This is then reduced to apparent place.

(c) The observed $\Delta\alpha$ and $\Delta\delta$ are applied to the apparent place of the reference object, obtained in (b).

Whereas all types of micrometer described below are capable of

297

determining the rectangular coordinates (α and δ) of an object with more or less accuracy, the filar micrometer may in addition be used to obtain its polar coordinates (position angle and distance). The use of such coordinates is restricted to double stars; they are related to $\Delta\alpha$ and $\Delta\delta$ in the following manner:

If θ=position angle of unknown object with reference to the comparison star (i.e. the primary, in the case of a binary),

r=their angular separation,

α, δ=RA and Dec of the reference object,

α', δ'=RA and Dec of the unknown object,

$\Delta\alpha=\alpha-\alpha'$,

$\Delta\delta=\delta-\delta'$,

$\delta_0=\dfrac{\delta+\delta'}{2}$,

then

$$\left.\begin{aligned}\sin\frac{\Delta\alpha}{2}&=\sin\frac{r}{2}\sin\theta\sec\delta_0\\[2mm]\sin\frac{\Delta\delta}{2}&=\sin\frac{r}{2}\cos\theta\sec\frac{\Delta\alpha}{2}\end{aligned}\right\}\quad\cdots\cdots\quad(a)$$

A simpler approximation, adequate when the two objects are very close together, or distant from the pole, or both, is given by

$$\left.\begin{aligned}\Delta\alpha&=r\sin\theta\sec\delta_0\\\Delta\delta&=r\cos\theta\end{aligned}\right\}\quad\cdots\cdots\quad(b)$$

Of the types of micrometer described in this section, namely:

Slade's (reticulated) micrometer,

Ring micrometer,

Cross-bar micrometer,

Filar micrometer,

Double-image micrometer,

Tilting-grating (interference) micrometer,

the first four employ graticules or webs located in the focal plane of the telescope; the last two employ other principles. Transit graticules and impersonal micrometers for use with meridian instruments are referred to in section 25.

18.2 Location of the graticule in the optical train

It is essential to the accurate use of the micrometer that the image of the star should lie precisely in the plane of the webs, i.e. the graticule must lie in the focal plane of the objective. If this condition is not satisfied parallactic displacement of the image will occur, its position relative to the webs depending to some extent on the position of the observer's eye. This parallax can be used to position the webs correctly: put an artificial star image on one of the vertical webs and move the eye laterally; if the star moves off the web in the same direction as that in which the eye is displaced, the graticule is too near the eye; if the direction of its displacement is opposite to that of the eye, it is too near the objective.

Only positive oculars are suitable for use with a micrometer. If the graticule is placed in the objective's focal plane and observed with a negative ocular, it will be seen with the eye lens only, whose aberrations will therefore be unbalanced by those of the field lens. It will show strong colour, due to chromatic inequality of magnification, and, if a web is not diametrical, distortion. With a positive ocular, on the other hand, both the aerial image and the physical web are observed with the same optical system.

Fixed cross-webs defining the centre of the field, on the other hand, may be placed in the plane of the diaphragm of a negative ocular, for although they will be invested with spurious colour they will not be distorted.

18.3 Slade's (reticulated) micrometer

The simplest and least accurate. A rectangular grid, with or without concentric circles, engraved on a glass diaphragm, is mounted in the focal plane of a positive ocular. The angular length of side of the divisions of the grid is determined by trailing a star of known Dec. The reticule can then be used to determine the approximate $\Delta\alpha$ and $\Delta\delta$ of any two objects which are in the field simultaneously:

(*a*) Rotate the eyepiece until the objects trail the webs. The number of squares and estimated fraction of a square, measured at right angles to the objects' diurnal motion, gives their $\Delta\delta$.

(*b*) The number of squares and estimated fraction of a square, measured in the direction of their diurnal motion, gives their $\Delta\alpha$.

For example (Figure 128): if object A is trailing the web XY, and the side of a grid square has been determined as $4'$, then

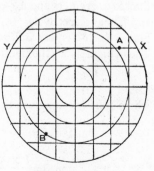

$$\Delta\delta = 18'$$
$$\Delta\alpha = 16'$$
$$= 1^m\ 4^s \text{ (assuming field on equator).}$$

FIGURE 128

18.4 Ring micrometer

A narrow, flat, opaque ring mounted on a glass diaphragm at the focus of the objective. Its inner edge is accurately circular; better still, both edges are accurately circular and concentric.

Its advantages are: simplicity of construction and use; no field illumination is required; no particular orientation or type of telescope mounting is required. Its chief drawback is that the measures made with it require a relatively complicated process of reduction as compared with those made with other forms of micrometer.

18.4.1 Preliminary to use: Before the micrometer can be used for the measurement of differential positions, the diameter (internal and external, or internal only) of the ring must be determined with the greatest possible accuracy.

(a) *Simple method*. Trail a star of known Dec across a diameter (estimated) of the ring. Take the mean of a large number of timings and convert to ' and " arc. Half this value is the required quantity, R, the radius of the ring. Since the exact position of the diameter can only be guessed at (the star is in fact probably trailing a chord) this method, even when a large number of measurements are averaged, gives a less accurate value for R than—

(b) *Accurate method*. Select two stars near the meridian (thus eliminating the effect of refraction on $\Delta\alpha$), of accurately determined Decs. The Pleiades (see Figure 129 and Table) will probably provide a choice of pairs whose $\Delta\delta$ is about three-quarters of the diameter of the ring, and whose $\Delta\alpha$ does not exceed about 15^s to 20^s. The effect of refraction upon $\Delta\delta$ must be allowed for, and may be taken as the difference of the refraction in Zenith Distance of the two stars, to be applied to $\Delta\delta$ (see section 26.3).

Suppose that:

Star S (α, δ) enters and leaves the inner edge of the ring at t_1, t_2 (Sidereal Time) respectively,

Star S' (α', δ') enters and leaves at t'_1, t'_2.

Then (Figure 130) the internal radius of the ring may be derived from:

$$R = \frac{\Delta\delta}{2 \cos A \cos B}$$

where A and B are given by

$$\tan A = \frac{7 \cdot 5(t'_2 - t'_1) \cos \delta' + 7 \cdot 5(t_2 - t_1) \cos \delta}{\Delta\delta}$$

$$\tan B = \frac{7 \cdot 5(t'_2 - t'_1) \cos \delta' - 7 \cdot 5(t_2 - t_1) \cos \delta}{\Delta\delta}$$

300

The Pleiades

Star	Mag	α (1950)			δ (1950)		
16 Tau	5·43	3ʰ	41ᵐ	49ˢ5	+24°	8′	1″5
17 Tau	3·81	3	41	54·1	23	57	27·7
18 Tau	5·63	3	42	10·4	24	41	2·0
19 Tau	4·37	3	42	13·6	24	18	42·8
4498	8·2	3	42	39·0	24	10	50·7
20 Tau	4·02	3	42	50·8	24	12	46·9
21 Tau	5·85	3	42	55·4	24	23	59·9
4506	6·46	3	43	3·9	24	22	24·3
23 Tau	4·25	3	43	21·2	23	47	38·9
4531	6·68	3	44	0·3	24	22	0·3
4535	8·0	3	44	19·1	23	34	23·7
24 Tau	? 6·7–8·1	3	44	22·4	23	57	46·9
4538	6·94	3	44	25·9	23	45	42·3
η Tau	2·96	3	44	30·4	23	57	7·5
4542	6·81	3	44	30·6	24	8	7·6
4567	6·56	3	45	31·1	24	11	36·3
4582	6·57	3	45	58·4	23	42	21·0
27 Tau	3·8	3	46	11·0	23	54	7·5
28 Tau	5·18	3	46	12·4	23	59	7·5
4588	8·25	3	46	13·7	23	44	8·3
4591	6·63	3	46	22·6	24	13	46·8
4603	6·11	3	46	45·2	23	33	39·8
4609	7·34	3	46	57·5	24	11	54·0
4611	6·68	3	46	59·6	23	41	53·2
4620	7·26	3	47	28·8	24	20	42·0

The four-figure numbers in the first column are the designations in Boss's *General Catalogue* (B. 483).

Take the mean of not less than five sets of transits. If extreme accuracy is justified, the probable error can be calculated (see section 29) and the radius of the ring expressed in the form

$$R = x'' \pm y''$$

The same procedure is used to give the external radius of the ring, if the outer edge of the ring is also to be used in the measurements.

18.4.2 Measurement of differences of position: Direct the telescope so that the diurnal motion of the unknown and the reference object will carry them across the ring; they may transit the ring on the same side of its centre, or one to the N and one to the S, but both should be as far from the centre of the ring as possible. Clamp the instrument slightly ahead of the *p* object and note the times of the four transits, as before. Then time of transit of centre of ring by $S = \frac{1}{2}(t_1 + t_2)$,

and „ „ „ „ $S' = \frac{1}{2}(t'_1 + t'_2)$.

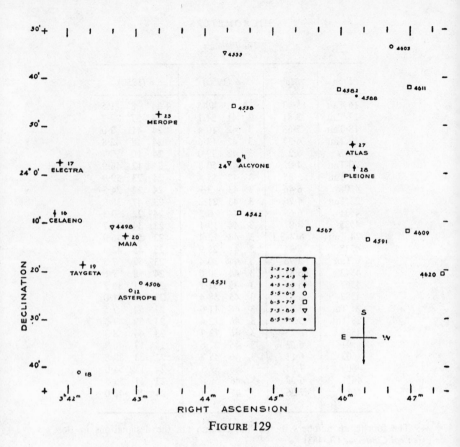

FIGURE 129

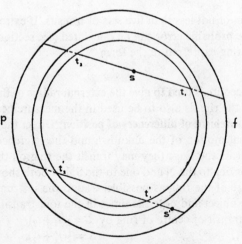

FIGURE 130

Hence
$$\varDelta\alpha=\frac{t_1'+t_2'}{2}-\frac{t_1+t_2}{2}$$

Increased accuracy will be obtained if the times of disappearance and reappearance of each star at both edges of the ring are taken, the mean of four observations being taken, instead of only two.

The derivation of $\varDelta\delta$ from the four (or, better, eight) observations depends upon the facts that (a) the intervals t_2-t_1 and $t_2'-t_1'$ are determined by the Dec of the centre of the ring, and (b) diametrical transit of the ring in Dec δ takes $\frac{R}{15}$ sec δ seconds (R being expressed in $''$ arc).

The difference in Dec between the centre of the ring and each of the two objects, S and S', is given by

$$d=R\cos\gamma$$
$$d'=R\cos\gamma'$$

where γ and γ' are such that

$$\sin\gamma=\frac{t_2-t_1}{\dfrac{R}{15}\sec\delta}$$

$$\sin\gamma'=\frac{t_2'-t_1'}{\dfrac{R}{15}\sec\delta'}\quad\ldots\ldots\quad(c)$$

Since the Dec (δ) of only one of the objects is known, δ' is taken as having the same value as δ, for a first approximation. Having derived d and d', the difference in Dec of the two objects is given by

$$\varDelta\delta=d\pm d'$$

The sign will be + when they are on opposite sides of the centre of the ring, − when on the same side. δ and $\varDelta\delta$ now being known, the second approximation to δ' is derived from $\delta'=\delta\pm\varDelta\delta$. This value is then introduced in equation (c) and its accurate value derived.

In all micrometer measurements the signs of $\varDelta\delta$ and $\varDelta\alpha$ depend upon whether the unknown object is N or S, p or f the reference object.

Times of transit may be determined by stop-watch, chronograph, or the ear-and-eye method (see section 28.18). A chronograph, though a great convenience, is by no means essential. If a MT clock or watch is used, the measured time intervals must be converted to ST intervals.

The simplicity of the ring micrometer favours both its use and its construction by amateurs. A large-scale ring (diameter about 6 ins) can be drawn with Indian ink and blacked in, then reduced photographically on to a thin glass slip; its reduced diameter will depend upon the focal length and field diameter of the ocular, but in most likely cases will be of the order of 0·5 to 0·75 ins. The slip is then mounted in the focal plane of a positive ocular. Alternatively, a thin metal ring may be accurately turned in a machine shop and cemented to a glass diaphragm with balsam.

18.5 Cross-bar micrometer

A grid consisting of two thick wires intersecting normally at the centre of the field, and a thinner third wire bisecting two of the angles

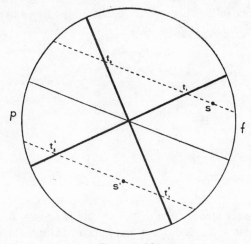

FIGURE 131

formed by the cross-bar, is mounted at the focus of the objective (Figure 131).

Its advantages are: simplicity of construction and use; no field illumination is required. Its limitation is that it cannot be used with an altazimuth mounting.

A cross-bar micrometer can quite easily be made at home, though a good deal of patience may have to be exercised before sufficient accuracy is obtained to ensure acceptable measures.* 10-amp fuse-wire is suitable for the cross-bar, and any thinner gauge for the third wire. Stretch the wire taut over a graved circle. Coat a 1-in brass ring with solder and lay it face downward on the wires so that their intersection lies at its centre: this is made easier if it has guide marks graved at 90° intervals. When in

* On the subject of determining the error of a cross-bar micrometer observationally, the reader is referred to B. 189.

position, warm those parts of the ring overlying the wires, so that the solder will soften and hold them. The outer parts of the wires are then cut to release the ring from the base.

18.5.1 Measurement of differences of position: Orient the micrometer so that the star travels along the third wire when the telescope is moved in RA. Clamp the telescope ahead of the two objects in such a position that S and S' will trail the field on opposite sides of the intersection of the bars. The four times t_1, t_2, t_1', t_2' are noted, and the intervals converted to ST if a MT clock was used.

Then
$$\Delta\alpha = \frac{t_1+t_2}{2} - \frac{t_1'+t_2'}{2}$$

$$\Delta\delta = \tfrac{1}{2}[(t_2-t_1)+(t_2'-t_1')] \, 15 \cos \delta$$

in seconds of time and $''$ arc respectively.

18.6 Filar micrometer

A system of webs, one or more of which is adjustable, placed in the focal plane and observed with a positive ocular. The adjustable web is displaced by means of a calibrated head actuating a screw, the amount of the displacement being read from the calibrations.

In its usual form (Figures 132) it consists of webs fixed normally to one another, the transverse web, TT, the fixed web, FF, and an adjustable micrometer web, MM, which can travel along TT whilst remaining parallel to FF. For greater precision in determining $\Delta\alpha$ by the method of transits, the single transverse web may be replaced by a grid consisting of an odd number of mutually parallel transverse webs.

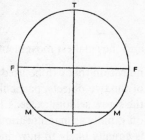

In addition, the entire micrometer is capable of rotation about the optical axis, its orientation being read off an engraved position circle with verniers. Any eyepiece

FIGURE 132

containing a web can be used to determine approximate position angles by fitting a protractor to the drawtube flange and an index of some sort to the eyepiece itself.

If the filar micrometer is fitted with a second adjustable web, it moves along FF and MM whilst remaining parallel to TT. The single adjustable web MM is sometimes replaced by a pair of parallel webs.

The specific advantages of the filar micrometer are: accuracy; flexibility of use (in particular, it can be used to derive polar coordinates, by

305

whose means the relative positions of the components of double stars are always expressed); it can be used (for deriving rectangular coordinates) with an altazimuth mounting.

Among its disadvantages may be mentioned: expense; it cannot be home-made with sufficient accuracy to justify the time and trouble that would be spent on it; for precision work on double stars it suffers from the disadvantages common to all web micrometers: (*a*) the closeness of the pair may necessitate the use of a magnification which is too high for the state of the atmosphere, resulting in ill-defined and moving images, with reduced accuracy in the measures; (*b*) the illumination either of the webs or of the field will render very faint *comites* invisible.

The limit of accuracy of the micrometer can easily be determined. The smallest fraction of a turn that can be read is first determined: if the graduated head has x divisions, and the eye can estimate the reading to $\dfrac{1}{y}$ of a division, then it can be read to the nearest $\dfrac{1}{xy}$ of a turn. Usually there are 100 divisions, the minimum practicable reading being $0 \cdot 001R$ to $0 \cdot 002R$, where R is the value of one complete turn, according to the size of the drum. The pitch of the screw is then measured by dividing its length by the number of turns. Suppose it to be m millimetres. Then if the focal length is F mm, one turn of the screw is equivalent to θ, where

$$\tan \theta = \frac{m}{F}$$

and the smallest movement of the web that can be read is $\dfrac{\theta}{xy}$. The magnification that can be used is dictated largely by the angular separation of the two objects, particularly their separation in Dec, since the run of the screw seldom exceeds 15′ or 20′. The accuracy of the measures falls off rapidly as magnification is reduced, though gross over-magnification is equally likely to introduce errors by accentuating atmospheric flaws.

18.6.1 Preliminary to use: (*a*) *Determination of the zero reading of the micrometer head.* Take the mean of 10 readings when *MM* and *FF* are judged to be exactly superimposed;* this reading is referred to below as R_0. The zero reading should be checked at regular intervals to guard against errors resulting from wear of the screw.

(*b*) *Determination of the zero reading of the position circle.* Allow an

---

* Whenever adjusting the web, whether on to a star or on to another web, it is advisable to make the last movement of the head in the direction that increases the tension of the springs, thus taking up any slackness and avoiding a fruitful source of error.

equatorial star near the meridian to trail across the whole field on MM, using a LP ocular (as low as $\times 10$, if possible). Take the mean of 20 readings of the circle, each taken to the smallest fraction of a division practicable. If this mean is called P_1, the zero reading of the circle is given by

$$P_0 = P_1 + 90°$$

This value also should be checked at regular intervals.

(c) *Determination of the value of one turn of the micrometer head.* If the micrometer is in constant use, this should be carried out at approximately six-monthly intervals.

(i) By measurement of $\Delta\delta$ of two stars of known Declinations: The Pleiades fulfil the following requirements: accurately known positions; near the same hour circle; similar magnitudes; $\Delta\delta$ equal to from 50 to 100 turns of the screw, which, since this is more than a normal field diameter, necessitates the existence of intermediate stars to act as stepping-stones. The seeing should be average or better, and the observations made near culmination. The procedure is to measure $\Delta\delta$ of the selected stars, first from N to S and then back again, on seven separate nights; the mean of these readings of the micrometer head is then taken. $\Delta\delta$ must be derived from the known Decs of the two stars at a given epoch by the application of their proper motions during the interval between that epoch and the date of observation, precession, the reduction from mean to apparent places, and correction of the apparent places for refraction (see sections 30, 26.3). Comstock's formula is accurate enough for determining the refractions, the difference between which is added to the difference between the Decs of their apparent places.

(ii) By means of transits of a circumpolar star: With the position circle reading P_0, and using a LP ocular, observe Polaris just before upper or lower culmination. The telescope drive is stopped, with MM in the f half of the field just p the star; throughout the observations the telescope's position must not be altered. The time of transit of Polaris at the micrometer web is noted. The web is then advanced one turn, or convenient fraction of a turn, and the transit time at the web in this new position is noted. Observe in all 30 or 40 transits before, and an equal number after, culmination. Repeat on several nights, with the telescope on opposite sides of the pier on alternate nights. In the reduction of each night's observations, take the mean of transits 1 and 41 (assuming there to have been 40 on each side of the meridian), 2 and 42, 3 and 43, etc, and with these 40 mean values of t set up 40 equations of the form

$$(m' - m)R = \sin(t' - t_0)\frac{\cos\delta}{\sin 1''} - \sin(t - t_0)\frac{\cos\delta}{\sin 1''}$$

where R =value of one turn of the micrometer head,

t_0 =ST of transit of star in position S_0,

m =micrometer reading when star is in position S,

m' =micrometer reading when star is in position S', (positions S_0, S, and S' being successively assumed),

t =ST when star is at S,

t' =ST when star is at S',

δ =Dec of the star ($+89°$ 2' 19" (1952.0) in the case of Polaris).

If a Mean Time clock is used, $(t'-t_0)$ and $(t-t_0)$ must be converted to ST intervals.

The value of R so derived, after the mean of the values obtained on the several nights is taken, must be corrected for refraction by means of

$$\Delta R = -R \tan 1'' \cot (\delta - \phi) r$$
$$\Delta R = -R \tan 1'' \cot [(180° - \delta) - \phi] r$$

for upper and lower culmination respectively, where ϕ =latitude of the telescope, r =mean refraction.

This is a special and more accurate case of the simpler method where the transit of any star is timed at MM, separated a large whole number of turns of the micrometer head from FF, and again at FF. Then

$$R = \frac{15(t_2 - t_1) \cos \delta}{n}$$

where R =the value of one turn of the head in " arc,

n =the number of turns by which MM and FF are separated,

t =is expressed in seconds of time.

18.6.2 Methods of use: The filar micrometer may be used to measure either $\Delta\alpha$ and $\Delta\delta$, or (more accurately) the position angle θ, and angular separation r, of two objects. For the former, either altazimuth or equatorial mounting is suitable, and the position circle is not used; but for the measurement of polar coordinates a clock-driven equatorial is required.

It is convenient for S and S' to be reasonably close to one another—in the case of double stars this is always so. Except in the method of direct measurement (A (ii) below)—where the separation must be less than the field diameter in both RA and Dec—it is less important that $\Delta\alpha$ should be less than the field diameter than that $\Delta\delta$ should be thus restricted.

A. MEASUREMENT OF $\Delta\alpha$ AND $\Delta\delta$

   (i) *Single adjustable web: any form of mounting*

   1. Clamp the telescope with the micrometer so oriented that S trails

MM. *MM* and *FF* are then coincident with parallels of Dec, and *TT* with an hour circle.

2. Clamp the telescope *p* the field containing *S* and *S'*, adjusting it in Dec so that *S* trails *FF*.

3. Adjust *MM* by means of the micrometer head, so that *S'* trails it. If the micrometer reading is R_1,

$$\Delta\delta = R_1 - R_0$$

4. Note the times at which *S* and *S'* transit *TT*. If they are *t* and *t'*, then

$$\Delta\alpha = t' - t$$

If a Mean Time clock is used, the usual correction must be applied to convert $(t' - t)$ to a ST interval. If the micrometer is fitted with a grid of transverse webs, take the means of the times of transit of *S* and *S'* at each.

5. Repeat the procedure several times (preferably not less than five) and mean the results.

(ii) Single adjustable web: clock-driven equatorial

1. Adjust *FF* parallel to the equator by trailing *S* along it.

2. Start the driving clock and measure $\Delta\delta$ by bisecting the images of *S* and *S'* by *FF* and *MM* respectively. Then

$$\Delta\delta = R_1 - R_0$$

3. Rotate the micrometer through 90° and again bisect the images of *S* and *S'* by the fixed and micrometer webs. Then, if R_2 is the new reading,

$$\Delta\alpha = (R_2 - R_0) \sec \frac{(\delta + \delta')}{2}$$

4. Rotate the micrometer through a further 90° and remeasure $\Delta\delta$ by bisecting the images of *S* and *S'* by *MM* and *FF*.

5. Take the mean of the two determinations of $\Delta\delta$.

(iii) Two adjustable webs: clock-driven equatorial

$\Delta\alpha$ and $\Delta\delta$ are measured as in (i) except that $\Delta\alpha$ as well as $\Delta\delta$ is measured by the displacement of *MM* from its zero position.

Although practice and care can achieve fairly reliable results with a hand-operated equatorial by this method—provided *S* and *S'* are not too remote in RA from each other, and that the instrument is supplied with good slow motions—a clock drive is nevertheless required for complete satisfaction with the measures to be assured.

B. MEASUREMENT OF θ AND r

Position circle, single adjustable web, clock-driven equatorial

1. (*a*) Rotate the micrometer so that S and S' are bisected by MM. Read the position circle (P_1). Take the mean of four independent readings, rotating the micrometer back and forth through roughly a right angle between each observation of the set. Then

$$\theta = P_1 - P_0$$

(*b*) An alternative method of measuring θ, not recommended for very close or unequal doubles, consists of separating MM from FF by a distance equal to roughly twice the diameter of the spurious disc of the primary, and then rotating the micrometer until it is judged that the line joining the centres of the discs is parallel to the webs lying on either side of them.

2. Rotate the micrometer through 90° from P_1, so that the line joining S and S' is now parallel to TT.

3. Bring S to the intersection of TT and FF by means of the slow motion in RA.

4. (*a*) Bisect S' with MM. Note reading of the micrometer head (R_1). Take the mean of several (three or four) independent measures of R_1. Then

$$r = R_1 - R_0$$

(*b*) An alternative, and more accurate, method of determining r is to measure the double distance: bring MM to the other side of FF and bisect the images of S and S' by MM and FF respectively. Note the micrometer reading (R_2). Take the mean of three or four measures of R_2. Then

$$r = \tfrac{1}{2}(R_2 + R_1)$$

The necessity of involving the zero reading in the expression for r is thus obviated; this is usually described as the method of crossed-zeros.

r and θ having been determined, $\Delta\alpha$ and $\Delta\delta$ could if required be derived from (*a*) or (*b*), p. 298.

18.7 Illumination of field or webs

The webs of a micrometer, or of a transit eyepiece, may be made visible at night by three means:

(*a*) silhouetting them against an illuminated field;

(b) illuminating the webs themselves, so that they stand out bright against a dark field;

(c) silhouetting them against an unilluminated field.

Method (a) is the most satisfactory with bright stars. It cannot be used when the observed star is fainter than, or only just as bright as, the minimum field intensity necessary to make the webs visible. It has been found that by using a red field illumination stars may be observed which would otherwise be invisible by this method. The illumination is provided by a light source in the telescope tube which either shines on an opaque white screen, whence light is diffused towards the eyepiece, or which is directed towards the OG, whence light is returned to the ocular by reflection from its air/glass surfaces. The light source must be fairly bright, and it is a great convenience if a variable resistance is included in the circuit, so that the strength of the field illumination can be adjusted as required.

Method (b) is applicable to fainter stars than method (a). The source or sources of illumination within the tube are so arranged that, while they illuminate the webs, no direct light can fall upon either the objective or the ocular; otherwise the field also will brighten. The webs may be illuminated from the side by a dim, screened bulb situated in their plane, facing which is a slip of plane mirror which ensures that they are equally illuminated on both sides, or they may be illuminated from the OG side by white opaque screens which are themselves illuminated by a bright screened source. A variable resistance overcomes the difficulty that if the webs are sensibly brighter or fainter than the star, the accuracy of the measures will be impaired. In the observation of very faint stars, it is helpful to try and align the image with a finite stretch of the web, rather than to restrict the attention to the actual point of the web with which it is in contact. When the highest degree of accuracy is sought, one source of systematic error can be removed by making the colour of the illumination the same as that of the star being measured, if the latter is at all marked.

Method (c) is confined to those cases where the object is so faint that it would be lost if either (a) or (b) were used. Spider web cannot be made visible at night if neither it nor the field is illuminated, and (c) relies upon the use of metal wires, which are not only opaque but are also much thicker than spider web. It is the least accurate of the three methods.

For the measurement of the positions of points on a bright surface, no illumination is, of course, required. When the detail is faint, however, the close juxtaposition of a precisely defined web often makes it invisible, or so difficult to see that accurate measures are impossible.

311

18.8 Materials for the graticule

The webs must have the three characteristics of (a) fineness, (b) uniform diameter, (c) elasticity, which ensures straightness under tension and less trouble from breaks.

Of the various materials that have been tried, spider web is the most generally satisfactory: (a) its diameter is only a few μ, (b) is extremely uniform, (c) adequately elastic. It is hygroscopic, but owing to the ease of replacement, this is not important.

Textile threads fail on (a) and (b), metal and quartz threads on (a) and (c). Quartz nevertheless has the advantage of being unaffected by temperature, humidity, or the lapse of time. Quartz or glass threads must be silvered.

Certain types of reticulated ocular contain a glass diaphragm on which lines are engraved with a diamond. When such a diaphragm is illuminated, the light being excluded from direct entry of the ocular, the engraved lines show bright on a dark field.

Almost any species of spider will provide web that is satisfactory for micrometer graticules, though some are preferable to others in respect of (b). That most commonly used is the garden spider, *epeira diademata*. For collection of the web, prepare a two-pronged fork, the separation of whose prongs is greater than the diameter of the graticule mount. Stand on a chair and shake the spider out of the box in which it has been kept. Wind the web on to the fork, taking care to keep each winding well spaced from its neighbours; the weight of the spider will keep the web sufficiently taut. It is useful to have several freshly varnished forks on hand, as the spider will often produce more web than can be wound on a single fork. After the web has been collected, the forks can be stored in a dustproof, and preferably airtight, box until required.

18.9 Mounting web and wire graticules

(a) *Single cross-webs*. Lay the end web of the fork across the graticule with the web lying in the graved lines marking its correct position; seal each end with a spot of varnish, and when this is dry cut the web free from the fork with a razor blade. Repeat the procedure with the second web.

(b) *Graticule of parallel webs*. Here it is convenient to construct a four-sided metal frame, larger than the graticule mounting, and pivoted at its corners. Lay the webs on the frame, as described above, the frame being locked rectangular. Unlock the frame and push it gently into the shape of a trapezium, thus bringing the webs closer together, while

312

maintaining their parallelism. Clamp the frame at the required degree of lopsidedness. Varnish the graticule mount and lay the webs across it. When dry, cut the projecting ends of the webs with a razor blade and remove the mounting frame.

(c) *For coarse graticules*—employing fine fuse-wire, for example—it is convenient to use an accurately machined mounting table, such as any machine-shop will make to specification quite cheaply. This consists (Figure 133) of a steel plate, drilled and plugged at four points (A—D),

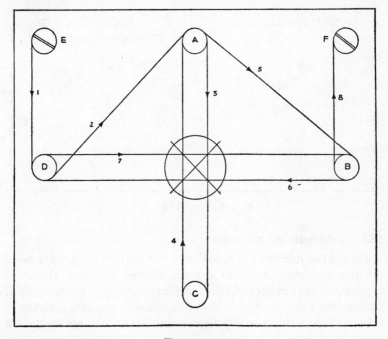

FIGURE 133

drilled and threaded at two points (E, F), and engraved with a circle to mark the position of the graticule mount, as well as with two diameters to mark its centre and to act as a check on the accuracy with which the wires are placed. The direction of winding is indicated by arrows and consecutive numerals. The engraved diameters should be exact diagonals of the central square formed by the cross-wires.

The same mounting table can be used for a single cross-wire by choosing a different winding sequence (Figure 134).

The wire having being wound, tautened, and secured by the screws E and F, it is fixed to the ring by drops of solder, and the ring cut free.

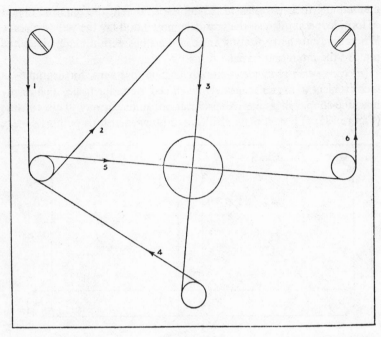

FIGURE 134

18.10 Double-image micrometer

The last two micrometers to be described have been designed to avoid the disadvantages inherent in web micrometers generally. These disadvantages, as already emphasised, are attendant upon the necessity of illuminating either the field or the webs, and also upon the difficulty of bisecting accurately a possibly ill-defined stellar or nebulous image by a well-defined web.

In the double-image micrometer the image of each star of the pair to be measured is divided by one of several available methods—a divided OG or Barlow lens, or a polarising prism mounted on the optical axis between the objective and the ocular. By separating the components of a divided lens, or by moving the prism along the axis, the two images of each star may be separated or closed; furthermore, the direction of their motion depends upon the orientation of the double-image device.

To measure θ and r of a binary, the micrometer is rotated until the direction of separation of the secondary images coincides with the line joining the primary images. The orientation of the divided lens or prism, read from a position circle, then gives θ (Figure 135(b)). The separation

314

of each pair of images is increased until S_2 and S_1' coincide (Figure 135(c)). the angular separation of the primary images, r, is then a function of the separation of the components of the divided lens or of the displacement of the Wollaston prism along the optical axis.

The double-image micrometer can be used on nights of such inferior seeing that a filar micrometer would be useless. For whereas the operation of the latter relies upon the simultaneous bisection of the two stellar

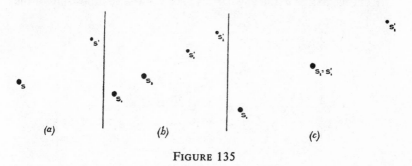

FIGURE 135

images—both suffering independent atmospheric distortion and movement—by clearly defined and motionless webs, the double-image micrometer involves the juxtaposition of two images, S_2 and S_1', whose 'boiling' and distortion are identical.

The decreased brightness of the secondary images amounts to very little in practice, since it is compensated by the fact that no web or field illumination is required. The main source of inaccuracy lies in the difficulty of exactly superimposing, with precise concentricity, two stellar images of different apparent sizes.

See further: B. 188, 190, 191.

18.11 Tilting-grating micrometer

18.11.1 Theory: When a screen containing two narrow slits, one near each end of a diameter of the objective and perpendicular to that diameter, is placed between the objective and a star, interference between the two halves of the divided beam changes the observed diffraction pattern of the stellar image: instead of being a single spurious disc it is spread out into a short series of fringes—alternate bright segments and dark interspaces—the axis of this elongation of the image being parallel to the line joining the slits.

If the two slits are multiplied to form a grating of parallel bars whose interspaces all act as individual slits the same phenomenon occurs,

315

though the images are brighter than before. To account for the formation of the fringes it is simpler to consider the case of a pair of slits at opposite ends of a diameter of the objective.

A parallel beam from a point source off the axis is represented (Figure 136) as falling upon the objective, the whole of which is diaphragmed except for two widely spaced slits. After refraction the two rays converge to a point X in the focal plane FG. The linear distance of X from the axis is d, the angular distance θ. θ is also a measure of the

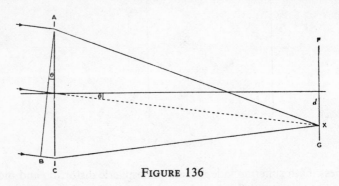

FIGURE 136

distance, BC, between the slits measured in a plane perpendicular to the wavefront AB.

When $BC = \lambda(n + \frac{1}{2})$, where n is an integer and λ the wavelength of the radiation, crests and troughs of the two converging wave systems will combine at X. Destructive interference will occur, with the formation of a dark interspace. It can be seen that there will be a series of positions for X along FG, each corresponding to a different value of n. At all intermediate positions, however, Y, the distance between the slits (measured normally to the wavefront), will be $n\lambda$, a whole number of wavelengths; the two systems will reinforce each other at all positions Y, with the production of a bright band at each.

The spacing of the fringes is a function of the wavelength, λ, and the linear separation of the slits, which may here be taken as equal to the aperture, D:

$$\frac{d}{F} = \frac{\frac{1}{2}\lambda}{D}$$

or

$$d = \frac{\lambda F}{2D}$$

If, now, the field contains two stars, two parallel linear interference patterns of bright and dark fringes will be seen. The distance between

316

them (*a* in Figure 137) can be reduced by rotating the grating so as to approximate the line joining the slits to that joining the stars. When the two patterns are overlapping, the relative displacement of one to the other, *b*, depends upon the angular separation of the two stars, *r*. The resolution of the individual pattern, *c*, varies with the separation of the slits. Hence the adjustment of the slit separation which is required in order to make the bright fringes of one pattern coincide with the dark interspaces of the other, so that the latter vanish, is a function of *r*:

$$r = \frac{\lambda}{2D} \quad \text{(measured in radians)}$$

or $\quad r = \frac{206,265\lambda}{2D} \quad \text{(measured in '' arc)}$

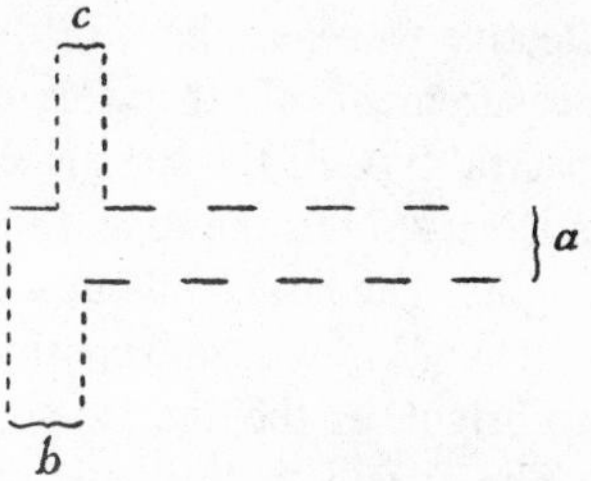

Since the fringes will only disappear completely if the stars are equally bright, it is advisable to make a set of measures with the fringes overlapping followed by a second set with the images in close juxtaposition.

FIGURE 137

When employed to measure the angular diameter of a disc image, the corresponding expression is

$$\text{diameter} = \frac{206,265\lambda}{D} \quad (\text{'' arc})$$

and the mode of operation consists in adjusting the grating or slits until the images just touch, taking the mean of a set of readings.

18.11.2 Performance: The limiting factor in the use of this type of micrometer is illumination. A single pair of slits occludes all but a small fraction of the incident light. For this reason a grating consisting of parallel bars—they and the interspaces being of equal width—is employed; each interspace functioning as a slit, and the width of each bar being the effective slit aperture. Even so the light loss is considerable (the 50% of the incident light passed by the grating being divided between three, five, or more secondary images) and with small instruments the utility of the micrometer is narrowly limited. Small refractors can deal with only the brighter doubles (which have already been sufficiently observed), though a 4½-in has yielded results with binaries whose *comites* are as faint as mag 7.

The chief sacrifice made by substituting a grating for a pair of slits is in the resolution of the image pattern (since the value of *c*, Figure 137, depends on the slit separation), with consequent uncertainty regarding the precise overlapping of the fringes. For this reason, primarily, the grating micrometer is rather less accurate than a filar micrometer used with the same instrument. Increasing the illumination rather than the resolution of the images has the effect of bringing fainter and more disparate doubles within the scope of the micrometer, without notably increasing the accuracy of its measures of brighter pairs.

A further source of inaccuracy is the necessity of introducing the effective wavelength, λ, into the expression for r. The precision of the measurement of r therefore depends upon the accuracy with which the spectral type of the stars is known; and a further difficulty arises when, as is very often the case, they are of markedly different effective wavelengths. The latter difficulty can only be satisfactorily obviated by using a filter of known transmission, and this is impracticable unless the stars are bright, or the aperture considerable, or both.

The tilting-grating micrometer has the same advantage and disadvantage as characterise the double-image micrometer: namely, that the two images to be compared are identically affected by atmospheric turbulence, whereas the exact point at which they coincide is difficult to judge.

With the skill that comes from practice the grating micrometer can be used to give results not markedly inferior to those of the filar micrometer, though in a more restricted field. In unfavourable cases—where one or both of the components is faint, where their magnitude disparity is great, or where their spectroscopic types are widely dissimilar —it is less reliable.

With this proviso, a home-made grating micrometer used with a small refractor may be expected to give measures accurate to 1% or 2% when the mean of a set is taken; individual measures will be wider of the mark. Good seeing is essential. Disc images may be rather more accurately measured: a $4\frac{1}{2}$-in refractor has given a mean value of a set of measures of the Martian diameter at opposition which is accurate to about 0·5%. Jupiter, Venus, and Mercury are also within the range of this instrument; Uranus, Saturn, and Neptune are too faint.

18.11.3 Construction and operation: In its simplest form the micrometer consists of a pivoted grid mounted in front of the objective, capable of being rotated about the optical axis and of being tilted about an axis in its own plane and parallel to the bars; the controls of both adjustments are carried back to the observing position.

If a position circle is incorporated, position angles can be derived from the position of the grating when the two image patterns coincide.

Slit separation is reduced by tilting the grating away from the plane perpendicular to the optical axis, and varies as the cosine of the angle of tilt except at the widest angles.

Individual measures being subject to rather wide random errors, sets of not less than six should be taken. To avoid eye fatigue, each individual adjustment should be made as quickly as possible. Each set should include measures made with the grating tilted in opposite directions. It is advisable to use the highest magnification compatible with sufficient illumination of the image; with small refractors this will seldom be found to exceed about 10D.

B. 192–4 contain details of home-made micrometers of this type.

SECTION 19

SPECTROSCOPES

19.1 Prismatic spectra

A ray of polychromatic light in traversing a prism undergoes two modifications, deviation and dispersion. A monochromatic ray, on the other hand, suffers deviation only. Figure 138 represents a section through a prism taken in a plane perpendicular to its base; AB and AC are the refracting surfaces, and A the refracting angle. The following relations are worth noting:

refracting angle: $\quad\quad\quad\quad A=r+r'$

total deviation: $\quad\quad\quad\quad \omega=(i+i')-(r+r')$

$$A+\omega=i+i'$$

and $\quad\quad\quad\quad\quad\quad \omega=(i+i')-A$

whence is seen that ω is minimal when $i=i'$. This is the position of minimum deviation, in which the ray passes through the prism symmetrically:

$$AB=AC$$

$$i=i'=\frac{\omega+A}{2}$$

$$r=r'=\frac{A}{2}$$

$$\sin i=\mu \sin \frac{A}{2}$$

$$\mu=\frac{\sin \dfrac{A+\omega}{2}}{\sin \dfrac{A}{2}}$$

The refractive index of a material $\left(\dfrac{\sin i}{\sin r}\right.$ when one of the media is air) is inversely proportional to the wavelength of the radiation with which

320

it is measured. Hence ω is a function of λ, and the measure of the differential deviation of radiations of different wavelengths is the dispersive power, or relative dispersion, of the prism. Using the suffixes F,

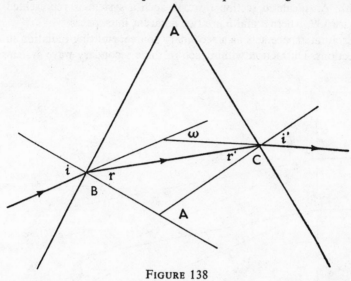

FIGURE 138

C, and D to denote quantities measured in the wavelengths of the F and C lines of hydrogen and the D line of sodium (roughly, the limits and centre of the visible range) respectively:

angular length of spectrum (when A is small):

$$\omega_F - \omega_C = (\mu_F - \mu_C)A$$

relative dispersion for F and C:

$$\frac{\omega_F - \omega_C}{\omega_D} = \frac{\mu_F - \mu_C}{\mu_D - 1}$$

and

$$\omega_F - \omega_C = \frac{\omega_D}{\nu}$$

where ν is termed the constringence of the prism (see section 6.1).

The position of minimum deviation is one of great importance, having the following characteristics:

    (*i*) maximum purity of spectrum is obtained for a given slit width;

    (*ii*) maximum light transmission;

    (*iii*) minimum contribution of astigmatism by the prism;

    (*iv*) minimum displacement of the spectrum results from any accidental displacement of the prism.

321

19.2 Diffraction spectra

A diffraction grating may be considered as a system of parallel, equidistant, opaque strips separated by transparent interspaces of the same width. An idealised section through such a system is represented in Figure 139, where a and b are two adjacent interspaces.

Each interspace acts as a secondary source, emitting radiation in all directions. Diffraction within each of these secondary wave systems is

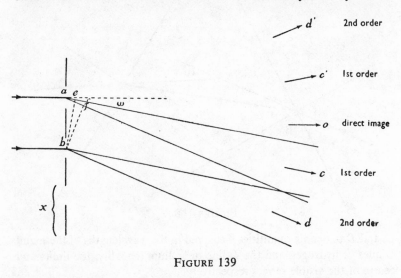

FIGURE 139

such that, in the case of incident light normal to the plane of the grating, reinforcement will occur when

$$x \sin \omega = n\lambda \quad . \quad . \quad . \quad . \quad . \quad . \quad . \quad . \quad (a)$$

where n is an integer, and maximum destructive interference when $n + \frac{1}{2}$ is an integer. Where the incident light is not normal to the grating, but inclined to the normal at an angle ϕ, the grating equation takes the form

$$x(\sin \omega - \sin \phi) = n\lambda$$

In the direction O will be formed, not a spectrum, but an integrated image of the slit, since all points on the wavefronts advancing in this direction are in phase on a plane surface. In the directions c and c' spectra of the 1st order will be produced, providing these directions are such that $ae = \lambda$, $(n = 1)$, i.e. the wavefronts advancing towards c and c' are in phase on a plane surface, but the fronts from adjacent secondary sources are one complete wavelength behind one another. Similarly, the 2nd-order spectra will be formed in the directions d and d', provided

322

$af = 2\lambda$, ($n=2$), i.e. wavefronts from adjacent interspaces are two complete wavelengths out of step with one another. And so on for spectra of higher orders. At positions intermediate between those just considered no image whatever will be observed, since for such positions n is not an integer, and destructive interference occurs. It will be noticed that the short-wave end of each spectrum is that which is nearest the direct image.

In practice, of course, the different orders are brought under observation most conveniently, not by shifting the eye or the camera, but by rotating the grating, if it be of the reflecting type.

By substituting the actual values of x and λ in equation (a), and taking successive integer values for n, the location of the various spectra for normal incident light can be calculated: taking $\lambda = 5700$Å (yellow), and $x = 1 \cdot 81 . 10^{-4}$ cms (14,000 rulings to the inch), we derive:

n	ω	
0	0°	direct image
±1	±18° 21'	1st-order spectra
±2	±39° 2'	2nd-order spectra
±3	±70° 50'	3rd-order spectra

Spectra in successive orders are increasingly deviated, dispersed, and faint.

Figure 139, though adequate for the explanation of principle, is misleading as to the spatial disposition of the different orders of spectra. They are distributed in a succession of planes, each one normal to the line joining its centre to the centre of the grating. Nor are the different orders separated as suggested by the Figure. Equation (a), p. 322, shows that the position of a wavelength λ in any order is given by

$$\sin \omega = \frac{n\lambda}{x}$$

Hence the wavelengths λ_1, λ_2, λ_3 will coincide in the spectra of the three orders n, $2n$, and $3n$, if

$$n\lambda_1 = 2n\lambda_2 = 3n\lambda_3$$

The result of this, so far as it affects spectra of the first four orders, is shown in Figure 140, where it can be seen that the overlapping of spectra becomes increasingly acute as n increases.

It is thus desirable to use spectra of low orders so as to obviate the necessity of colour filters or plates of restricted colour sensitivity. On the other hand, if n is small, high dispersion (see section 19.3.1) can only be obtained by reducing x, i.e. increasing m, the number of rulings per cm.

The surplus field light which is encountered with a grating, and which

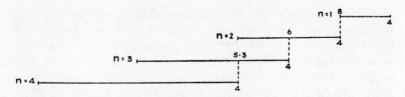

FIGURE 140
Each spectrum extends from 4000Å to 8000 Å. Wavelengths are
shown in units of 1000Å

reduces contrast in the 1st and 2nd-order spectra, can be eliminated by
means of an ocular mask covering the two segments of the field above
and below the image itself.

19.3 Efficiency of a spectroscope

Three characteristics of the performance of a spectroscope are
relevant to a consideration of its efficiency:

its dispersion (analogous to the magnification of a telescope);
its resolving power (analogous to that of a telescope);
the brightness of its spectrum (analogous to the light grasp of a
telescope).

19.3.1 Dispersion:
The dispersion of a spectrum may be defined as
the rate of change of deviation with wavelength. If we write ω and ω'
for the deviations of two rays, and λ, λ' for their respective wavelengths,

then $\omega - \omega' = \Delta\omega$, $\lambda' - \lambda = \Delta\lambda$, $\dfrac{\lambda' + \lambda}{2} = \bar{\lambda}$, and we have for the wavelength λ

$$\text{angular dispersion} = \frac{\Delta\omega}{\Delta\lambda}$$

expressed in, e.g., degrees or radians per Ångström. The linear dispersion in the focal plane of the spectroscope depends upon the focal
length, f, of the camera lens forming the image (or the magnification of
the view telescope):

$$\frac{\Delta l}{\Delta\lambda} = f \cdot \frac{\Delta\omega}{\Delta\lambda}$$

whence
$$\Delta l = f . \Delta\omega$$

expressed in, for example, mm per Ångström.
If the refractive indices of the prism for λ and λ' are μ and μ', and we

write, as before, $\mu - \mu' = \varDelta\mu$, and $\bar{\mu} = \dfrac{\mu + \mu'}{2}$, then when the prism is set at minimum deviation we have:

$$\varDelta\omega = \varDelta\mu \cdot \frac{2 \sin \dfrac{A}{2}}{\left(1 - \bar{\mu}^2 \sin^2 \dfrac{A}{2}\right)^{\frac{1}{2}}}$$

and

$$\frac{\varDelta\omega}{\varDelta\lambda} = \frac{2\dfrac{\varDelta\mu}{\varDelta\lambda} \sin \dfrac{A}{2}}{\left(1 - \bar{\mu}^2 \sin^2 \dfrac{A}{2}\right)^{\frac{1}{2}}}$$

which is the expression for the angular dispersion of a prism of refracting angle A, for a wavelength $\bar{\lambda}$ whose refractive index is $\bar{\mu}$. Angular dispersion at a given wavelength is thus a function of the refractive index, position (minimum deviation, giving minimum dispersion, is here assumed), and refracting angle.

If we put $A = 60°$, the last equation reduces to

$$\frac{\varDelta\omega}{\varDelta\lambda} = \frac{\varDelta\mu}{\varDelta\lambda} \cdot \frac{1}{\left(1 - \dfrac{\mu^2}{4}\right)^{\frac{1}{2}}}$$

To compare the average angular dispersions of 60° prisms of typical crown and typical flint glass, substitute the values of μ for the wavelengths of the F and C lines as given in section 6.1:

	Crown	Flint
$\varDelta\mu$	0·008	0·018
$\varDelta\lambda$	1702Å = 1702 × 10⁻⁸ cm	
∴ $\dfrac{\varDelta\mu}{\varDelta\lambda}$	$\dfrac{0\cdot008}{1702 \times 10^{-8}} = 470$	$\dfrac{0\cdot018}{1702 \times 10^{-8}} = 1058$
∴ $\dfrac{\varDelta\omega}{\varDelta\lambda}$	$\dfrac{470}{\left(1 - \dfrac{1\cdot519^2}{4}\right)^{\frac{1}{2}}} = 724$	$\dfrac{1058}{\left(1 - \dfrac{1\cdot625^2}{4}\right)^{\frac{1}{2}}} = 1814$

Hence the average overall dispersion of the flint is roughly two and a half times that of the crown. In both cases the dispersion is about seven times as great in the ultraviolet as in the far red.

The angular dispersion provided by a grating is independent of N, the total number of rulings illuminated, being dependent solely on the separation of the rulings and the order of the spectrum:

$$\frac{\Delta\omega}{\Delta\lambda} = \frac{mn}{\cos\omega} \quad \cdots \cdots \cdots \quad (b)$$

where m = number of rulings per cm,
n = order of the spectrum.

Since $m = \frac{1}{x}$ (where x is the distance separating the centres of adjacent rulings) this can be rewritten

$$\frac{\Delta\omega}{\Delta\lambda} = \frac{n}{x\cos\omega}$$

whence it can be more readily seen that the dispersion of a grating depends, ultimately, on nothing but ω—i.e. on the position of the spectrum—since an increase in x is automatically compensated by a larger value of n of the spectrum that is formed there.

If f is the focal length of the view-telescope objective or camera lens forming the image, the linear dispersion is given by

$$\frac{\Delta l}{\Delta\lambda} = f \cdot \frac{n}{x\cos\omega}$$

One can deduce from expression (b) that a diffraction spectrum, unlike a prismatic spectrum, is normal, since $\cos\omega \backsimeq k$, and therefore $\omega \propto \Delta\lambda$. Also that the dispersion of high-order spectra is greater than that of low-order spectra—whence, also, they are fainter.

19.3.2 Resolving power: The resolving power of a prism is its ability to resolve into visually perceptible intensity-maxima the different wavelengths of which the incident radiation is composed. If, with the slit at minimum width, λ and $\lambda + \Delta\lambda$ are just resolved, then the prism's resolving power in that region of the spectrum is given by

$$R = \frac{\lambda}{\Delta\lambda}$$

R is a function of the refractive index and of the difference between the longest and shortest light paths through the prism, and is independent of the prism's refracting angle. If the whole face of the prism is illuminated, the difference in light path is equal to b, the length of the prism's base, and

$$\frac{\lambda}{\Delta\lambda} = b \cdot \frac{\Delta\mu}{\Delta\lambda}$$

Thus b is related to prismatic resolution in a manner analogous to that of D with telescopic resolution.

For average optical glass, $\dfrac{\Delta\mu}{\Delta\lambda}\backsimeq10^3$, where λ is expressed in cms. Hence, for a prism,

$$R\backsimeq10^3b$$

when b is also measured in cms.

Analogously to the magnification and resolving power of a telescope, an increase of prismatic dispersion leads to increased separation of the spectral lines, but since it involves a corresponding increase in their width, resolution is unaffected.

Thus, determining the resolving power of a prism, we have:

(*i*) Slit width: maximum theoretical resolution being obtained with an infinitely narrow slit. Hence the purity and brightness of a continuous spectrum are inversely related to one another, in so far as they are dependent upon slit width. For a given value of f_{col}, spectral purity will vary inversely with slit width; for a given slit width it will vary directly with f.

(*ii*) Effective base length of prism; or, in the case of a prism train, the sum of the bases; or, in the case of reflection back through the same prism, the base length multiplied by the number of such reflections. To resolve the 6Å-apart D lines, a base length of not less than 1 cm of dense flint is required.

(*iii*) Refractive index: hence a flint prism has a greater resolving power than a crown prism, other factors being the same.

(*iv*) Wavelength: as with dispersion, resolution with a prismatic spectroscope is seven or eight times as great in the ultraviolet as in the extreme red.

The resolution of a normal spectrum, produced by a grating, is similarly measured in terms of λ and $\lambda+\Delta\lambda$, two wavelengths which are just resolved:

$$R=\frac{\lambda}{\Delta\lambda}=nN$$

where $N=$ total number of rulings illuminated,
    $n=$ order of the spectrum.

nN then expresses the wavelength at any point in the spectrum divided by the least difference of wavelength that is perceptible there.

Hence with a given grating, the greater the area that is illuminated, the greater will be the resolution. To avoid impossibly large gratings, the finer the engraving the better; the commonly employed scale is about 15,000 per inch. Gratings thus possess a signal advantage over prisms in this respect: the resolving power of the latter can only be increased by increasing the thickness of glass traversed—a procedure which is limited by the absorption becoming prohibitive (apertures of 3 ins are near the maximum that can be usefully employed)—whereas that of a grating can be increased by increasing the number of rulings per inch, or the number of inches illuminated, or both. Since about 1000 rulings will resolve the D lines in the 1st-order spectra, a 4-in grating, normally engraved with 15,000 per inch and fully illuminated, will have a 1st-order resolution equivalent to that of a dense flint prism train of base about 60 cms.

As in the case of a telescope, the visibility of resolved detail assumes the use of adequate magnification ($\geqslant M_r$). This condition may, and should, be fulfilled with small spectroscopes; but in the case of high-dispersion spectrographs, to increase the linear dispersion to an extent that would separate the images of adjacent resolved wavelengths would so dilute the intensity of the image that exposures would have to be impracticably long.

It should be stressed again that incomplete illumination of the prism or grating will result in wasted resolving power. Hence the collimator lens must be large enough, and must be filled with light from the slit.

19.3.3 Brightness of the spectrum: The brightness of the spectrum of a source of unit intensity is dependent on:

(*a*) *Size of objective*

(*i*) Point objects: Owing to the increased size of stellar images with increased D, owing again to their increased susceptibility to atmospheric turbulence, the brightness of the spectrum of a given spectroscope varies more nearly with D than with D^2, as would be expected from theory. Increasing the aperture of the collimator (and therefore its focal length, to preserve the relation $D/F=d/f$) will increase the purity but not the brightness of the spectrum, which in the case of stellar images is determined solely by D.*

(*ii*) Extended objects larger than the slit width, but smaller than the angular aperture of the telescope: Here the brightness of the spectrum is determined by the effective aperture of the collimator, irrespective of the fact that the intensity of the focal plane image of such an object varies inversely as F^2 for unit telescopic aperture. Provided only that

* The increase in F necessitated (with a given spectroscope) by an increase in D will also widen the spectrum of an extended source, and the dispersion; this springs necessarily from the fact that F and M are directly proportional.

equality is maintained between the focal ratios of the telescope and collimator, variation of F has no effect upon the brightness of the beam emergent from the collimator, and therefore none upon that of the spectrum. This can easily be demonstrated:

Putting F and D for the focal length and aperture of the telescope,
      f and d for the focal length and aperture of the collimator,
         x and a for the diameter and area of the focal image, dependent upon F,
      b for its brightness,
then for maximum efficiency,

$$\frac{D}{F} = \frac{d}{f}$$

If, now, F is halved, so that

$$\frac{D}{F'} = \frac{2d}{f}$$

only one-quarter of the radiation from the slit will be intercepted by the collimator, and $b' = \frac{b}{4}$. But at the same time $x' = \frac{x}{2}$, hence $a' = \frac{a}{4}$ and $b' = 4b$. By the combination of these two factors, $b' = b$, and the brightness of the image will remain unchanged.

If f is also halved, to reinstate the relation

$$\frac{D}{F'} = \frac{d}{f'}$$

all the radiation from the slit will indeed be intercepted by the collimator, but the effective width of the slit (its angular width at distance f) will have been doubled, and to achieve the same resolution as formerly its linear width will have to be halved, so that b' will again be reduced to b.

(*iii*) Extended objects whose angular size exceeds the angular aperture of the telescope: These are best observed by doing without the telescope altogether, when the brightness of the spectrum will be directly proportional to d.

(*b*) *Width of slit*

The significant measure of slit width is its angular width as seen from the centre of the collimating lens, i.e. from a distance f. Effective slit width can therefore be reduced by a linear reduction or by an increase of f.

With normal slit widths and average seeing, from 60% to 90% of the light from a point source transmitted by the objective passes the slit; in the case of a large instrument, this figure is often reduced to 50% or less owing to the expansion of the theoretical spurious disc by atmospheric turbulence.

The brightness of a bright-line spectrum is independent of slit width, and dependent on (*a*) and (*c*) solely: opening the slit merely widens the lines.

(*c*) Transmission factor of the optical train

Light loss in the optical train of a high-dispersion spectrograph may be as high as 45% to 48%. Assuming that 50% of the incident light has passed the slit, this gives a transmission factor of about 27%. With smaller instruments of lower dispersions, it is very much higher.

In addition to the inevitable light loss by absorption in, and reflection at, prisms and lenses, the transmission factor of the instrument will be needlessly reduced if:

> the focal ratios of objective and collimating lens are not the same, in which case a peripheral ring of one or the other will be out of commission;
> any part of the optical train is not large enough to pass the whole of the beam delivered to it;
> it is, on the other hand, needlessly large;
> the prism or prisms are not in the position of minimum deviation; even at minimum deviation from 10% to 20% of the light incident upon a 60° prism is lost by reflection: the denser the glass the greater this loss.

Loss by absorption is also greater in flint than in crown; whilst the light loss from this cause is increased by either increasing the density of the glass or the length of the light path through it, this additional loss is not borne equally by all wavelengths, but more heavily by the actinic than by the longer wavelengths.

(*d*) Dispersion

Doubling the dispersion while retaining the original width of a continuous spectrum reduces its brightness to one-half its former value. But if the doubling of the dispersion is achieved by an overall magnification of the spectrum, the brightness will be reduced to one-quarter. The brightness of an emission spectrum would be unaffected, since increasing the dispersion merely increases the separation of the monochromatic images.

(*e*) Grating spectra

The brightness of a normal spectrum is, within limits, a function of the closeness of the rulings (brightness increasing as the number of

330

rulings per cm increases) and of the size of the grating, while being inversely proportional to the order of the spectrum.

19.4 Spectroscope construction: optical requirements

1. Slit in focal plane of telescope objective.

2. Slit in focal plane of collimator (which then transmits a parallel pencil to the prism or grating).

3. Slit parallel to the prism edge or the grating rulings.

4. Focal ratios of telescope objective, view-telescope objective and collimating lens the same ($F/D=f/d$), to utilise their full aperture.

5. Prism just large enough to accept the whole of the parallel beam from the collimator. (If too small, light grasp is being wasted; if much too large, unnecessary light loss through absorption.)

6. Prism in position of minimum deviation for the central wavelength.

7. Camera lens or view-telescope objective to reconverge the beam from the prism or grating (unless the latter is concave) to a focus at the photographic plate or the eyepiece.

19.5 Diffraction gratings

Reflecting gratings are engraved with a diamond on speculum metal. Transmission gratings are engraved on glass. Both are expensive.

Replica gratings are collodion impressions of the original, and are comparatively cheap—comparable with prisms of equal aperture, and about one-tenth as costly as an original. Replicas are mounted on plane or concave glass, silvered or unsilvered, and may be used as either transmission or reflecting gratings. Their definition is inferior to that of original gratings when the slit is 'closed'; but when widened—for the observation of prominences, for example—there is nothing to choose between them.

Concave reflecting gratings have the advantage that they both form and focus the spectrum, thus eliminating the view telescope or camera objective with its unavoidable additional light loss.

Most modern gratings have from 14,000 to 15,000 rulings to the inch or about 30,000. For maximum efficiency the separation of the rulings should be comparable with the wavelength of light, i.e. of the order of 1/50,000 in. Concentration of most of the light into a particular spectrum (usually of the 1st order) is determined by the form of the lines ruled with the diamond.

The outstanding advantages offered by gratings are (i) normal spectra i.e. dispersion uniform throughout the spectra of a given order; (ii) choice of dispersions, that of the 2nd order being approximately twice that of the 1st, of the 3rd three times that of the 1st, etc.

The greatest enemy of gratings is dust and, in the case of reflecting gratings, tarnishing and corrosion of the reflecting surface. They must always be kept under airtight cover when not in use; dusting, as infrequent as possible, should be done with a sable-hair brush, the strokes being very light and made in a direction parallel with the rulings.

19.6 Direct-vision prisms and gratings

By suitable choice of refracting angles and refractive indices, it is possible to design two-prism combinations which (*a*) deviate an incident polychromatic ray without dispersing it, thus giving an achromatic prism,* or (*b*) disperse it without deviating it, thus giving a direct-vision prism. Such prisms may be regarded as a halfway stage from the plane parallel plate (which displaces the ray laterally but neither deviates nor disperses it) to the ordinary prism.

In the case of (*b*), nil residual deviation is arranged for a mean wavelength (usually in the region of the *D* lines of sodium), other wavelengths diverging on either side of the selected undeviated wavelength.

By means of two prisms (Figure 142), nil deviation can only be achieved with lateral displacement. To make the emergent *D* ray not

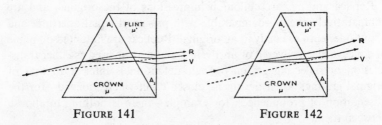

FIGURE 141   FIGURE 142

simply parallel to, but coincident with, the incident ray, a similar but reversed unit must be added, producing a three-prism combination (Figures 143, 144) known as an Amici prism. Any odd number of components may be used, each additional pair increasing the dispersion without introducing deviation. Three- or five-prism combinations are the usual d/v spectroscope arrangement, the first and last components being always of crown glass. Total internal reflection is avoided by cementing the components. The convenience of d/v for the centre of the

* Achromatism, in this sense, indicates that the emergent rays of different wavelengths, though separated, are mutually parallel (Figure 141).

spectrum is to some extent offset by the heavy absorption in such combinations, though this is of little account in solar work.

The refracting angles, refractive indices, and constringences of the

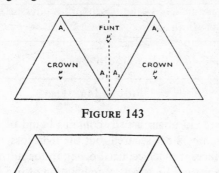

FIGURE 143

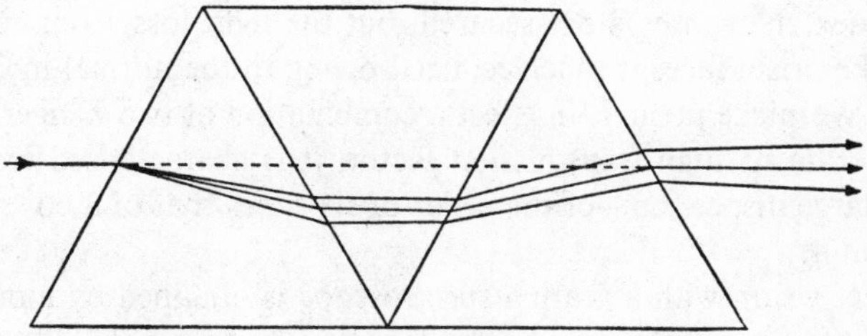

FIGURE 144

two (crown and flint) components are related to the dispersion of the combination as follows:

Given that $\omega_D=0$

he n $\dfrac{A_1}{A_2}=\dfrac{\mu_D'-1}{\mu_D-1}$

and $\omega_C-\omega_F=(\mu_D-1)\left(\dfrac{1}{\nu'}-\dfrac{1}{\nu}\right)$

where the constringence ν, is given by

$$\nu=\dfrac{\mu_D-1}{\mu_F-\mu_C}$$

Two types of d/v combination using one liquid component are illustrated in Figures 145 and 146. The first is known as a Zenger prism: it consists of two right-angle prisms, cemented at their hypotenuse faces and set with the base of the first component normal to the incident beam. If the materials of which they are

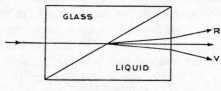

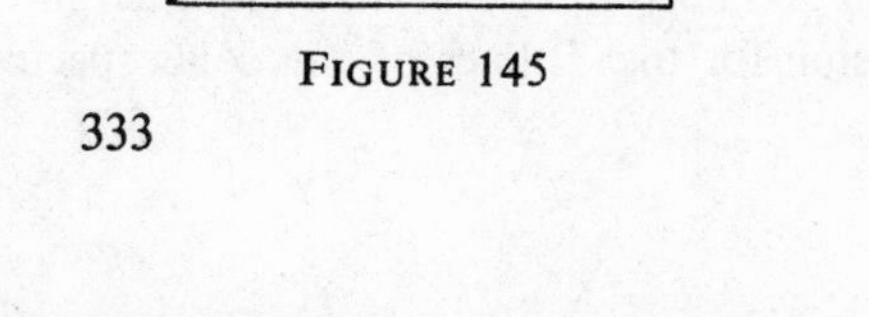

FIGURE 145

333

made each has the same value of μ_D but different dispersions, λ_D will be transmitted undeviated, while other wavelengths will be deviated and

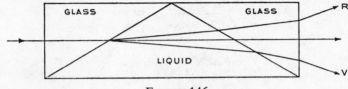

FIGURE 146

dispersed. The second prism usually contains liquid of suitable refractive index. Not only is d/v secured, but the light loss from reflection from the prism faces is much reduced owing to the normal incidence.

The Wernicke prism is in effect a combination of two Zenger prisms. Apart from its high transmission factor, it is characterised by abnormally large dispersion—of the order of six times that of a 60° prism of dense flint.

Direct vision with a grating spectroscope is obtained by mounting a replica grating on the hypotenuse face of a prism of suitable A and μ

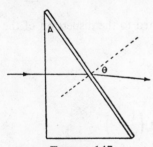

FIGURE 147

(Figure 147). The value of A to give zero deviation for the centre of the 1st-order spectrum depends upon the number of the grating's rulings per cm, and upon the refractive index of the prism. The angle θ, for any wavelength λ, is given by

$$\lambda = x(\mu \sin A - \sin \theta)$$

where x = separation of adjacent rulings; and if λ is to be undeviated, we have

$$\lambda = x(\mu - 1) \sin A$$

Substituting the values, $\lambda = 5700\text{Å}$, $\mu = 1 \cdot 6$, $x = 1 \cdot 81 . 10^{-4}$ cm (14,000 rulings per inch), we have approximately

$$A = 32°$$

In practice a 30°/60° prism is usually employed in small spectroscopes of the type sold for the observation of prominences.

D/v grating spectroscopes have the double advantage that their spectrum is normal, while zero deviation for any desired wavelength can be obtained merely by juggling with μ and x.

A Thorp replica grating mounted in this way gives sufficient dispersion for the D lines of the solar spectrum, or the lines in the spectra of bright A-type stars such as Sirius to be separated without further magnification.

19.7 Direct-vision and slitless spectroscopes

The slit is often omitted from small d/v spectroscopes. These commonly go by the name of 'star spectroscopes' or 'eyepiece spectroscopes'; the McClean and Zollner d/v spectroscopes are examples. They are employed either in place of the telescope's ocular, or as an attachment to be screwed on to the ocular.

Having no slit they can be used only on point sources if the spectrum is not to be hopelessly impure. The stellar image is in effect a very short length of a slit whose width is equal to the diameter of the spurious disc. For this reason good seeing is important, or an impure spectrum will result from the enlargement of the spurious disc.

Since absorption lines are impossible to see in a spectrum which, whatever its dispersion, is a mere line of colour, the spectrum must be widened in some way.* This is usually effected by means of a cylindrical lens (or a cylindrical curve may be incorporated in the prism itself) which draws out the spurious disc into a linear streak in a direction perpendicular to that of the dispersion, the amount of widening achieved in this way being inversely proportional to the radius of the curve. Effectively, therefore, such spectroscopes incorporate a slit whose width is equal to the diameter of the spurious disc of the star under observation.

The McClean spectroscope employs a concave cylindrical lens mounted close to the prism on the side furthest from the eye, and is screwed into the drawtube in place of the ocular. The Zollner employs a convex cylindrical lens mounted between the prism and the eye, the spectroscope being screwed on to an already focused LP ocular ($M=20D$ is recommended) in the manner of a suncap. Espin's star spectroscope consists of a d/v prism without either a slit or a cylindrical lens. It is mounted on the objective side of an ordinary ocular, the dispersion varying with the distance of the ocular from the prisms.

Horne and Thornthwaite used to manufacture a useful 'universal' eyepiece spectroscope, with d/v prism, cylindrical lens, slit, and achromatic viewing lens. For work on the Sun, the cylindrical lens is omitted; for the telescopic observation of stellar spectra, the slit is omitted; and for meteors, aurorae, the Zodiacal Light and the like, the 'star' spectroscope is used without the telescope.

* Over-widening of the spectrum, however, will reduce the visibility of the details again, owing to excessive dilution.

Inevitably both dispersion and spectrum width are small, and spectroscopes of this type are not suitable for photography, though in conjunction with quite small telescopes they will show the main spectroscopic type characteristics of stars down to about mag 3 or 4. A Zollner used with a 3-in or 4-in refractor, for example, shows the prominent hydrogen emissions of the B-type stars, the fluted spectra of the brighter M-type, the major A-type lines, and some lines can be glimpsed in the spectra of bright M-type stars when the seeing is good.

More powerful spectroscopes can also be used without slits in stellar work, and for the photography of stellar spectra often are. The usual practice in this case is to place a negative collimating lens a short distance on the objective side of the focal plane, so that a parallel pencil is delivered to the prism train, and thence to the camera objective in the usual way.

On the other hand, not all 'eyepiece' spectroscopes are slitless. The function of the slit is to increase the purity (and therefore resolution) of the spectrum, since it can be made narrower than the spurious disc, and also because wanderings of the primary image due to turbulence have no effect upon the definition of the photographic image. Spectrum width can be obtained, in photographic work, by trailing the image along the slit during exposure: the slit is oriented EW, and the driving clock set to run a little slow or fast. For visual work a cylindrical lens must still be resorted to: set in front of the slit it draws out the spurious disc into a line which must be made coincident with the slit. Maintaining this coincidence requires a very accurate drive and smoothly working slow motion in RA, and with small-dispersion 'eyepiece' spectroscopes little is to be gained by incorporating a slit.

19.8 Short-focus spectroscopes

The secondhand market always carries a wide variety of spectroscopes, both prismatic and diffraction, which are more powerful than the d/v 'eyepiece' spectroscopes, but are still compact and light enough to be used with telescopes of not more than moderate aperture.

The type often described as a 'student's' spectroscope consists of a single, usually 60° prism, with collimator and view telescope. A more elaborate form, giving greater dispersion, employs a train of three or more prisms, and should incorporate a pivoting arrangement whereby any required wavelength may be refracted at minimum deviation and the view telescope made parallel to its emergent ray when it is in this position.

In the type known as the constant-deviation spectroscope the latter provision is not required, the collimator and view telescope being set permanently at 90° to one another. Different wavelengths are brought to the axis of the view telescope by means of a constant-deviation prism (Figure 148); this may be either a single block of glass of the shape

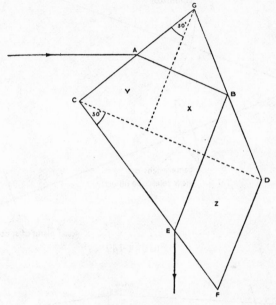

FIGURE 148

shown, or may be made up of a right-angled prism, *X*, and two 30° prisms, *Y* and *Z*. As such a prism is rotated, successive wavelengths become incident in turn at 45° to the face *GD*, and are consequently deviated through a total of 90°. It is to the action of the 30° prisms that the dispersion of the combination is due, and the latter is therefore equivalent to a single 60° prism set at minimum deviation.

Typical layout of a small reflecting grating spectroscope is shown in Figure 149. Diffraction by a transmission grating, cemented to the hypotenuse of a 30° prism (giving nearly d/v, collimator and view telescope being inclined at only 11°), is made use of in Sellers' simple but extremely effective prominence spectroscope.* The optical layout is shown diagrammatically in Figure 150. It was designed for ease of construction at home, and for use with small telescopes in the 2-in to 6-in

* Further referred to in *O.A.A.*, section 1.12, devoted to the spectroscopic observation of the chromosphere and prominences.

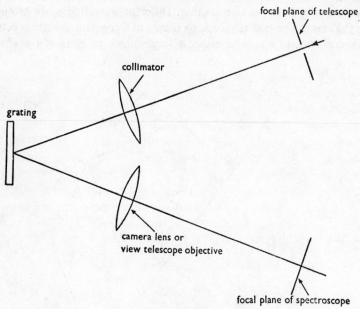

FIGURE 149

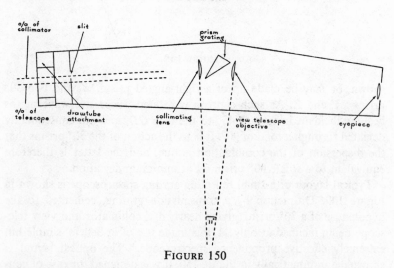

FIGURE 150

range; for constructional details, the reader should refer to B. 198. The rotation of the spectroscope about the optical axis of the telescope, allowing the offset slit to travel tangentially round the limb of the solar image, is effected by rotating the drawtube to which it is attached; correct focus is maintained during this operation by means of a sleeve round the drawtube, whose length is such that one end of it bears against the end of the telescope tube and the other against the flange at the ocular end of the drawtube when the focus is correct.

By modifying this design to allow the view telescope to pivot about an axis passing through the centre of the grating, a useful all-purpose solar or stellar spectroscope is obtained.

An even simpler arrangement for the observation of the prominences, though less powerful than Sellers', consists merely of a d/v prism with slit, offset in the drawtube by an amount equal to the semidiameter of the solar image in the focal plane.

Still another type of prominence and general solar spectroscope is that designed by Evershed for Messrs Hilger; it is light and compact, but of good dispersion and extremely convenient to use. Its basic elements are two d/v prisms and a plane mirror (Figure 151). The prisms

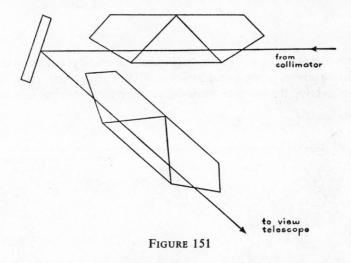

from
collimator

to view
telescope

FIGURE 151

are fixed, different wavelengths being brought to the centre of the field by rotating the mirror in the plane of dispersion. Means of rotating the whole spectroscope are provided for prominence work. Hilgers also make a three-prism, two-mirror version of this spectroscope, and in both cases supply either curved or straight slits.

Two ingenious designs of prominence spectroscope are described by Rees Wright in B. 199; they are extremely compact, and make use of image rotators (section 8.11) so that the spectroscope remains fixed relative to the telescope during observation. The first (Figure 152) employs a transmission grating cemented to a 30° prism; a right-angled prism and a totally reflecting prism are inserted in the optical train between the objective and ocular of the view telescope in the interests

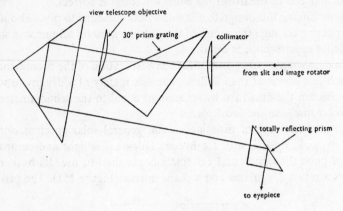

FIGURE 152

of compactness. The second design (Figure 153) employs two constant-deviation prisms, giving a total deviation of 180°, and a totally reflecting prism to bring the image to a convenient observing position. Like the

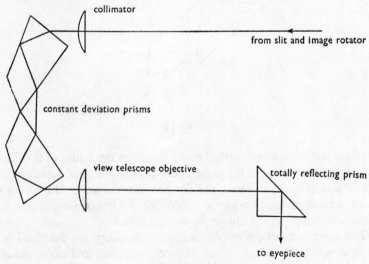

FIGURE 153

340

Evershed spectroscope, they provide for the scanning of the spectrum, any required wavelength being brought to the centre of the field by rotating, respectively, the prism-grating or the constant-deviation prisms.

Prominence spectroscopes in general should be used with fairly high magnifications, of the same order as those used for visual work on the Sun. Some designs, as we have seen, employ an image rotator to rotate the solar image on the slit plate, as being in many ways simpler than rotating the slit (and with it the rest of the spectroscope) round the solar image.

Spectroscopes of the type mentioned in this section are capable, when used in conjunction with even quite small telescopes, of showing a considerable amount of detail in the Fraunhofer spectrum, the characteristic differences between the spectra of spots and photosphere, and the chromospheric reversals and prominences, but their dispersion is generally too great for them to be suitable for work on objects other than the Sun.

19.9 Home-made short-focus spectroscopes

The basic specifications for any spectroscope are listed in section 19.4. Gratings are, on the whole, preferable to prisms, since they are lighter and easier to mount; their superior dispersion in the long wavelengths is also an important advantage, especially for prominence work involving, as it does, the widening of the slit in the position of the *Hα* line.

The only part of the spectroscope that is liable to cause practical difficulty is the slit, and the amateur building his first spectroscope should not be surprised by several failures, and should be prepared to scrap these as being valuable experience which will, at the third or fourth attempt, enable him to produce a satisfactory slit.

The vital part of the slit is the small section of the jaws which defines the pencil passing to the collimator: the spectrum itself is composed of innumerable images of this aperture, each emission or absorption line being one such image. The following are the essential points:

(*i*) The edges of the two jaws must be absolutely straight and absolutely parallel.

(*ii*) The edges of the jaws must be sharp, to prevent light being reflected from them into the collimator; they must be bevelled almost to a knife-edge, the bevelled side being that facing the collimating lens.

(*iii*) The two jaws must lie in the same plane, so that when closed they come precisely together and do not overlap.

(*iv*) If the slit is adjustable, the parallelism of the jaws and their location in a single plane must be maintained at all slit widths.

Slit jaws may be made from $\frac{1}{16}$-in brass or aluminium sheet, which can be cut with scissors. Each edge that is to form the jaws is first

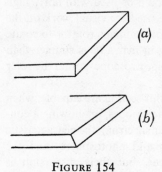

(*a*)

(*b*)

FIGURE 154

worked on a fine-toothed file till it is absolutely straight, perpendicular to the sides of the jaw, and, in section, accurately square (Figure 154(*a*)). This work is finished off on an oil-stone, and examined with a magnifying glass; any serrations or other imperfections must be ground down. Then, starting with the file and finishing on the stone, one edge is worked down to an increasingly deep bevel, this operation being stopped when it is just short of becoming a knife-edge, i.e. of interfering with the edge already perfected (Figure 154(*b*)). The work is examined with a powerful magnifying glass; the bevel on the two jaws should be at least approximately equal.

For a fixed slit (Figure 155) cut two 1-in diameter semicircles of the metal sheet and work the diameters to the bevelled jaw edges. Cut two rings of fine-grain $\frac{1}{8}$-in cork, outer and inner diameters 2 ins and $\frac{1}{8}$ ins. The central hole is the stop limiting the height of the spectrum*; it should be bevelled on the side facing away from the slit, and finished very carefully with fine-grade glasspaper. One cork ring is coated with carpenter's glue and one of the slit jaws laid on it, with the edge that is to form the slit edge lying diametrically across the central aperture. The second jaw is then laid approximately in position and moved up to the first with a needle or sharp pencil point; with the slit supported over a bright light on a sheet of glass, and

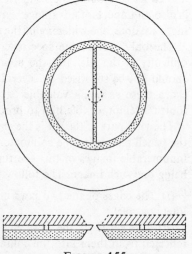

FIGURE 155

* The diameter of this aperture should be about one-third of the diameter of the focal image of the Sun.

examined with a magnifying glass or ocular, the position of the second jaw is adjusted until a very fine and parallel-sided streak of light is visible. A 2-in/1⅛-in ring of the same sheet of metal from which the jaws were cut is then glued to the cork outside the jaws; when it and the jaws are set firm, a second cork ring, of the same dimensions as the first, is glued over the jaws and the metal ring.

A simple arrangement for an adjustable slit is described by Sellers (B. 198). Another is illustrated in Figure 156. The slit plate, *B*, fixed

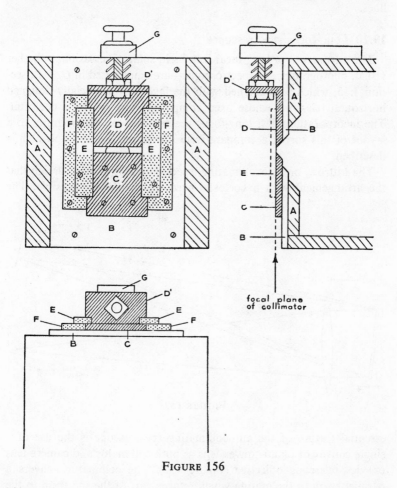

FIGURE 156

jaw, *C*, adjustable jaw, *D*, slides for the adjustable jaw, *E*, and bases for the slides, *F*, must be cut from the same sheet. Clearance of a few thousandths between *E* and *D* can be obtained by packing tin foil between *E* and *F*. All screws must be countersunk.

Not only should HP examination against a bright light show the slit to be accurately rectangular at all widths, but when closed it must exclude light completely along the whole of its length.

Slits must always be kept scrupulously clean, and be treated with great care. Specks of dust between the jaws will produce dark lines running the length of the spectrum, i.e. at right angles to the absorption lines.

19.10 Long-focus spectroscopes

The advantages of long focal lengths in solar observation—whether visual, photographic, or spectroscopic—are emphasised in *O.A.A.*, section 1.15, which is concerned with the layout of a specially designed horizontal telescope whose fixed components are fed by a coelostat. The incorporation of a long-focus spectroscope or spectrograph in a layout of this sort is a comparatively simple matter, and one such is described.

The Littrow, or autocollimating, spectroscope is another design that the arrangement with a coelostat-fixed image renders possible. The

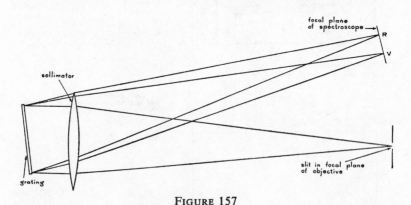

FIGURE 157

essential feature of the autocollimating spectroscope is the use of a single convex or plano-convex lens as both collimator and camera lens or view-telescope objective (Figure 157). The collimator delivers a parallel beam to the grating which returns part of the spectrum to the collimator; this, now in its capacity of view-telescope objective, forms an image of the spectrum at the same distance from itself as the slit, but, owing to the grating being slightly tilted, at a convenient distance above or below the slit. The collimating lens should be provided with a few inches of horizontal adjustment for focusing; the grating can be mounted on an altazimuth fork, altitude adjustment varying the position of the image at the photographic plate or eyepiece, and azimuth adjustment bringing spectra of different orders, or different regions of a given spectrum, to the observing position. There should be provision for these adjustments to be made from the observing position.

The 150-ft Sun telescope at Mt Wilson incorporates a 75-ft Littrow spectrograph (12-in OG of 150 ft F, 10-in collimating lens of 75 ft F, 8-in grating) yielding a dispersion of 7 mm per Å in the 4th-order spectra: a linear dispersion from the B to K lines of about 67 ft. But even the 20-ft, 5-in objective suggested in *O.A.A.*, used with an 8-ft, 2-in collimating lens, will provide a solar spectrum which is a revelation to anyone whose experience has been confined to telescope attachments.

Another way to secure a longer focal length than is possible with an eye-end telescopic attachment is in effect to turn the whole telescope into a spectroscope, just as in prime-focus photography the telescope is

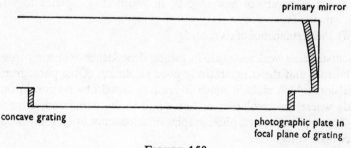

primary mirror

concave grating

photographic plate in focal plane of grating

FIGURE 158

turned into a camera. The use of an objective prism achieves this. Another method is Butterworth's use of a concave grating with a Herschelian reflector (Figure 158). Much greater convenience and adaptability are, however, offered by a fixed-image arrangement, as in the Sun telescope, allowing the fixed mounting of the spectroscopic apparatus.

19.11 Objective prisms

See *O.A.A.*, section 15.9.2.

345

SECTION 20

PHOTOGRAPHY

20.1 Introduction

The power of photography as a weapon in the astronomer's hand springs from:

(a) its precision and impersonality, which adapt it to positional work, in which it is comparable with visual methods;

(b) the cumulative effect of light upon the photographic emulsion, adapting it to the investigation of very faint objects, in which it is superior to visual methods;

(c) its wholesale character, which, together with (b), adapts it to the discovery of new objects, in which it is superior to visual methods;

(d) the permanence of its records.

Its outstanding weaknesses are its greater dependence upon atmospheric turbulence and the comparatively poor resolution of the photographic emulsion, which make it much inferior to visual observation in those fields where the resolution of fine detail is the paramount requisite.

The applications of photography in astronomy usually divide into three main categories:

(a) photographing the Sun, Moon and planets;

(b) photographing faint extended objects, e.g. nebulae, comets, and star fields;

(c) meteor photography.

20.2 Image scale

In each of the categories mentioned above, different techniques or apparatus are used, one significant factor being image scale.

The scale of the image in the focal plane of a lens, whether camera or telescope objective, is given by equation (a), p. 52:

$$h = \frac{F\theta}{57 \cdot 3}$$

346

where θ is measured in degrees, and F and h in the same linear unit, whence

$$\theta = \frac{57 \cdot 3}{F}$$

when $h = 1$, i.e. the scale of the primary image is $\dfrac{57 \cdot 3}{F}$ degrees/inch. To fix ideas, the diameter of the primary focal image of the Sun or Moon $\simeq 0 \cdot 009 F$ ins.

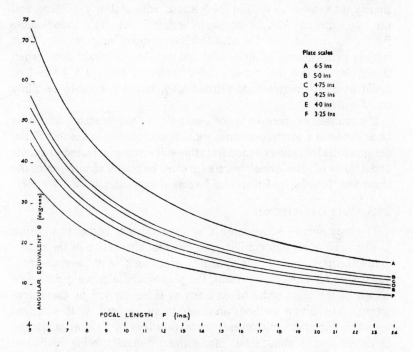

Plate scales

A 6·5 ins
B 5·0 ins
C 4·75 ins
D 4·25 ins
E 4·0 ins
F 3·25 ins

FIGURE 159

The above graph shows, for the range of F from 5 to 24 ins, the angular values of six selected values of h. These h-values are the lengths and breadths of three commonly used plates, viz:

$$3\tfrac{1}{4} \times 4\tfrac{1}{4} \text{ ins} \qquad 4 \times 5 \text{ ins} \qquad 4\tfrac{3}{4} \times 6\tfrac{1}{2} \text{ ins}$$

from which it can be seen, for example, that a quarter-plate ($3\tfrac{1}{4} \times 4\tfrac{1}{4}$ ins) used in a camera of 8 ins focal length will cover about $23° \times 30°$; to cover the same area with a half-plate, a focal length of about 12 ins would be required. A focal length of about $7\tfrac{1}{2}$ ins would give a plate scale very nearly equal to that of Norton's maps.

The field of acceptable definition will be less than this by an amount depending upon the quality of the lens, the focal ratio at which it is used, and the purpose for which the photograph is taken; it must be determined by trial, comparing the contents of the acceptable field at different relative apertures with a star chart or atlas.

The linear size of the field of tolerable definition is primarily a function of focal ratio, and within a certain range is largely independent of focal length. Thus the linear area of acceptable definition is approximately the same with a 20-in $f/4\cdot5$ lens as with a 10-in $f/4\cdot5$ lens, and not four times as great. As the focal length is increased the linear usable field does indeed increase also, but very much more slowly. This relative independence of linear field and focal length breaks down when things are carried to extremes: for instance, the $f/4\cdot5$ 10-in lens could cover a quarter-plate satisfactorily, but a $f/4\cdot5$ 1-in certainly could not.

If any material increase in the linear size of the tolerable field is to be achieved at a given focal ratio without enormously increasing F, the design of the lens must be modified; this will involve increased thickness and density of glass, which in turn involves increased absorption of the short wavelengths, and therefore increased exposures.

20.3 Plate measurement

The only normal occasion on which the amateur wishes to measure positions from a photographic plate is as an alternative to the micrometrical determination of the position of a comet. With care, and provided a stellar nucleus is present, the photographic method can yield results of the same order of accuracy as those derived by the micrometer, and the two methods are in fact equally used. With sufficient magnification and a micrometer screw, positions of a stellar image can be determined to about $0\cdot005$ mm without difficulty, being equivalent to about $2''5$ on a plate exposed in a camera of 20 ins F. The particular advantage of the photographic method lies in its greater applicability to faint comets. Times of both the beginning and end of the exposure must be stated.

Plate-measuring machines are expensive, but with ingenuity one may be improvised, or projection used instead. In principle, any device incorporating controlled motion in two mutually perpendicular directions—such as, e.g., a microscope mechanical stage, or a double-slide plate carrier—could be adapted to plate measurement. With such makeshifts, however, error is liable to be introduced by the fact that insufficient attention was devoted in their construction to the precision of the mutual perpendicularity of the two motions. A more satisfactory plan is to use a filar micrometer for its magnification and its micrometer screw. With a single-web micrometer, it or the plate must be mounted in a frame which can be rotated to allow the second coordinate to be measured, and the machining of the frame must be of a high degree of precision to ensure that the two positions of the web relative to the plate are accurately perpendicular to one another. Measurements with such adaptations are liable to be rather laborious, and the field is small; but for the measurement of cometary positions—of its nature, not a very frequent operation—with short F cameras these disadvantages are not serious.

Alternatively, plates may be measured by projection on to a rectilinear grid (mm or 0·1-in squares). It is important that the planes of the grid, of the plate, and of the projecting lens should be accurately parallel. This can be checked by projecting a photographically reduced circle or square on to a screen on which a similar figure has been drawn; the grid is then, after the necessary adjustments have been made, placed on this screen. It is also wise to confine measures to the central region of of the field. (See B. 214.)

Reductions can be conveniently made by Comrie's method of dependences using three comparison stars (B. 207); see also B. 221 for Merton's two-star method, a modification of Kaiser's interpolation method, and discussion of these papers, B. 208. Schlesinger describes a much shortened method in B. 233, and B. 4 has a useful introductory treatment of the subject. See also B. 244, 245.

20.4 The photographic emulsion

The light-sensitive emulsion consists of a crystalline silver halide (commonly bromide), further sensitised by an organic dye, which is suspended in gelatin on a base of film or glass. The photographic speed, or sensitivity, of the emulsion is related to the size of the silver salt particles: the grain of modern emulsions ranges for the most part from $0·3\mu$ to $1·5\mu$. The graininess of the final image is also influenced to some extent by the development.

The precise mechanism whereby the latent image is formed by the action of light on the silver crystals is still imperfectly understood. However that may be, the particles of the latent image are reduced to dark metallic silver by the action of the developer. Those halide crystals which were not exposed to light are then removed by the fixing solution, in its simplest form sodium thiosulphate or hyposulphite ('hypo').

The term optical density, or just density, referred to developed photographic emulsions, is defined as follows:

If the intensity of light falling on a plate or film is I_0 and the intensity of light transmitted through it is I, then the density of the exposed area (denoted by D to distinguish it from aperture) is

$$D = \log \frac{I_0}{I}$$

Thus for a totally clear area, $\qquad\qquad\qquad\qquad\qquad D=0$
for an area transmitting $0\cdot1$ of the incident light, $\qquad D=1$
for an area transmitting $0\cdot01$ of the incident light, $\quad D=2$, etc.

If a graph of emulsion density produced by a given exposure (in logarithmic units) is plotted, the characteristic D–log E curve results,

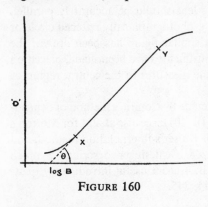

FIGURE 160

sometimes called the H and D curve after the pioneers Hurter and Driffield (Figure 160). This curve indicates that the photographic emulsion is far from being the perfect light detector. Ideally, the curve would be a straight line, and density would be directly proportional to exposure. In fact a maximum density, D_{max}, is reached, usually with a value of 3 or 4. At the other end of the scale, even the emulsion which has had no exposure at all will have a certain density, the fog level, notwithstanding any density due to the emulsion support, film or plate.

The transition region between the fog level and the straight line portion of the curve is called the 'toe'. The slope of the straight line is the contrast, usually referred to as gamma (γ), given by tan θ. The term 'contrast index' is becoming used—this is measured taking into account the effects of the toe. In order to differentiate faint details, emulsions with high contrast should be used. Such emulsions are generally slow, and have a desirably fine grain.

An ideal emulsion would obey what is known as the 'reciprocity law', which states that if exposure is defined to be the product of the intensity of light and the duration of exposure to it, then equal exposures will give rise to equal image density regardless of the actual intensity and duration values. In practice it is found that exposure of an emulsion

350

to light of high intensity for a short time will in general produce an image of greater density than exposure to a low intensity for a long period, even though the exposures be calculated to be equal. This effect is a failure of the reciprocity law as applied to photographic emulsions, and is commonly known as 'reciprocity failure'. For a photographic grain to be made developable, it must absorb a certain number of light 'quanta'. However, if the light intensity is low, it becomes probable that the thermal motions of the molecules within the grain will cancel the effect of the first grains to arrive, and it may be that the required number to sensitise the grain may never be achieved. It is possible to reduce the effect of reciprocity failure by cooling the emulsion with dry-ice or liquid nitrogen during exposure, thereby reducing the thermal agitation of the molecules in each grain.

A less effective way of overcoming reciprocity failure is to give the emulsion an overall pre-exposure before use, in order to produce a certain amount of fog. This will reduce the number of additional photons required to make the grains developable. This has the effect of rendering visible the fainter images which might not otherwise have been developed. It has most effect upon the lower density part of the D–log E curve.

A chemical method of achieving higher emulsion speed is hypersensitization, in which pre-soaking the emulsion in liquid ammonia, or more simply but less effectively, in distilled water, has the effect of removing anti-fog chemicals which suppress some of the sensitivity but produce more acceptable pictorial results. The emulsion must be dried quickly and used as soon as possible. This method is most frequently used with infra-red sensitive plates, which are normally particularly slow.

The technique of baking the emulsion at 50°C–70°C for 12–48 hours is also widely used professionally, and creates more sites within the emulsion in which photons can be trapped.

When an emulsion leaves the manufacturer, it is not necessarily at its highest sensitivity—if it were, it would be foggy and unpredictable. Treatment to bring the emulsion to its optimum performance is left to the experienced user, and no guidance is generally given by the manufacturer. Some of the methods described above could be used in combination, but in general only one is used at a time.

20.5 Magnitude thresholds and exposure limits

The effectiveness of an emulsion is influenced by so many variables that it is not feasible to give exact values for the faintest magnitude detectable with a given aperture. It is better to detail the various factors and allow the astronomer to experiment on the basis of these.

If the stars were seen against a perfectly black ground, and if a perfect detector were used, there would be no limit to the magnitude reached with any aperture other than the exposure time. In practice the limits are set by the inefficiency of the emulsion, and by the general brightness of the sky background which tends to drown out the fainter stars' images. The problems connected with the emulsion itself are summarised at section 20.4. Exposures cannot be prolonged indefinitely because fogging will occur and will eventually reduce the ratio between the image and background densities (the signal to noise ratio) to a level at which the image becomes indistinguishable. There are two kinds of fog present: sky fog, which is entirely due to the exposure, and emulsion fog which is in turn made up of chemical fog and any pre-exposure fog. The emulsion fog normally has little effect on the detect-ability of the image, but merely raises the base level above which the density of subsequent exposures is measured.

We have seen in section 1.6 that whereas the image intensity of an extended object is proportional to $(D/F)^2$, that of a point object is proportional to D^2. We can apply this property to the sky background itself, to show that the larger the value of the focal ratio of the objective, the smaller will be the amount of sky fog recorded. In practice the star images themselves are extended to a degree depending on the seeing conditions, which will slightly reduce the image contrast. Thus an instrument working at $f/8$ would in general have a fainter limiting stellar magnitude than one of $f/3$ having the same aperture. It should, however, be remembered that the faster instrument would perform better in the recording of extended objects such as nebulae.

A variety of factors influence the degree to which sky fog will be present on a given occasion, among them:

(a) Diffuse auroral light and sunlight reflected from interplanetary matter, which largely accounts for the brightness of the moonless night sky, usually taken to be equivalent to about 13·5 mag per square minute of arc, though this varies rather widely from night to night, from one part of the sky to another, and with the geographical position of the observing site.

(b) Distance of Sun below horizon. If less than about 10° fogging of the plate is rapid. Severe trouble is encountered by the cometary photo-grapher, for instance, when his object is near perihelion.

(c) Distance and phase of the Moon.

(d) Azimuth distance of the field from the Sun.

(e) Altitude of the field.

(f) Clarity of the atmosphere.

(g) Presence of artificial lights. These, aggravated by the products of combustion suspended in the atmosphere, were responsible for the retreat of the Royal Observatory from Greenwich to Herstmonceux. Again, diffusion of the lights of Los Angeles from the 'smog' layer which normally overlies that city reduces the maximum exposure time of the $f/5$ instruments at Mt Wilson from 80 mins to about 30 mins, as was demonstrated during the war when Los Angeles was put under a partial blackout.

Moonlight being markedly polarised (about 60% at 90° from the Moon), the use of a crossed polaroid filter in front of the plate has been tried with good effect. A 15-min exposure with the 10-in $f/4·5$ Franklin-Adams instrument at Johannesburg produced a fog density of $1·04$ with the filter in the parallel position and $0·51$ with it crossed, though the filter factor of $1/3$ would involve trebling the exposures for this advantage to be fully utilised.

Although in practice the threshold magnitude appropriate to the equipment and materials to hand may be discovered only by experiment, it can be said that using a 35 mm. camera with the standard lens of 45 mm. or 50 mm. focal length at the common widest aperture of $f/2.8$, in conjunction with a fast film (speed of 400 ASA or more) it is possible to record images of stars fainter than the eye can see using exposures of only a few seconds duration.

20.6 Equivalent exposures

The intensity of a true point image is, for a given exposure, directly proportional to D^2. Thus the exposure E' with an aperture D' which is equivalent to the exposure E with an aperture D is given by

$$E' = \frac{D^2}{(D')^2} . E$$

A 4-sec exposure with an aperture of 6 ins is equivalent to a 1-sec exposure with an aperture of 12 ins, irrespective of F in the two cases. By the same token, a 4-sec exposure with a 6-in lens is equivalent to a 4-sec exposure with any other 6-in, irrespective of their focal ratios.

In the case of an extended image the photographic speed (the degree of darkening of the emulsion with unit exposure) is again a function of D^2, but in addition is inversely proportional to F^2 since the area of the image increases as F^2. Photographic speed in this case is therefore a function of $\left(\dfrac{D}{F}\right)^2$, and equivalent exposures of $\left(\dfrac{F}{D}\right)^2$. The exposure E' with a lens D', F', which is equivalent to an exposure E with another lens D, F, is given by

$$E'=\left(\frac{F'}{D'}\right)^2\left(\frac{D}{F}\right)^2.E \quad . \quad . \quad . \quad . \quad . \quad . \quad . \quad (a)$$

When equivalent exposures with a single lens used at different focal ratios are under discussion, $F=F'$, and equation (a), above, simplifies to

$$E'=\left(\frac{D}{D'}\right)^2 E$$

F being the same in each case, D and D' are given by the respective f/numbers of the two stops. Thus if a plate has been exposed for 1 hr with the full aperture of a $f/5\cdot6$ lens, the equivalent exposure with the lens stopped down to $f/8$ is given by

$$E'=\left(\frac{8}{5\cdot6}\right)^2\times60 \text{ mins}$$
$$=122 \text{ mins}$$

If the focal length is doubled, therefore, the length of the equivalent exposure is quadrupled. Similarly, doubling the exposure is equivalent to decreasing the focal ratio by $\sqrt{2}$; or in the case of a series of stops to be used with a particular lens (F therefore constant) to increasing D by $\sqrt{2}$. The following exposures at their respective focal ratios are therefore equivalent:

$f/1$	1^s	$f/8$	1^m	4^s
$f/1\cdot4$	2	$f/11$	2	8
$f/2$	4	$f/16$	4	16
$f/2\cdot8$	8	$f/22$	8	32
$f/4$	16	$f/32$	17	4
$f/5\cdot6$	32			

It will be clear from equation (a), that, for example, a 2-in lens working at $f/2\cdot8$ is faster than the 200-in at $f/3.3$. Where a large lens or mirror gains over a smaller one—even though the latter's focal ratio, and therefore photographic speed, may be the same—is, of course, in its greater resolution and (because of correspondingly larger F) image scale. Thus a $f/15$ 2-in and a $f/15$ 10-in will provide equally dense images of the Moon with exposures of the same length, but one will be 1·3 ins in diameter while the other will be less than 0·3 ins.

20.7 Photographic resolution

The resolving power of the photographic emulsion is much inferior to that of the eye, and this is not connected (as is often supposed) with the size of the grain.

Grain diameter, depending upon the speed of the emulsion and the developer, is of the order of $0·3\mu$ to $1·5\mu$. But owing to diffusion in the emulsion the diameter of the image of a point source may be anything from about 20μ (0·0008 ins) if under-exposed, to 1000μ (0·04 ins) if grossly over-exposed. Taking 50μ as a rough figure for the diameter of a 'normally' exposed stellar image, two such images will be resolved on the plate if their separation is not less than about 25μ. Substituting 0·001 ins in equation (*a*), p. 52, we have, for the threshold of photographic resolution,

$$R_p = \frac{243}{F} \, '' \text{ arc}$$

(*F* being measured in ins), and also, equation (*g*), p. 45,

$$R_v = \frac{4''56}{D}$$

(*D* also expressed in ins). Now $\frac{243}{F} < \frac{4·56}{D}$ if $\frac{F}{D} > 53$, approximately, which is to say that at all normally employed focal ratios photographic resolution is inferior to visual. For the photographic plate to compete with the eye in this respect, focal ratios larger than about $f/50$ must be used. The Cassegrain is the only form of optical system where such ratios can be practicably achieved, but exposures of extended objects would be prohibitively long. For focal ratios of 15 and 5 (typical, respectively, of refractors and Newtonian reflectors) the values of R_p are respectively 3 and 10 times those of R_v. Given a by no means abnormal degree of atmospheric turbulence these figures may easily be doubled.

It is for this reason that all the lunar and planetary detail visible in photographs taken with the Mt Wilson 100-in reflector can be seen visually with a 10-in, and that these photographs taken with the world's second largest telescope are inferior to Lyot's Pic du Midi lunar plates. The corollary of this is that the ideal objects for photography are those which tax light grasp rather than resolving power.

The desirability of large F-values to increase plate-scale and resolution, and at the same time large relative apertures to reduce exposures of extended objects, unfortunately tends in the direction of taking the ideal instrument out of the amateur's hands altogether, for the two desiderata can only be satisfied by increasing both F and D.

For lunar and planetary photography, it is, however, possible to achieve some improvement by the technique of superimposing several negatives each deliberately underexposed to give a lower density than normal. Resolution of the emulsion varies directly with contrast and inversely with the square root of the density, so the optimum is with high contrast and low densities, which have associated low granularities. Thus the exposure should be made to give a density on the upper part of the toe of the D–log E curve, just before the straight line region, with a density of about 0·2. This is also the region of maximum light detecting efficiency of the film. Tolerating a less dense exposure means that a shorter exposure time (where seeing is troublesome) or slower, finer-grain film may be used.

Superposing a number of such exposures will give a printable density which is less grainy and shows more detail than a single exposure. The effectiveness of the technique increases only as the square root of the number of exposures used, the optimum being about six. It should be remembered that this technique depends for its success on the accurate registration together of the multiple negatives.

20.8 Emulsions and developers

The choice of the best emulsion and its development for a specific job are matters in which experience alone brings confidence. The popularity of the 35 mm. film size camera means that this is now the most widely used material, whereas twenty years ago the glass plate was nearly always used for serious work even from the amateur point of view. The currently available materials can be divided into four main categories:

(a) 35 mm. width film with sprocket holes for miniature cameras, supplied in cassettes for easy loading.

(b) Roll film, most commonly 120 size, for larger cameras. This is backed with a detachable paper roll with numbers on the reverse side so that the film can be wound on an appropriate distance.

(c) Glass plates, usually about 2 mm. thick, in recognised sizes of which those most often used are 5×4 ins, $4\frac{3}{4} \times 6\frac{1}{2}$ ins (half plate), and 10×8 ins. The astronomer may also find the $3\frac{1}{4} \times 3\frac{1}{4}$ in. size useful.

(d) Sheet film (sometimes called cut film or flat film), available in the same sizes as glass plates.

Of the above, glass plates are the most expensive and, for amateur purposes, the least widely used. Kodak have now in fact ceased general production of plates in Britain, though their production by Ilford continues. The advantages of plates are their flatness and dimensional stability: sheet film can easily curl if not held properly, although most plate holders can be modified to hold sheet film as flat as will be required for purposes other than astrography. In professional work, however, dimensional stability is often of paramount importance since the image positions will be measured accurately, possibly over a span of many decades, and the glass plate will reign supreme in such work for years to come in the absence of an alternative flat and stable film base.

The Kodak Spectroscopic series of plates are almost exclusively used in professional astronomical photography. These plates have specific colour sensitivity ranges and are thus most suitable for photometry and the examination of features in various wavelength ranges. Some emulsions are manufactured so as to have little reciprocity failure, and are thus suitable for recording faint sources. The cost of these plates, however, generally prohibits their use by amateurs.

The spectral sensitisation of the emulsion is of great significance when a selection is to be made for a specific purpose. An unsensitised emulsion is responsive to blue and ultra-violet light only, an ortho-chromatic one is sensitive to green light as well, and a panchromatic emulsion is sensitive over most of the visible spectrum in addition to the ultra-violet. Emulsions with infra-red sensitivity in addition to panchromatic sensitivity are also available.

An exhaustive list of the properties and availabilities of emulsions suitable for astronomical work is not feasible, but the following is offered as a guide. Emulsions are listed in approximate decreasing order of speed.

(a) 35 mm. films.

Kodak 2475 Recording Film
Kodak Tri-X
Ilford H.P.4
Kodak High Speed Ektachrome (Colour)
Kodak Plus-X Pan.
Ilford F.P.4
Kodak Infrared Film

Ilford Pan F
Kodak Panatomic-X
Kodachrome II (Colour)

All the above emulsions are panchromatic, with the obvious exception of the two colour films.

(b) Roll films.

Kodak Royal-X Pan
Kodak Tri-X
Ilford H.P.4
Ilford Selochrome
Kodak Verichrome Pan.
Ilford F.P.4
Kodak Panatomic-X

All of the above emulsions are panchromatic.

(c) Sheet films.

Kodak Royal-X Pan
Kodak Panchro-Royal
Ilford H.P.4
Kodak Tri-X Pan.
Kodak Tri-X Ortho.
Kodak Plus-X Pan.
Ilford F.P.4
Ilford Commercial Ortho.
Kodak Commercial Ortho.

Except for the three orthochromatic emulsions, the above are panchromatic.

(d) Plates.

Ilford H.P.3 Plate   (panchromatic)
Ilford Selochrome Plate   (orthochromatic)
Ilford Astra III Plate   (panchromatic)
Ilford Zenith Plate   (unsensitized)
Ilford Special Rapid Plate (unsensitized)
Ilford R.40 Rapid Process Panchromatic Plate

The Ilford Astra III and Ilford Zenith plates have reduced reciprocity failure.

The characteristics of fast and slow emulsions are, in summary:

Fast plates are more sensitive, permitting shorter exposures with minimum loss of definition through atmospheric turbulence; they fog more rapidly, produce lower contrast, and have a coarser grain which may limit possible degree of enlargement.

Slow plates are less sensitive and require longer exposures. They run a lower risk of fogging, have higher contrast, with better consequent discernability of detail, and have a finer grain allowing greater enlargement without loss of detail.

Processing of emulsions is an art in itself, and largely empirical. Plates and sheet films are processed in dishes which should be considerably larger than the plate or film itself, and contain an adequate depth of the solution. Roll and miniature films are processed in spiral holders in small tanks; once the film is loaded, the rest of the process may be carried out in the light. Very rarely is it possible to inspect a film with a safelight, so all processing should be done by the time-temperature method. Particularly when developing, continuous agitation of the dish or tank is advisable.

All solutions should be at the same temperature. The penalty for ignoring this rule is reticulation, which is often mistaken for graininess. Although raising the processing temperature accelerates development, the development becomes less controllable. A temperature of 20°C (68°F) is a good standard.

Fine-grain developers suffer from the disadvantage that they reduce the speed of the emulsion and increase the effective exposure necessary to give a satisfactory density. Medium fine-grain developers, with a high sodium sulphite content and less alkaline than the normal fast hydroquinone types of developer, are generally preferable except where the highest speed possible is required, and if much subsequent enlargement is envisaged they are essential. The detailed technique of development depends on the type of work being undertaken, and there is always considerable room for experimentation. The development times recommended by manufacturers are generally intended to give the most pleasing pictorial results, with a minimum of fog and fine grain. It is usually possible to use longer times with increased film speed resulting from the consequent full development.

Polaroid film is now quite popular, and offers the astronomical photographer some advantages. The process is physically slightly different from conventional photography in that printing paper is squeezed into contact with the exposed negative material and the chemicals, and the image is transferred onto the paper by chemical diffusion. As only a thin emulsion layer is needed to produce a normally dense image on paper, a markedly slower, less grainy emulsion can be used. Polaroid film is widely available with a speed rating of 3000 ASA, and the process takes only a few seconds. The technique is comparatively new and cannot yet yield the same high quality that conventional photography has achieved, but it offers many interesting possibilities.

It is not necessary to use a Polaroid camera for this work; indeed, these have such small apertures and large focal ratios that they are most unsuitable. A device for holding the film and carrying out the processing can be bought or made.

20.9 Halation

Backed plates are essential for stellar work, as for extended images if these are bright. With unbacked plates fainter and fainter stars will, with increasing exposure, be ringed by halation owing to reflection from the rear surface of the emulsion support. Bright extended images, such as those of the Sun or Moon, will be overlaid by a halation haze which will reduce contrast and may completely submerge faint detail. The backing absorbs the wavelengths to which the emulsion is sensitive, and is decolorised during development.

If backed plates of the type required cannot be obtained, it is not difficult to back them oneself (though in the case of panchromatics—where one has to work in the dark—this is hardly feasible). Melt 1 lb of white sugar over a low flame and add 8 oz of burnt sienna; stir thoroughly together and add absolute alcohol in the proportion of $\frac{1}{2}$ oz per pint of the mixture. This stock mixture can then be applied to the plates by brush, when diluted to the consistence of a thick paste with water. After exposure and before development, the home-made backing must be removed with wet cottonwool.

20.10 Focusing and squaring-on

The surface of the emulsion must lie accurately at the photographic focus of the objective, and at the same time be accurately perpendicular to the optical axis. Insufficient attention given to these adjustments is one of the common causes of disappointing results.

What the correct focus is depends upon the equipment, and can be determined by various methods. A reflector being perfectly achromatic,* the visual (yellow) and actinic (ultraviolet–blue) foci coincide, and the camera can be focused by either visual or photographic methods. This is virtually true, also, of a photovisual refractor. The correct photographic focus of a visually corrected objective (see Figure 25) can also

---

* Though unequally efficient in different wavelengths; see section 6.

be determined visually provided that the strongly actinic wavelengths are eliminated by a yellow filter and orthochromatic, isochromatic, or, better still, panchromatic plates are used. The disadvantage of this procedure is that such filters have factors of from 1·5 to 3, involving increased exposures with an instrument whose photographic speed ($f/15$ or thereabouts with a typical refractor) is already slow. The use of a correcting lens, placed in the convergent pencil to alter the objective's colour correction, bringing the blue to the minimum focus, suffers from the same drawback; the $2\frac{1}{2}$-in correcting lens used 39 ins inside the visual focus of the Lick 36-in, for example, introduces a light loss of 10%.

20.10.1 Visual determination of focus: (*a*) *Reflector or photovisual.* Load a plate carrier, from which the back has been removed, with either an exposed plate on which a cross has been scratched in the emulsion, or a sheet of very fine ground glass of the same thickness as a plate (ground side nearest the objective), on which a cross has been drawn in pencil and covered with a microscope slip cemented by a drop of balsam. The aerial image now has to be brought into the plane containing the cross: with a high-power (not less than × 10) adjustable magnifying glass placed in contact with the back of the focusing screen adjust its focus so that the cross is sharply defined, the telescope or camera being pointed at the day sky. At night, and with the same adjustment of the focusing lens, adjust the setting of the plate in relation to the objective so that the aerial image of a moderately bright star or of the Moon is brought to the same plane as the cross; when this is achieved, a side-to-side movement of the head will result in no parallactic shift of the image relative to the cross.

(*b*) *Visually corrected objective.* The procedure is as above, with the additional provision of a yellow filter and the subsequent use of ortho-, iso-, or panchromatic plates. If an ocular is used to enlarge the image, the filter should be placed in contact with the field lens; if the plate is exposed at the primary focus, the filter should be as near to the plate as possible. The filter will itself change the focal length slightly, owing both to its modification of the effective wavelength, and to the lengthening of the focus (by about one-third of the total thickness of glass in the filter) dependent on the two plane parallel glass slips protecting the gelatin.

20.10.2 Photographic determination of focus: (*a*) *Star trails.* Expose five trails of a mag 2 star situated near the equator and the meridian. These trails should be distributed around the visually determined focus; in the case of a visually corrected objective a yellow filter must be used (see section 20.10). It is essential that the adjustable element (lens or plate-holder of a camera, drawtube of a telescope) should be fitted with a rigid and accurately divided scale, so that its position corresponding to any particular trail can be repeated as required. The trails can either be spaced diametrically across the field (the suitable exposure time depending upon the plate-scale and the Dec of the star; the first trail, on the f side of the field, should be given double the exposure of the others for identification purposes), or, preferably—since it keeps the trails more nearly central in the field—the instrument can be raised or lowered in Dec and moved back in RA between each exposure; identification of the trails in this case is facilitated if each is exposed for a constant interval longer than its predecessor. The two procedures would produce plates represented schematically below:

—— —— —— —— ——— and

The trails are examined on the negative with a HP magnifying glass. If one trail is narrowest, and that on either side of it is equally wide, then it was made at the focus. If one trail is narrowest, but the trails on either side of it are unequally wide, then the focus lies between that trail and the narrower of its neighbours but near the former. If two adjacent trails are equally narrow the focus lies between them. If there is a progressive narrowing of the trails from one end of the series to the other, the focus may lie beyond the series of plate positions used, and a further series of exposures must be made. If the objective–plate distance was varied between exposures by more than about 0·25 mm (as certainly it would have been unless the position of the focus was already known fairly precisely) a second series of exposures must be made, at about this value of Δf, centred upon the best position as shown by the trails of the preliminary series.

(b) *Star images.* The general procedure is the same as in (a), except that the images are not trailed. Equal exposures are given, and the telescope readjusted in Dec between each exposure; for identification purposes, this displacement is doubled between the first and second exposures. The weakness of the method is that any errors of guiding will enlarge the images and thus confuse the issue. On the other hand, if the instrument is to be used for photography at all it ought to be safe to assume that it has an efficient drive and slow motions—otherwise its unsuitability for the job renders the whole procedure of accurately determining the focus rather beside the point. The outstanding advantage of the method is the detailed information that the appearance of the images in different parts of the field provides regarding both focus and squaring-on. This is discussed fully in B. 210, to which the reader is referred for further treatment of the subject.

In the case of a portrait lens it will probably be found that there is a certain amount of latitude in the focusing, there being no material change of focus over a small range of movement of the plate or objective along the optical axis. In these circumstances it is best to keep the distance objective–plate to a minimum, but if this is exaggerated it will be found that the focus at the centre of the field becomes inferior to that in an annular zone surrounding the central region.

20.10.3 Hartmann method: An application of the principle of the Hartmann test provides a simple photographic method of determining the focus of a lens. The objective is covered by a mask in which two circular holes are punched; these should be equidistant from the centre and should lie on the same diameter; their diameters should not be less than about 0·04 ins. Two photographs of a mag 1 star are then taken, one well inside the focus and the other outside. Then if s_1 and s_2 are the separations of the two images in the two photographs, and f_1 and f_2 the corresponding lens–plate distances, the focal length is given by

$$F = f_1 + \frac{s_1}{s_1+s_2}(f_2-f_1)$$

20.11 Photography at the telescopic focus; image amplification

The obvious position for the photographic plate is in the focal plane of the objective—the prime focus. With a refractor, the dark slide carrier—a short drawtube terminating in a grooved plate fitted with one or more spring clips to hold the dark slide absolutely firm in all positions —replaces the ordinary ocular drawtube. With a reflector, a similar arrangement can be used at the Newtonian focus, or the flat can be removed and the plate mounted on a special spider at the prime focus. The latter arrangement results in image reversal; also the smallest possible plate must be used, to reduce the central obstruction to a minimum. Larger plate-scale can be obtained at the Cassegrain focus, if this is accessible.

When the primary focus is occupied by the plate, guiding (see section 20.13) is most conveniently carried out by an off-centre ocular mounted beside the plate. If a separate guide telescope is used, its focal length should be as large as possible. Quite an inferior long-focus lens can be used for the objective, and it and the ocular need not be enclosed in a tube, though adequate protection against dew must be provided. Amplification of the guiding image by a Barlow increases the observer's comfort and therefore, in the long run, the accuracy of his guiding. For the same reason a diagonal is often valuable.

Without amplification images at the prime focus are too small for effective examination unless telescopes of much larger F than are normally found in amateur hands are used; the primary image of the Moon, for example, does not attain to a diameter of 2 ins until F reaches 18 ft. The image can either be amplified before it falls on the plate, or an enlargement can be made from the unamplified negative later. Generally the former is preferable. If the required detail is below the limit of resolution of the emulsion no amount of subsequent enlargement from the negative will reveal it: such a procedure is exactly analogous to trying to reach detail or separate points which are beyond the resolving power of an objective by increasing the magnification.

Taking the resolving power of the normal eye as 2′ to 3′ (section 24.5), which approximately equals 0·01 ins at a distance of 10 ins, and taking the minimum photographic 'point' image as 0·0008 ins (section 20.7), an enlargement of about 12 times is the maximum feasible; and since 20μ is an optimistic figure for the photographic resolution, it may be expected to be smaller. Thus at a focal length of 8 ft the image of Jupiter at opposition is 0·02 ins in diameter. All detail smaller than 1/50 of the diameter of the disc will be submerged by emulsion diffusion. For subsequent enlargement to reveal smaller detail than this, the image must be amplified before it reaches the plate.

The use of an amplifying lens is simply a means of obtaining the advantages of long F without the mechanical, financial, and other difficulties that go with it; for the scale of a plate exposed at the primary focus of a 40-ft telescope is the same as that of a plate taken with a 4-ft telescope and a ×10 amplifying lens. Amplification, on the other hand, increases the practical difficulties. Increased exposures are required in proportion to M^2 (for amplification M), with increased trouble from atmospheric turbulence. Guiding errors are also exaggerated in proportion to M.

Either a positive or a negative amplifying lens may be used, though the latter is to be preferred, since the outer parts of the field are imaged by rays passing through the peripheral regions of a positive lens, resulting in marked deterioration of the definition towards the edges.

A negative lens is also smaller than the equivalent positive lens. Cemented negative lenses of the Barlow type have been used very successfully, but even better results can be obtained with telenegative lenses (see section 8.8). These are preferable to ordinary ocular on account of their smaller light loss.

If there is to be no subsequent enlargement, the fastest plates available should be used. If the image is to be enlarged after development, it is necessary to experiment with plates of different speeds to find the fastest one whose grain does not become obtrusive at the degree of enlargement required. (It should be remembered that if enlargement reaches the stage at which the grain becomes visible, it has far exceeded the stage at which the resolving power of the emulsion is exhausted.) A rather cumbersome way round the difficulty is to enlarge in steps, enlarging to about half the size at which the grain appears, photographing this enlargement, and enlarging again from the new negative.

20.12 Aberrational tolerance of a photographic reflector

We have seen (section 20.7) that the diameter of a stellar image due to photographic diffusion varies between wide limits. For a well-exposed image 60μ may be taken as a rough figure, and aberrations may be disregarded so long as they do not enlarge the image beyond about this size. That is to say, coma and astigmatism can be ignored if their combined effect is to produce a star image whose longest dimension ξ is less than 60μ. Consider the case of an $f/5$ 12-in reflector. From equation (a), p. 52, the plate-scale is $7\cdot4\mu$ per $1''$. Hence the maximum tolerable value of ξ is $8''\!.1$. From Figure 19 it can be seen that $\xi=8''\!.1$ occurs at a distance of about $17'$ from the axis. Therefore the usable field, limited by coma and astigmatism, is about $34'$ in diameter; and correspondingly larger or smaller for over- or under-exposed images respectively.

A grossly over-exposed image, however, even though the aberration may be masked by the photographic diffusion, cannot be accurately measured. A fainter or less exposed star, lying outside the region of tolerable definition, will be asymmetrically comatic, and similarly not susceptible of accurate measurement. A still fainter or more underexposed star may have only the brighter part of the image recorded, in which case its position can be measured with reasonable precision. Finally, a star which with a given exposure would just be recorded at the centre of the plate will be lost altogether towards the edge of the field, owing to the dilution of its light throughout the enlarged aberrational image.

Similarly with other focal ratios and linear dimensions—see Table and Figure 161 below. It should be re-emphasised that these figures are necessarily very approximate, owing to the assumptions involved in their derivation. They are indicative of the order of magnitude, however, and furthermore show very clearly how the reduction of F with D constant, or the increasing of D with F constant, results in the rapid reduction of the size of the usable field.

Diameters of fields (' arc) limited to photographic tolerance by coma and astigmatism.

		D (ins)			
		6	9	12	15
f/ratio	$f/3$	41	28	21	17
	$f/5$	63	44	34	27
	$f/7$	82	58	44	36
	$f/9$	110	80	64	52

20.13 Guiding

Given a bright enough object and a fast enough lens, exposures may be so short that the images are not blurred by diurnal motion even when the camera is stationary. But in the greater part of astronomical photography the camera must follow the object in its diurnal motion. This necessitates an equatorial mounting with an efficient clock drive and serviceable slow motions in both directions. Extraordinary feats have indeed been achieved from time to time in the way of guiding by hand (i.e. without a mechanical drive), exposures of half an hour or more having been made; but one is left wondering why such indubitable feats—like that of the man who climbed to the top of the Empire State Building twice a day for seventeen years—were thought worth performing.

Accurate guiding—which demands practice and efficiently operating equipment—is one of the first conditions of successful astronomical photography. If the clock, the setting of the polar axis, or the slow motions are individually or collectively faulty, guiding becomes very tiring and the resulting plates correspondingly unsatisfactory. Comfort while guiding is much increased by the use of a diagonal; it is also advisable, before starting on a long exposure, to try out the limits of the observing position to see that they are feasible.

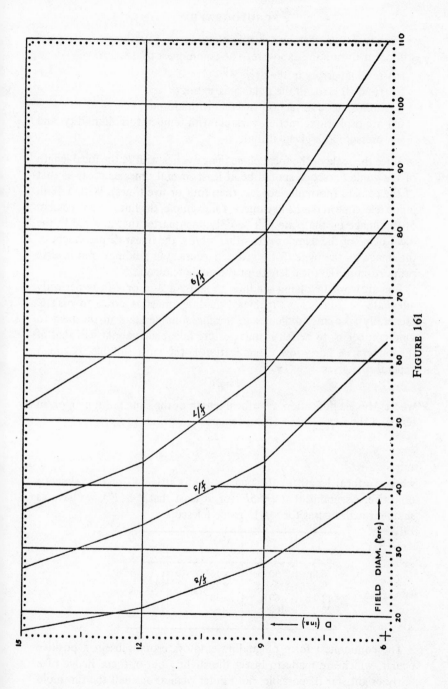

FIGURE 161

Even with a good clock drive, displacements of the image on the plate will occur during the exposure. The commonest causes are:

(a) residual errors in the drive;

(b) residual error in the polar axis setting;

(c) atmospheric refraction variable with altitude;

(d) atmospheric refraction variable with temperature, humidity, and pressure at a given altitude.

Since the scale of the focal plane image varies with F, the focal length of the guide telescope should be at least several times as great as that of the camera (normally not less than four or five times). With a 48-in guide telescope and a 12-in camera, for example, the linear displacement of the image on the plate will be only one-quarter that allowed in the focal plane of the telescope. In other words, the linear displacement of the image on the plate (for a given inaccuracy in guiding) is in inverse proportion to the focal length of the guide telescope.

For all work involving guiding on stars, a 3-in or 4-in is perfectly adequate—a 3-in being capable of guiding on stars down to mag 6. Generally speaking, difficulty in finding suitable stars in the field to guide on begins to be felt if the aperture falls below about half that of the camera. A fairly low magnification is the most suitable. A rough general rule gives

$$M = 3F'$$

where M = magnification, F' = focal length of the camera (in ins), or, in terms of the focal length of the ocular,

$$f = \frac{F}{3F'}$$

where f = focal length of the ocular, F = focal length of the telescope objective. Assuming that $F \simeq 5F'$ (or at least that $F \nless 5F'$), we have as sample rough figures for a $f/15$ guide telescope:

F'	F	M	f
6 ins	30 ins	×18	
9 ,,	45 ,,	×27 $\left.\right\} = 9D$	$\left.\right\}$ 2·5 ins
12 ,,	60 ,,	×36	
20 ,,	100 ,,	×60	

The commonest form of guiding employs cross-webs in a positive ocular, which are made to bisect the slightly out-of-focus image of a fairly bright star (if possible, not fainter than about half the threshold magnitude of the instrument). To facilitate quick correction of runs-off, the ocular should be oriented so that one web is coincident with the hour circle through the star, and the other with the Dec circle; this is done by trailing the star with the clock drive disconnected.

A rather simpler arrangement to use, being more sensitive to run-off, is a photographically reduced circle located in the plane of the ocular and illuminated. The image of the guide star is enlarged until it just does not fill the circle, and placed at the centre of the circle. The narrow ring of dark sky separating the image and the ring, which should be of equal width in all position angles, provides a very sensitive criterion for guiding purposes. It is essential that each run-off should be corrected with all possible speed, as well as being halted while still small.

When photographing an object with proper motion, the object itself must, of course, be the guide object. In the case of star-like objects, such as asteroids, this presents no difficulty apart from the additional manual operation of the slow motions and the possible faintness of the image. But comets, unless they have a sharply defined nucleus, present a different problem, methods of meeting which are described in *O.A.A.*, section 16.3.

A third method of guiding employs a double-slide plate carrier (similar to the mechanical stage of a microscope) fitted with an ocular for the examination of a stellar image just outside the field covered by the plate; the necessity of a guide telescope is thus obviated altogether. Even without a double slide the same avoidance of a separate guide telescope can be obtained, guiding then being achieved by movement of the whole camera-telescope. In really large instruments the advantage of the double-slide plate carrier is that it obviates the disparity of displacement of the camera and guide telescope images due to unequal flexure in the two instruments. For the same reason it is particularly useful with reflectors, slight shifts of whose specula introduce image displacements which are not duplicated in the field of the guide telescope.

20.14 Photographic seeing

Exposure of the plate when the atmosphere is prohibitively turbulent is one of the commoner causes of disappointment in astronomical photography. Where exposures can be short—the Sun, for example, and under certain circumstances the Moon—it is not only possible but essential to watch for and utilise the fleeting moments of better than average seeing. The view system (whether reflex camera or separate telescope) should have the same magnification as the photographic, since even a small amount of turbulence will distort the magnified image. Even where exposures of some length are necessitated it is still essential to exercise some choice of when to expose the plate; many nights are totally unsuitable for photography; on others the seeing improves or deteriorates either spasmodically or continuously throughout the hours of darkness.

It should be borne in mind that photographic and visual seeing are to some extent different. Visually, turbulence may either confuse the image while introducing little movement, or it may cause bodily movement of a reasonably clearly defined image (see section 26.7), or both may be present. Visually, movement is much less troublesome than confusion; photographically they are equally undesirable.

20.15 Photographic lenses

Doublets, the simplest type of camera lens, consist of widely spaced achromatic pairs. The Petzval portrait lens was the prototype of these lenses. Characteristic features of portrait lenses generally are small focal ratio—$f/2$ to $f/5$ being commonly employed—with consequent high photographic speed and small plate-scale. They are therefore ideally suited for wide-angle work such as patrol photographs, star fields, comets, etc; Barnard's superb photographs of the Milky Way, for example, were taken with a 6-in portrait lens of 31 ins F ($f/5$). Conversely portrait lenses are useless for angularly small objects such as the planets or Moon, unless an amplifying lens is also used; for such work the telescopic focal plane is the position for the plate.

Portrait lenses of less than about 2 ins aperture are of limited but real usefulness; a 3-in or 4-in working at $f/3$ to $f/4$, on the other hand, is valuable, while a 6-in will give very fine results with comets and for wide-field work generally. Though their fields of tolerable definition are in general considerably smaller than those of anastigmats of the same aperture and focal ratio they are also very much cheaper, and a given outlay will secure a larger and faster portrait lens than anastigmat.

The original Petzval doublet consisted of a plano-convex doublet furthest from the plate, and an air-spaced meniscus and biconvex nearest the plate. It suffers from distortion and a falling off of illumination towards the edge of the field. The famous 'Astrographic' doublet* consists of a positive and a negative unit, both cemented, and is characterised by flat field, good spherical correction, negligible coma and astigmatism, but some residual secondary colour.

* Not to be confused with the type used for the International Astrographic Chart.

Doublets generally are rather bulky and heavy. Small ones are, owing to reduced absorption, faster than large ones of the same focal ratio. Useful fields of 10° to 12° may be expected.

The old-fashioned orthoscopic rectilinear lenses, consisting of meniscuses arranged in two achromatic components (the positive distortion of one being compensated by the negative distortion of the other), have been superseded by the modern anastigmat, in which freedom from astigmatism is secured (see section 5.7). An anastigmat of 4 or 5 ins aperture and 20 ins F is a first-rate piece of equipment, but probably ruled out on the score of expense; there is nevertheless plenty of work to be done with less powerful equipment.

Since the war it has been possible to buy quite cheaply ex-R.A.F. aircraft cameras and lenses, most of which are triplets; these are an excellent investment, even though the falling-off of definition towards the edges of the field is rather severe with some types. The Zeiss, Aldis, and Ross 'Xpres' lenses are good examples of this type. (See also B. 222.)

Other types of lens that have been put to good use in astronomical work are telephoto lenses of many sorts (anastigmats with an additional negative component to increase the effective focal length and plate-scale), and projection lenses, which can occasionally be picked up cheaply and are excellent for such extended objects as comets.

20.16 Dewcaps, shutters and plate-holders

Dewcaps are even more essential for camera than for telescope objectives, since they reduce extraneous light (therefore retarding fogging) as well as dew. Particularly is this function important when, for example, the Moon is above the horizon or when, during total solar eclipse, sunlight is diffused through the nearby sky outside the totality zone.

The most satisfactory type of shutter depends upon the nature of the work undertaken. For long and approximately timed exposures (e.g. stellar) a hinged flap mounted on the dewcap and operated from the eye end of the telescope (by Bowden cable, electromagnet acting against a spring, or double string) gives good results. The practice of removing by hand a cap covering the objective is not to be recommended; even when eased off as carefully as possible it is difficult to avoid vibration, and during an interval at the beginning and end of the exposure the telescope is not guided. For planetary and lunar photography, where the exposures though comparatively short need to be precisely timed, Bowden cable or electric operation of the dewcap cover works satisfactorily, a metronome being useful for giving the required exposure time.

Short and precisely timed exposures, as of the Sun, require a good mechanical shutter. Focal plane are to be preferred to compur shutters, and electrical to bulb or plunger operation.

It is essential, particularly when a bright object is being photographed, that both the shutter itself and its operating mechanism should be vibrationless. Shutters should always be tested out at the beginning of each observational spell.

Although curved (concave) plate-holders would increase the useful field with refractors, this is by no means essential (cf. Schmidt and Maksutov cameras), and a double dark slide is the most suitable form of plate mounting. Lateral movement of the dark slide in the holder allows multiple exposures of such objects as planets to be obtained on one plate. This is not only a saving of time and money, but also ensures uniform development and hence surer comparability of the images.

It is a good plan to form the habit of drawing the plate slide as soon as the object is correctly placed in the field, or the guide star on the cross-webs, and immediately before opening the shutter. Before removing the dark slide from the carrier, make sure that the plate slide is fully home.

It has already been mentioned (section 20.8) that most plateholders can be adapted to hold sheet film adequately flat. One positive method of ensuring the flatness of sheet film is to employ a partial vacuum to draw the film against a flat perforated surface forming the back of the film-holder. The use of a glass pressure plate in front of the film is not to be recommended.

20.17 Camera: construction and mounting

A strong, well-made box of seasoned wood, or a metal tube, is preferable to the bellows construction of commercial folding cameras. Besides being rigid in construction it should be as light as possible, and of course completely dust- and light-proof; matt black finish internally. The squaring-on of the plate in a home-made camera is a matter of importance, and may present some practical difficulty. The back of the camera, to which the runways for the dark slide are attached, should be fitted with a screw adjustment so that it can be tilted in two mutually perpendicular planes. Star images on a focusing screen, at opposite ends of two mutually perpendicular diameters of the field, are then examined with a high-power magnifying glass; when they have been made as uniform as possible (great precision is difficult owing to the comatic and astigmatic distortion) final adjustment is made from stellar exposures (see section 20.10.2).

An alternative method, which is easier to operate, employs the curtain-ring dodge already described in section 13.15 when dealing with the squaring-on of anastigmats. Load the open plate carrier with a sheet of plane glass and observe it through a bright ring from behind, the cover being in place over the objective. Then adjust the setting of the dark slide runners to make the reflections of the ring at the glass sheet and at the objective concentric.

Complete rigidity between objective and plate, and between camera and guide telescope, is essential, and should be achieved with the minimum weight consistent with this.

Cameras which are small compared with the telescope are conveniently mounted on the telescope tube in a continuation of the Dec axis. They may be bolted to a wooden base which is itself bolted to the tube cradle, or may be fixed to the tube by metal bands, screwed tight, and widely spaced wooden U-shaped distance pieces.

For larger cameras, the best position is probably the opposite end, from the telescope, of the Dec axis. This arrangement has several advantages: the weight of the camera is subtracted from, instead of being added to, the counterpoise, thus reducing the load on the bearings; and it ensures good clearance between the camera and telescope axes— if these are too close together the telescope dewcap may obtrude on the field of a short-focus, very wide-angle lens. If, however, the telescope is mounted under a dome it entails a very wide shutter opening.

Short-focus cameras (even ordinary commercial cameras) can be mounted on a small equatorial head, with a small telescope of the 'pocket' variety fixed to it for guiding (a diagonal will probably be required to provide a practicable observing position); for meteor work, even the telescope can be dispensed with, providing the drive is reasonably accurate. With lightweight apparatus of this sort the most unpretentious forms of drive give good results: an ordinary spring-escapement clock, for instance, running at sidereal rate and connected to the driving circle by a tough rubber band or chain. The driving circle can be mounted at the upper end of the polar axis, and the camera mounted on it in pivots to provide adjustment in Dec. (See also section 14.6.)

For photographic even more than for visual work the mounting must provide:

(a) Firmness and solidity; an iron pedestal or masonry base is infinitely preferable to a tripod, even with stretcher bars. Sources of vibration (see also section 14.25) include wind, traffic, the driving motor, and the operation of the slow motions.

(*b*) Accuracy of the setting of the polar axis, so as to eliminate as much slow-motion adjustment as possible and to reduce the rotation of the plate about the guide star.

(*c*) Efficient (instantly and smoothly acting) slow motions.

20.18 Colour- and cine-photography

Colour films are now of quite sufficient speed to permit their use in astrophotography, though their use is usually limited to their pictorial value. A colour film has three photographic layers, with other layers acting as colour filters. The details vary from film to film, each having its own properties, which makes evaluation rather more tricky than in the case of monochromatic film. Colour reversal films, producing transparencies, are the most popular since the image is contained on a small and consequently cheap piece of film. Negative films require prints to be made on colour paper, which, apart from adding to the expense, introduces further problems of colour balance.

The processing and printing of some types of colour materials can be undertaken by the amateur; in the case of the negative materials the colour balance of the final print is often judged by eye, trial and error. A grey calibration patch on the film, negative or reversal, using the same exposure time as the sky photograph is invaluable in assessing the correct balance.

The colours reproduced by photography are on the whole accurate, though certain anomalies may be noted. In particular, it is found that features with emission line spectra can vary in colour. For example, the Orion nebula, M42, appears red in photographs, due to the presence of a strong line of the hydrogen spectrum which falls in a sensitive region of the red layer of the emulsion. The eye is more responsive to the green lines of doubly ionised oxygen, which fall in a region of relatively poor emulsion sensitivity. Anomalous response of colour emulsions to light of low intensity will also arise from the tendency of the three layers to have different reciprocity failure characteristics.

Any amateur who has built for himself a long-focus Sun telescope* might turn his attention to cine-photography. The very fine McMath-Hulbert films of solar prominences were made with a 50-ft tower telescope (10½-in aperture), and interesting results could certainly be obtained with a 20-ft Sun telescope.

20.19 Three common faults

Among the commonest causes of faulty and disappointing plates

* Such as that described in *O.A.A.* section 1.15.

(they have been mentioned already, but bear repetition) are:

(*a*) Inefficient slow motions. A faulty drive may be largely offset by smooth and efficient slow motions, free from backlash and lag; but if the slow motions themselves are inefficient, decent exposures are almost impossible to obtain.

(*b*) Insufficiently precise adjustment of the plane of the emulsion to the actinic focal plane.

(*c*) The exposing of plates when the atmosphere is too turbulent for photography.

SECTION 21

THE MEASUREMENT OF RADIATION INTENSITIES

21.1 Introduction

The human eye, though relatively insensitive as a judge of absolute brightness, can be trained to estimate the brightness difference of two sources to within 0·1 mags. This fact is the basis and justification of the visual methods of observing variable stars,* which are valuable where considerable variations or differences of brightness are concerned. For the measurement of small brightness differences, however, some form of photometer is necessary. Photocells of various types, for example, can provide consistent determinations of differential magnitudes to 0·01 mag.

It would be worthwhile, before proceeding to a description of photometric methods, to review the frequently unrecognised difficulties inherent in the theoretically simple operation of determining precisely the apparent magnitude of a star. The apparent brightness of the image will be to some extent modified by, among others, the following factors:

(*a*) Transmission factor and wavelength sensitivity of the telescope: a function of time in the case of a reflector, as well as differing from one instrument to another.

(*b*) Wavelength sensitivity of the eye, photocell, or photographic plate; that of the eye, at least, varies widely from individual to individual, as well as from time to time with a given individual.

(*c*) Transmission factor and selective absorption of the atmosphere: both vary with place and time, depending on local meteorology, geographical position and altitude of the observing station, and the altitude of the star.

(*d*) Colour of the star.

(*e*) To these must be added, if we are concerned with magnitude as an

---

* These are described in detail in *O.A.A.*, sections 17.2, 17.3.

376

index of the actual conditions at the star; transmission factor of inter-stellar space along the light-path.

In view of this host of disturbing factors, mostly uncontrollable, it is hardly surprising that many published magnitudes are infected with errors whose size there is no means of estimating.

Those types of photometer that are applicable to astronomical work can be broadly classified under four heads: extinction, equalisation, photographic, and energy conversion.

21.2 Extinction photometers

The principle is to reduce the apparent brightness of the source until it falls below the threshold of visibility. The amount of the adjustment of the extinguishing mechanism necessary to perform this is a measure of the star's brightness.

Disadvantages

(a) The visibility of the star does not remain steady and continuous right down to the extinction point; close to the limit of vision it will be glimpsed intermittently. The precise point of extinction is therefore difficult to determine. In an effort to get round this difficulty some observers make their measurements from the point of 'minimum steady visibility', but it is doubtful if this can be much more consistently recognised than the point of total extinction. Alternatively a device such as a stepped wedge may be used, so that the star's magnitude may be certainly assigned to a magnitude range between two definite values, instead of being of an unknown degree of uncertainty.

(b) The value of the extinction reading varies with the seeing, the clarity of the atmosphere, the ZD of the star, the brightness or otherwise of the background, and the personal equation of the observer. With all these variables involved, series of observations by a single observer cannot be assumed to be truly consistent, nor those by different observers to be strictly comparable.

(c) Results obtained by different extinction methods are not comparable without the introduction of an adjusting factor, owing to different rates of decrease of background brightness and image brightness: the wedge extinguishes the ground before the stellar image, whereas with reduction of aperture the field darkens relatively more slowly.

(d) If the image does not always fall on the same region of the retina, results obtained at different times will not be comparable; and keeping the eye directed precisely at an image which is close to the limit of vision is nearly impossible. More uniform results are obtained if the eye is directed at another, non-vanishing, point in the field, always at the same distance and position angle from the image being measured. This procedure keeps the error well below 0·1 mags, whereas discrepancies as large as 0·5 mags can be encountered when attempts are made to keep the vanishing image on the macula.

(e) Within a certain range, the extinction point is affected by the magnification.

(f) Extinction achieved by reducing the aperture produces greatly enlarged spurious discs with the small apertures at which the brighter stars disappear,* which again makes the extinction point difficult to determine. With a reflector, this may be overcome by substituting an unsilvered flat for the ordinary diagonal; or, with reflectors or refractors, a known preliminary reduction of brightness can be obtained by polarisation or other means.

(g) Eye fatigue, brought on by the length of time the adjustment requires, is probably not a factor of great importance (provided the observations are made with reasonable expedition), and in any case applies to some extent to all types of visual photometer.

For all these reasons extinction methods have now been discarded by all serious workers in photometry, and are accordingly dealt with only summarily here.

21.2.1 Optical wedge: An optically worked, tapering wedge of glass, neutrally and uniformly tinted, and mounted against a reversed wedge of plain glass so that the outer surfaces of the combination are parallel; or a plane parallel glass slip upon which a metallic film of graded thickness has been deposited; or a graded gelatin filter (colloidal carbon or granular silver) of the type supplied by such firms as Ilford, Kodak, or Dufay-Chromex.

The light from a star, transmitted by the wedge (which is mounted against the ocular), is then reduced by an amount which is proportional to the distance of its image from the end of the wedge; this distance can be read from a scale. When the wedge constant (the magnitude difference of two stars whose scale readings at extinction differ by one division) has been determined, scale readings are directly convertible to magnitudes or differences of magnitude.

21.2.2 Reduction of aperture: An objective iris diaphragm operated from the eye end of the telescope by means of an endless screw is a convenient arrangement, since it allows an accurate reading of the scale

* Sirius, just before the extinction point is reached, resembles a small planet, so exaggerated is its spurious disc.

to be made, and at the same time is quickly accomplished, thus reducing eye strain. In the case of bright stars, however, it will be found necessary to employ also a neutral wedge so that extinction can be obtained without too gross a reduction of D.

Knobel's equilateral triangular stop, which maintains the shape of the aperture (and hence the diffraction pattern) with only three leaves, has the additional advantage of being fairly easily home-made. A (Figure 162) is a brass plate which carries the stop and which is clamped

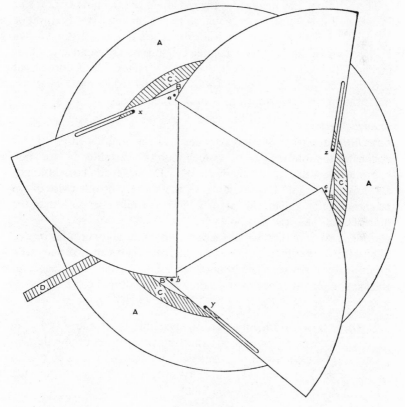

FIGURE 162

to the dewcap, OG cell, or tube mouth. B is a raised brass rim screwed to A. Its outer edge is machined, and it is bored for screws at three equidistant points, a, b, and c. C is a brass ring, machined to revolve smoothly round the rim B. It carries projecting pins at the three points x, y, and z. Secured at a, b, and c, but not so tightly that they cannot revolve on these pivots, are the three triangular leaves of the shutter. The base of each is an arc of a circle centred at the point of pivot. Three slots, one cut in each leaf, run over the pins x, y, z. Rotation of the arm D, actuated from the observing position, will cause the leaves to open or close, varying the dimensions of the triangular aperture. An engraved scale on the base-plate can be read from the edge of the arm D.

21.3 Equalisation photometers

The principle here involved is the reduction of the intensity of a star image until it is judged to be equal to that of another star (natural or artificial) in the field. The magnitude difference of the two stars is then $2 \cdot 5 \times$ the log of the ratio by which the brightness has been decreased. If the magnitude difference is great, a preliminary known reduction to near equalisation when the brighter is observed—by reduction of aperture, polarisation, rotating sector, etc—may be required.

Advantages

(a) Providing that the two stars are near one another in the sky, it eliminates the following factors which modify the results of extinction photometry: seeing, atmospheric clarity, ZD, background brightness.

(b) With training, the estimation of the equality or otherwise of the brightness of two stellar images is a more accurate operation than the judgment of the extinction point.

(c) The eye is naturally more sensitive to the equality or otherwise of two point images than to the exact difference between two unequally bright images. An equalising photometer thus has the advantage over the usual visual methods of estimating the magnitudes of variables.

Disadvantages

If there is a natural comparison available, well and good. But more often than not there is no other star of suitable magnitude near enough to be included in the same field. Some designs of photometer (e.g. Danjon's) get round this difficulty. Others rely on an artificial star, which, however, suffers from the following drawbacks:

(i) It never exactly resembles the natural star in appearance, and is often—even if filters are used—of dissimilar colour; all of which makes accurate and consistent comparisons difficult.

(ii) The consistency of a series of observations depends upon the invariability of the standard source providing the artificial star image.

(iii) The comparison of different observers' results is complicated by the fact that their standard sources cannot be directly compared.

21.3.1 Optical wedge: If the wedge is adjusted so as to cross half the field, occluding one but not the other of two stars in the field, it can be used as an equalisation photometer to reduce the intensity of the brighter image until it is judged to be equal to that of the fainter. Results so obtained are much to be preferred to those obtained by the extinction method, but a serious source of error is the greater darkness of the half of the field covered by the wedge.

Unfortunately, so conveniently placed a comparison star of known magnitude is not always available—though in the case of well-observed variables it very often is. Two ways round the difficulty suggest themselves: to feed the image of a more remote star into the field by reflection, or to employ an artificial star. The former is the method used by, e.g. Pickering's meridian photometer (section 21.3.6) and Danjon's cat's-eye photometer (section 21.3.7); the latter by the Zöllner and Flicker photometers (sections, 21.3.2, 21.3.8).

21.3.2 The Zöllner photometer: A typical layout is illustrated schematically in Figure 163. The image of the observed star is formed

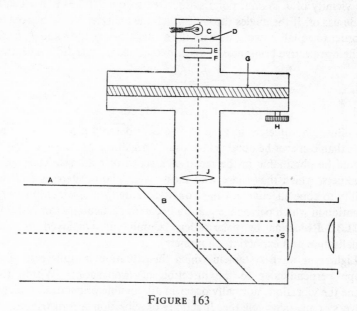

FIGURE 163

by the telescopic objective at S, the flange A fitting into the telescope drawtube. B is an unsilvered plane parallel plate which reflects two images of the artificial source into the field, one from each surface; that from the front surface, s, is used in making the comparisons, the other

381

being out of focus. C is a low-candlepower lamp which illuminates a pinhole in the metal-foil sheet D. Pinholes of a range of different sizes (from, perhaps, 0·1 to 0·3 mm, though this depends upon the design of the instrument) should be available, so that the appearance of the natural star under different seeing conditions can be reproduced as closely as possible. This is a point of some importance, since dissimilarity in the appearance of the images makes the estimation of equality more difficult and is one of the commonest sources of inaccuracy with a Zöllner-type photometer. Variation of the appearance of the artificial image can also be obtained by varying the distance of the ground-glass screen, E, from the pinhole. The function of F, a light blue filter, is to reduce the yellowness of the artificial star. G is the optical wedge, operated by the milled head H. The associated scale should be graduated to at least 0·05 ins in divisions large enough for tenths to be estimated. J is the projection lens.

Reduction of the brightness of the artificial star can be accomplished in other ways. For instance, by means of an iris diaphragm situated in the vicinity of J. Again, polarisation (see section 21.3.3) is frequently made use of. If the angles through which the analyser has to be rotated in order to equalise two stars with the artificial image are θ and θ', b and b' their respective brightnesses, and Δm their magnitude difference, then

$$\frac{b}{b'} = \frac{\cos^2 \theta}{\cos^2 \theta'}$$

and
$$\Delta m = 5(\log \cos \theta' - \log \cos \theta)$$

Zöllner photometers in general give considerably more accurate results than can ever be obtained by visual estimation; at least $\pm 0·05$ mag should be obtainable in the mean of a series of readings. Their chief weakness, which they share with most visual photometers, is their inability to cope effectively with strongly coloured stars, which present a situation in which personal equation becomes particularly rampageous.

21.3.3 Reduction by polarisation: Commonly employed in both equalisation and extinction photometry.

Light is a wave system in which the vibration is transverse—in a plane perpendicular to the direction of propagation. Within this plane the vibrations normally occur in all possible orientations. Such a wave system, in which the directions of vibration are distributed at random, is said to be unpolarised. If, conversely, the vibrations occur predominantly or solely in one direction, then the light is said to be partially or completely plane polarised.

If an unpolarised beam is passed through a filter of some polarising material (see section 21.3.4), the emergent beam will consist of waves vibrating in a single direction. The direction, in the filter, parallel to the direction of vibration of the transmitted light, is termed the polarising axis of the filter. If the vector a represents the amplitude of a vibration inclined at an angle θ to the polarising axis, then the component of a lying in the polarising axis is given by

$$a' = a \cos \theta$$

Also the intensity of a vibration is a function of the square of its amplitude. Hence, writing I_θ for the intensity of the transmitted radiation and I for the intensity when θ is zero,

$$I_\theta = I \cos^2 \theta$$

Hence the action of the polarising filter upon an incident beam of unpolarised light is not simply to transmit those vibrations whose direc-

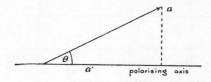

FIGURE 164

tion is exactly parallel to the polarising axis, but also to transmit, of vibrations inclined to it, their components in the polarising axis.

If, now, a second filter, known as the analyser, is placed in the path of the polarised ray transmitted by the first, it will behave in an entirely analogous fashion, but now θ is the angle between the polarising axes of the two filters, and I is the intensity of the beam transmitted by the analyser when these axes are parallel. Whereas a single filter will transmit (neglecting absorption) about 50% of an unpolarised incident beam irrespective of the orientation of its axis (the directions of vibration in the unpolarised beam being random), the second filter will transmit anything from 100% of the polarised beam incident to it (when $\theta = 0°$) to 0% (when $\theta = 90°$); see Figure 165. In practice these theoretical figures are reduced by reflection at each surface of the filter, and also by absorption: the transmission factors of different Polaroid filters range from 25% to 42%.

21.3.4 Polarising agents: (a) *Transmission polarisers.* Polaroid, a synthetic product commercially produced within the last twenty-five years, is a development of 'herapathite' (iodoquinine sulphate), discovered as long ago as 1852. It derives its polarising action from the linear

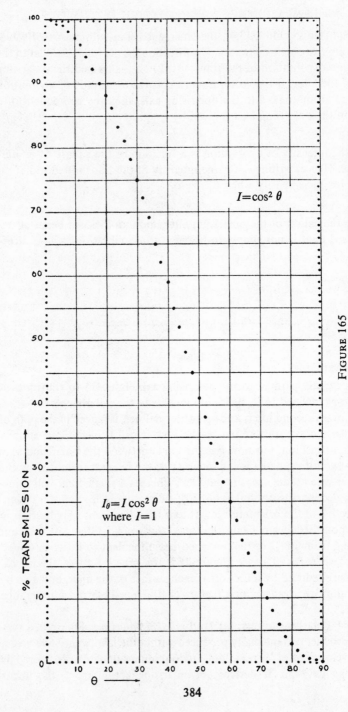

FIGURE 165

Percentage transmission of two polarising filters for different inclinations (θ) of their axes

$I = \cos^2 \theta$

$I_\theta = I \cos^2 \theta$
where $I = 1$

% TRANSMISSION

θ

arrangement of innumerable microscopic herapathite crystals embedded in a protective, stable, and transparent medium. It produces 90% polarisation at 5000Å, is opaque to ultraviolet, and comparatively so to the shorter visible wavelengths; the infrared is transmitted unpolarised.

Various commercially produced plastic polarisers derive their action from the linear arrangement of large and complex molecule chains. An example is iodinised polyvinyl alcohol.

(b) *Double-refraction polarisers.* Many naturally occurring crystals (e.g. calcite) and some artificial materials (e.g. cellophane) have the property of refracting an incident ray in two directions, each of the refracted rays being completely polarised in mutually perpendicular planes. One ray arises from normal refraction, and is known as the ordinary ray. The other, the extraordinary ray, obeys neither of the laws of refraction; μ_e, as measured by it, may be either greater or less than μ_o, measured by the ordinary ray. The ordinary and extraordinary images may be distinguished by rotating the crystal, when the ordinary image will remain stationary while the extraordinary image will revolve round it. That the two rays are plane polarised with their axes of polarisation inclined to one another at an angle of 90° is proved by observing the two images through an analyser: when this is rotated the images will alternately appear and disappear, both never being fully visible or invisible simultaneously.

An ingenious use of the doubly refracting calcite crystal is the type of prism devised by Nicol in 1828. Two calcite crystals are cut to the shape

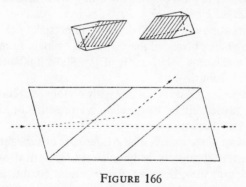

FIGURE 166

illustrated in Figure 166, polished, and cemented together at their shaded faces with Canada balsam. The refractive index of the balsam is intermediate between μ_o and μ_e, and by adjusting the angle between the balsam interface and the incident ray it is possible to eliminate one of the rays by internal reflection (its angle of incidence at the interface exceeding the critical angle) while the other is transmitted.

The advantage of the Nicol prism, which secures its use for all critical work, including photometry, resides in the fact that it is the only known agent whose polarisation and transmission are independent of wavelength, i.e. its image is colourless. Its disadvantages are, firstly, cost, since sufficiently large calcite crystals are hard to come by, and secondly, for one ray to be eliminated while the other is transmitted, the range of angles that the incident ray may make with the first face of the prism is limited, i.e. its field is small.

(c) *Polarisation by reflection.* This finds its application less in photometry than in solar work, where it is employed in solar eyepieces.* If a ray of light is directed at a plane glass surface, the angle of incidence i_p being such that

$$\tan i_p = \mu$$

then the reflected portion of the ray will be completely polarised in a plane parallel to the reflecting surface. This special value of i is known as Brewster's angle; for flint ($\mu = 1 \cdot 615$) it is $58\frac{1}{4}°$, and for crown ($\mu = 1 \cdot 515$) $56\frac{1}{2}°$. Polarisation is zero when $i = 0°$ and $i = 90°$, and partial between these limits and Brewster's angle. The refracted portion of the ray is never more than partially polarised.

Even matt surfaces partially reflect as glass, or mirror-like, surfaces, and hence the light by which they are seen is partially polarised; this fact is the basis of the use of polaroid anti-glare spectacles.

(d) *Polarisation by scattering.* Particles in suspension, whether in a liquid or in the atmosphere, both scatter and partially polarise the light passing among them, provided they are small enough. In the case of the atmosphere, maximum polarisation of the Sun's light occurs along a great circle 90° from the Sun.

(e) *Internal strain.* Glass is not normally doubly refracting, but under great strain, caused either by mechanical pressure or by too rapid annealing, it may become so.

(f) *Conditions in the source.* The greater part of sunlight and of light from most artificial sources is unpolarised. Electrical and magnetic fields in the source may, however, under certain conditions cause it to emit partially polarised light.

21.3.5 The Pickering photometer: A generic name given to a series of polarisation photometers developed by E. C. Pickering at Harvard (B. 264–5). They share the characteristics of being restricted (like the

---

* For detailed description, refer to *O.A.A.*, section 1.6.

simple wedge photometer) to the direct comparison of two stars which can be contained in the same field, and of relying on equalisation by means of a double-image prism and a rotary analyser. They differ for the most part as regards the relative positions of the components in the optical train. They all involve a considerable light loss (the brightness of the equalised images being only 30% that of the primary image of the fainter star), and in an effort to circumvent this disadvantage one model incorporated a prism for reflecting the image of a bright comparison star into the field; its brightness was reducible by means of an adjustable cat's-eye diaphragm, the faint star then being viewed direct.

The general principle underlying all designs of the Pickering photometer is as follows. A doubly refracting prism, mounted near the telescopic focus, produces an ordinary and an extraordinary image of each star. By adjusting the prism's orientation and position along the axis, the ordinary image of either star can be brought close enough to the extraordinary image of the other for comparisons to be made conveniently. Either such pair of images is then equalised by rotating the analyser. If, then,

b, b' = brightnesses of the two stars,

$\Delta\theta = \theta - \theta'$, where θ = analyser reading when the two images are equal,

θ' = analyser reading when one image disappears,

the brightnesses of the two images of the first star are given by

$$b \sin^2 \Delta\theta$$
$$b \cos^2 \Delta\theta$$

and those of the second by

$$b' \sin^2 \Delta\theta$$
$$b' \cos^2 \Delta\theta$$

Of these, the second and third have been equalised,

$$\therefore \quad b \cos^2 \Delta\theta = b' \sin^2 \Delta\theta$$

$$\text{or} \quad \frac{b}{b'} = \tan^2 \Delta\theta$$

Therefore, from equation (e), p. 33,

$$\Delta m = 5 \log \tan \Delta\theta$$

Equalisation can be produced by four positions of the analyser, all of which are used.

21.3.6 The meridian photometer: Also developed at Harvard by Pickering; it was with an instrument of this type that the Revised Harvard Photometry was compiled. Though one hesitates to say of any instrument that it is not for amateur use—now that horizontal **Sun**

telescopes, spectrohelioscopes, Schmidt and Maksutov telescopes, and apparatus for long-wave solar study are all to be found in the hands of amateurs—it must be admitted that the meridian photometer comes very near to falling within this category.

It is a refinement on the types already described, in that for the sake of consistency the comparison star is neither an artificial star nor a casual neighbour, but in every case Polaris. Even this gain is not without its adverse side, however, for it means that the comparison star and stars in low Decs are subject to different atmospheric and meteorological modifications.

Light from Polaris and light from a star on the meridian (in which position all measurements are made) are fed into a horizontal tube, lying E and W, by two objectives; these are of equal aperture and focal length, and are inclined to one another at a small angle. The two images in the common focal plane of the objectives are then equalised as in the simple Pickering photometer. A more detailed description will be found in B. 266.

21.3.7 Danjon's cat's-eye photometer: Since the original description of this instrument is comparatively inaccessible (B. 251) it will be worthwhile treating it here in some detail, for it has outstanding advantages for the amateur. It is simple in construction and operation; it can be home-made quite cheaply by anyone used to handling tools; it introduces no false colour (cf. wedges); it employs natural stars for comparison (cf. Zöllner photometers), but these are not limited to the small area of sky covered by the telescopic field (cf. Pickering's photometer); the measures can be made quickly—as many as five a minute without difficulty; less trouble from ghosts is experienced than with polarisation photometers; and, finally, its results have been found to be of an accuracy comparable with those obtained by Stebbins with a photoelectric photometer.

The optical arrangements are extremely simple, and are illustrated in Figure 167(a). A and B are right-angled borosilicate prisms set for total internal reflection, B being mounted centrally in front of the telescope OG and A at the side of the tube. A is movable through a small arc about an axis in the plane of the diagram and perpendicular to the optical axis; B is similarly adjustable about an axis perpendicular to the diagram plane. By this means the image of a star several degrees from the axis of the OG can be reflected into the field; the controls are carried back to the eye end of the telescope. C is a cat's-eye diaphragm (Figure 167(b)) oriented so that the sides of its aperture are parallel to those of the prism faces. It is operated by remote control from the observing position through pinions engaging a ratchet on each section of the diaphragm, so that an equal and opposite movement is imparted to each. The length of the diagonal of the aperture is recorded by means of a scale mounted on one section of the diaphragm and an index on the other. The precision of the scale and the mounting and operation of the cat's-eye are determining factors in the success of the instrument. The scale is read by means of a small view telescope mounted on the tube

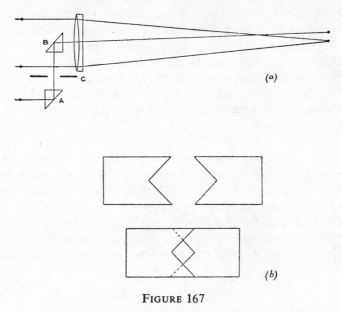

FIGURE 167

near the eyepiece, being illuminated at will by a low-amp bulb and a pocket battery, and viewed (since it lies parallel to the telescopic axis) via a totally reflecting prism mounted in front of it.

The zero reading of the scale is determined by pointing the telescope towards a bright light and reducing the aperture of the diaphragm until the image of the prism is completely extinguished. Let the mean of a large number of such determinations be x_0. If, now, $C=$ the magnitude difference of the two images of the same star for $(x-x_0)=1$ scale unit, the mag difference of two stars whose images are equalised at a scale reading of x units is given by

$$\Delta m = C - 5 \log (x - x_0)$$

Three ranges of mag variation by means of the cat's-eye (i.e. three independent values of the constant C) may be obtained by reversing one

389

or both of the prisms from the position of total internal reflection to that of external reflection at the hypotenuse face: C_{ii} when A and B are both reflecting internally (as in Figure 171); C_{ie} when the reflection is internal at A and external at B; C_{ee} when both A and B are reflecting externally.

These constants must be determined empirically with the greatest precision possible:

(a) *Two internal reflections.* A rotating sector of known angular aperture is set up in front of the instrument so that it occludes the OG but not prism A when the telescope is pointed at an artificial star some 50 yards away. From the known proportion of the open to the closed sectors the relative brightness of the two incident beams, and hence the mag difference of their respective images Δm, can be calculated. If x is the scale reading when the images are equalised,

$$C_{ii} = \Delta m + 5 \log (x - x_0)$$

(b) *One internal, one external, reflection.* Equalisation would now involve too great a reduction of the open/closed ratio of the sector for the accurate use of the latter alone. With both prisms set for internal reflection, reduce the used aperture of the OG by a diaphragm (whose area need not be known). Let x_1 be the scale reading at equalisation. Now reverse B and place the rotating sector (which can now have a relatively wide opening) in front of the OG diaphragm; let x_2 be the new equalisation reading. Then

$$C_{ie} = C_{ii} + \Delta m + 5 \log (x_2 - x_0) - 5 \log (x_1 - x_0)$$

(c) *Two external reflections.* The same differential method as in (b) is used, giving the value of $C_{ee} - C_{ie}$, whence

$$C_{ee} = C_{ie} + \Delta m + 5 \log (x_2 - x_0) - 5 \log (x_1 - x_0)$$

Danjon emphasises the danger of measures being invalidated by the deposit of dew on the exposed prism faces. To combat it he recommends that observations be made from inside an observatory rather than in the open air, that measures be taken to keep the temperature of the instrument slightly above that of its surroundings, and that the glass surfaces be gently wiped periodically; with borosilicate prisms he was unable to detect any systematic variation of C over long periods as a result of this practice.

21.3.8 Flicker photometer: Two further types of equalisation photometer must be mentioned briefly. The first, the flicker photometer, is a true equalisation photometer since it involves the adjustment of the brightness of one source until it is equal to that of another; this, however, is accomplished by a visual estimation not of equality but of the absence of 'flicker' in the images when viewed in rapid succession and in the same position.

The principle is illustrated in Figure 168. A is a circular mirror mounted centrally so as to revolve about the axis XY. Its appearance in plan is shown at A'. The convergent beam from the telescope objective is intercepted by this mirror near its edge. B is a fixed plane mirror at which the ray from an artificial star is projected via two Nicol prisms.

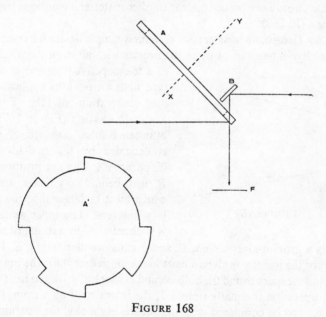

FIGURE 168

Both rays—from the polariser and from the objective—are brought to a focus at the same point in the focal plane F. If, now, A is rotated, the the images of the natural and the artificial star will be seen in rapid succession: the former when one of the mirror projections intercepts the cone from the objective, and the latter one-eighth of a revolution later, when mirror B is revealed. If the two images occupy the same position in the focal plane, they will be seen as a single stationary image, which, however, will flicker—i.e. fluctuate rapidly in brightness—so long as the images are of unequal brightness. By adjusting the orientation of the analyser the flicker can be reduced and finally eliminated. At this point, to which the eye is extremely sensitive, the artificial star has been brought to equal brightness with the natural star.

The outstanding advantage offered by this type of photometer is that owing to the rapid alternation of the images, their colours, if these are different, will blend, and the task of comparing the brightness of different-coloured images, which is so productive of systematic error, is very largely obviated.

The chief practical difficulty in the construction of a flicker photometer is likely to be the elimination of vibration set up by the motor driving the rotating mirror. This, however, should not be insuperable, and the photometer would appear to offer material advantages over the Zöllner. (B. 260).

21.3.9 Holophane lumeter: An extremely simple device for comparing the relative intensities of extended objects without even the assistance

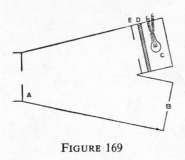

FIGURE 169

of a telescope. A photometer of this type finds an obvious application in the observation of the Zodiacal Light. It consists (Figure 169) of a bifurcated tube, one end of which is occupied by the non-adjustable diaphragm A. The eye position is at B, just behind a circular aperture which must be larger than the pupillary aperture. The inner surface of A is illuminated by a standard source, C, via a ground-glass screen, D, and a cat's-eye diaphragm, E. The interior of the lumeter is given a matt black finish except for the inner side of A and the space round the bulb behind D which are painted matt white.

Its operation is equally simple. If the intensity of an aurora, for example, is to be compared with that of the night sky, the instrument is directed so that the eye at B can see the aurora through the central aperture in A. E is then adjusted, so varying the illumination of A, until the intensity of A is equal to that of the aurora. At this point, providing the two are of the same colour (whence filters will have to be experimented with), the hole in A will disappear—i.e. the boundary between it and the diaphragm will no longer be distinguishable. The process is then repeated with the instrument directed at another part of the sky, and a second reading taken. The difference in the two screen intensities being proportional to the square of the diagonal of the cat's-eye in the two cases, the relative intensities of the aurora and of the sky can be calculated, either in terms of b/b' or of Δm.

21.4 Photographic photometry

Not well adapted to amateur needs owing to the complexity of the technique, the auxiliary apparatus required, and its comparative slowness.

The brightness of a stellar image as recorded photographically is a function of D^2,* plate speed, and exposure. Various methods of determining it are available:

(*a*) Diameter of the image: for a given plate and exposure, image diameter is a function of magnitude and can be calibrated by means of the NPS. Alternatively a test plate can be prepared, on which a series of exposures of a star is made. The ratio of the exposure times is geometrical, and the telescope is advanced slightly in RA between each. The resultant images now exemplify 1-mag differences very nearly; when the test plate and the plate containing the image of unknown magnitude are superimposed, and the images examined in close juxtaposition with a magnifying glass, the brightness of the unknown can be estimated to 0·1 of the difference between adjacent scale images.

(*b*) The density of the extrafocal images being a function of their brightness, this can be calibrated in terms of magnitudes and measured with a microphotometer, which makes an automatic record, via a photocell and galvanometer, of the image density. The method is applicable only to brighter stars.

(*c*) A coarse grating placed in front of the objective spreads the light from each star into a series of secondary images arranged on either side of a central primary image. The number and intensity of these secondary images can be calibrated in terms of stellar magnitude.

The most troublesome sources of error in photographic photometry are emulsion grain, local variations in the sensitivity of the emulsion, and non-uniform development.

21.5 Energy-conversion photometers

These rely on the conversion of the radiant energy of the star into a more accurately measurable form, namely electricity. Photoelectric cells fall into three main categories:

(*a*) photo-emissive, or Elster Geitel, cells—either vacuum or gas-filled;
(*b*) photo-voltaic, barrier layer, rectifier, or Sperrschicht cells;
(*c*) photo-conductive cells.

* Stellar image brightness in theory increases proportionately to D^2. In practice, the ratio is somewhat lower, since the image is a disc and not a mathematical point. Similarly, reducing F for a given value of D should theoretically have no effect upon stellar image brightness; in practice it has.

Photoelectric photometers generally are distinguished by their great sensitivity and by the fact that their systematic and random errors are smaller than those of any other type, an accuracy of the order of $\pm 0 \cdot 001$ mag being attainable. Their disadvantages are restricted applicability, cost, the complexity of the technique of their operation, and the desirability of large apertures. The threshold of Stebbins's calcium hydride cell, for example, was mag 9 with the Yerkes 40-in, and mag 7 with a 15-in. However, this restriction has been to a large extent relaxed as a result of the development of the multiplier type of tube, which with a 12-in telescope should reach to about the 10th magnitude.

21.5.1 Photo-emissive cells: These rely upon the characteristic of the alkali metals that when exposed to light they emit electrons from their surface, the rate of this emission being exactly proportional to the number of quanta in the incident radiation. The cell in its most primitive form consists of a photo-sensitive cathode mounted at the back of a glass tube whose inner surface, except for a small window opposite the cathode, is silvered. A ring anode is located centrally in the tube. The electrons emitted by the cathode when it is illuminated are collected at the anode, which is maintained at a constant potential by a H.T. battery. The current carried by the electron stream is directly proportional to the number of electrons, hence to the intensity of the radiation, and is read from a galvanometer included in the circuit; the latter may have to be at a considerable distance from the telescope, under stable physical conditions, thus involving two observers. For a given illumination the current is proportional to the applied potential up to about 25 volts, if the cell is evacuated; further increase is unaccompanied by an increase in the photoelectric current. But if the tube is filled with an inert gas (usually argon), gaseous amplification through collision ionisation will occur, and the current can be increased by raising the potential right to the point at which a glow discharge occurs. Gas-filled tubes are therefore preferred for most purposes. The external circuit should contain a resistance of about 50,000 ohms as a protective measure in the event of a glow discharge occurring. The potential can, however, be safely maintained at about 5 volts below the discharge level.

If the photoelectric current is below the threshold of the galvanometer a more sensitive meter may be used (such as a Lindemann electrometer), or the current may be amplified to a level at which it can be registered by the galvanometer; electronic amplifiers are thoroughly dealt with in B. 67. Nowadays the most usual practice is to employ the type of cell known variously as an electron-multiplier, multiplier phototube, or photomultiplier tube.* This is a vacuum tube which contains, in addition to the cathode and anode, a number (usually about ten) of dynodes. The electron stream emitted by the cathode is directed by electro-static force to the first dynode, where it produces more electrons; thence to the remaining dynodes in turn, being amplified by secondary emission at each stage. Finally it is collected by the anode.

The tube's dark current is of little importance when the illumination is strong, but in any application of the photocell to very low intensities it becomes a factor of the first importance, since it determines the threshold to which the tube may be used. Dark current arises mainly in three ways: (a) ohmic leakage across incomplete insulation within the cell or between it and the base; (b) thermionic emission arising from the fact that the cathode emits some electrons by virtue of its own temperature; (c) regeneration ionisation, which occurs in a multiplier tube at about 110 volts. Of these, (b) is in practice usually the limiting one and can be reduced by refrigerating the cell with melting ice, or virtually eliminated with dry ice (solid carbon dioxide, $-125°C$). With dry ice refrigeration a gain of as much as 5 mags may be achieved.

Another factor affecting the performance of the cell is photoelectric fatigue. This affects all types of cell to some extent, introducing an element of uncertainty regarding their colour sensitivity as well as their absolute response. Checking with a standard source from time to time is therefore an essential precaution.

The response of the cell being independent of both the shape and the size of the image falling upon it (provided only that it is completely contained by the cathode), it is convenient to let a slightly unfocused image of the star fall on the cell window, a diaphragm in the focal plane restricting the linear field to a few millimetres. Alternately, a Fabry lens may be used to image the objective itself. In any case the telescope, preferably a reflector, should not have a smaller focal ratio than $f/5$ or so, or difficulty will be experienced in keeping the widely diverging cone on the small photo-sensitive surface. A totally reflecting prism sliding in a direction perpendicular to the optical axis, just behind the diaphragm, can be used to check the accuracy of the finder, and ensure that the star's image is central. In addition, the focal plane of the finder should contain a glass diaphragm engraved with a circle whose diameter is equivalent to that of the photocell diaphragm.

Measures can be made on the unknown star and either an artificial or a natural comparison star. If the former, it is best to arrange for the

* The 1P21, manufactured by the Radio Corporation of America, is well adapted to photoelectric photometry. See B. 257–9, 267. B. 250 has a useful bibliography.

rapidly alternating exposure of the cell to the two sources, the intensity of the brighter being reduced by any convenient calibrated accessory until the flicker is zero, in this way eliminating the linearity or otherwise of the cell's response from the results. A reliable routine procedure when a natural comparison star (which should be of similar spectral type to the star being measured) is used is as follows:

(a) Measure the sky brightness by taking alternate readings with the telescope pointing at a starless field near the star to be measured, with and without the dark slide drawn. Note the time, and repeat hourly, or more frequently if the circumstances warrant it.

(b) Take a set of readings of the comparison star with the dark slide alternately open and closed. Note the time of the mid-point of the set.

(c) Repeat with the unknown star.

(d) Repeat (b) and (c) twice, and end with a final set of measures of the comparison star.

By plotting the sky brightness readings against time, the correct value to subtract from the star readings, at the time they were made, can be determined. The mean of each set of readings of the unknown star is compared with the mean of the preceding and following sets of the comparison star. Finally b/b' is converted to Δm.

The sensitivity ranges and other characteristics of some photo-emissive materials which are commercially employed are summarised below:

Caesium oxide (CsO). Has one of the widest ranges: 3400–10,500Å, maximum at about 7700Å, secondary maximum at 3500Å; high infra-red sensitivity. Unfortunately has a high dark current (average 10^{-12} amps), though this, being mostly due to thermionic emission, can be reduced by refrigeration. The caesium cell is not the most suitable for photometric purposes owing to its dark current and because the infrared must be eliminated by filters.

Potassium oxide (KO). Maximum sensitivity near 3500Å, falling off sharply on the short-wave side to zero at about 3000Å, and less sharply on the long-wave side to zero at about 7500Å or 8000Å. It thus covers the visible spectrum and part of the ultraviolet. One of the most satisfactory cells as regards linearity of response.

Potassium hydride (KH). Narrow sensitivity range: maximum at about 4300Å, falling off to zero at 6000Å, with an equally steep drop on the short-wave side. Notable for its extremely small dark current. Stebbins' Yerkes photometer (B. 268) was of this type.

Sodium. Individual cells usually have a range of about 2000Å between 2000Å and 5500Å. Only a slight response to visible light, therefore.

Cadmium. Increasingly sensitive down from 3200Å, limited only by absorption of the quartz cell window, i.e. sensitive to the ultraviolet only.

Thorium. 3600–2500Å.

Titanium. 3200–2500Å.

Tungsten. 2700–1700Å.

21.5.2 Photo-voltaic cells: These exploit the fact that the incidence of light causes the transfer of electrons across the rectifying boundary of two dissimilar materials such as copper and copper oxide. Photo-voltaic cells thus generate their own EMF, and no battery is required in the external circuit. The sensitivity ranges vary rather widely with different photo-voltaic combinations.

Advantages: (*i*) simplicity; (*ii*) obviation of the necessity for an external source of potential; (*iii*) greater current sensitivity than the photo-emissive cell.

Disadvantages: (*i*) amplification of the output is not practicable, and it must therefore be read with a galvanometer; by means of an electrometer, or amplifier and galvanometer, much smaller currents can be detected with the photo-emissive cell, which will therefore be preferred when low intensities are to be measured; (*ii*) subject to fatigue; (*iii*) response is never strictly linear.

Hence the photo-voltaic cell is the more convenient when there is plenty of light and the utmost precision is not demanded; but for most astronomical requirements the photo-emissive cell is to be preferred.

21.5.3 Photo-conductive cells: These exploit the fact that the conductivity of certain materials varies with the intensity of the radiation to which they are exposed. One such substance is lead sulphide (PbS), with a sensitivity maximum in the region of 20,000Å. B. 254 describes in some detail the construction and performance of a sulphide cell which in the infrared showed itself to be of much superior sensitivity to the most sensitive thermopiles (section 21.6). Another photo-conductive material is selenium, whose sensitivity is superior to that of the eye (both as regards contrast and absolute values) over a range of approximately 1800–15,000Å. Its main disadvantage is that its response is not linear, being initially a function of the intensity of the radiation but finally of the square root of the intensity.

21.6 Thermocouples and thermopiles

These—together with radiometers, bolometers, and other instruments of little interest to amateurs—are designed to detect and measure minute intensities of long-wave (heat) radiation.

They exploit a rather similar phenomenon to that made use of in the photo-voltaic cell: if two wires of dissimilar material are joined at both ends, a current (recorded by a galvanometer included in the circuit) will flow if there is a temperature gradient across the junctions; this current is proportional to the temperature difference, for a given junction. Commonly used materials for couples are iron and copper, iron and constantan, platinum and a platinum-rhodium alloy, antimony and bismuth, bismuth and a bismuth-tin alloy, copper and alumel (a nickel alloy containing a few per cent of manganese and aluminium and traces of silicon and iron), alumel and chromel (a nickel-chromium alloy). The thermojunction is covered by a small disc of blackened metal foil on which the stellar image is projected. Evacuating the tube containing the couple increases its sensitivity: it is doubled, for example, by reducing the pressure in the tube from 10^{-3} to 10^{-6} mm.

A thermopile consists of a number of thermocouples in series: the current produced by a given intensity of radiation is then that produced by a single thermocouple multiplied by the number of couples in the pile. Alternate junctions are maintained at a constant temperature, the remainder being exposed to the radiation which it is desired to measure.

Thermocouples are less sensitive than the eye within the visual range, but in the longer wavelengths they are incomparably more sensitive than thermometers, a good thermopile and a sensitive galvanometer being able to measure temperature differences of the order of 10^{-5} °C.

B. 575, 261–3 describe thermocouples in use at Mt Wilson and elsewhere. The technique of making couples is described in, *inter alia*, B. 271, 67.

21.7 Microradiometer

A development of the simple thermocouple. Suspended by a quartz fibre between the poles of a magnet in an evacuated cell is a small coil. Below and connected to it is the thermocouple. When radiation falls on the couple a current passes through the coil, which turns in the magnetic field. The amount of the rotation is determined by the strength of of the current generated by the couple, and therefore by the intensity of the radiation. The angular displacement of the coil can be conveniently measured by the deflection of a beam of light directed at a small mirror attached to it.

Microradiometers are relatively awkard to use, since their orientation cannot be disturbed or varied, and this to some extent offsets their great sensitivity, which is superior to that of the vacuum thermocouple.

The application of a home-made microradiometer to the amateur study of the lunar temperature during eclipse is described in B. 273–275.

21.8 Bolometer

Designed for use at the prime focus of a large instrument. As the temperature of a conductor is increased, so its resistance increases. The bolometer usually consists of a thin, blackened metal-foil strip, mounted in an evacuated cell. This is exposed to the star's image in the focal plane of the objective. It is connected as one arm of a Wheatstone bridge, and variations of the resistance of the strip are measured by a sensitive bridge galvanometer. Its sensitivity is comparable with that of thermopiles.

ACCESSORY INSTRUMENTS AND EQUIPMENT

22.1 Binoculars

A good pair of prismatic binoculars is an invaluable accessory. Not only does it provide some of the best views of the Milky Way that are to be had with any instrument, but it is frequently of use as a preliminary to the finder of the telescope, as well as in the observation of variables and comets.

The advantages of binocular over monocular vision (i.e. of both eyes using identical separate optical systems) are:

(*a*) faint objects are slightly more easily seen;
(*b*) fine detail is more quickly, if not better, seen ;
(*c*) it is less tiring;
(*d*) it provides—though not in the field of astronomical work—an increased parallactic effect.

The advantages of the binocular over the naked eye are:

(*a*) increased light grasp;
(*b*) increased resolution;
(*c*) increased magnification.

Its advantages compared with the telescope are:

(*a*) wider field;
(*b*) increased contrast.

Prismatic binoculars—in which the rays are turned twice through 180° by Porro prisms—are, with few exceptions, the only type suitable for astronomical work; so-called opera glasses are usually Galilean systems (see section 9.2).

Provision should be made for the independent focusing of each eyepiece; if, in addition, the two tubes can be adjusted together, so much the better. The connecting pieces between the tubes should also be hinged, so that the interpupillary distance can be exactly matched by the separation of the two axes. The two systems must provide identical magnifications, and their axes must be parallel. The tolerance of inaccuracy in the latter respect decreases as magnification increases, since an angle θ between the axes separates the images by an angle $\theta' = \theta(M+1)$. If the direction of separation of the images is parallel to that joining the eyes, the images are more easily superimposed than when the separation occurs at right angles to the line of eyes. The limits of θ beyond which the images can only be superimposed by conscious effort are respectively 1° and 10′, approximately. A useful pair of binoculars will secure about $1\frac{1}{2}$ to 2 mags beyond the naked-eye threshold.

Binoculars are usually described in terms of their magnification and aperture measured in mm, e.g. 7×50 ($M=7$, $D=5$ cms). Many varieties of useful instrument are always to be had on the secondhand market. Particularly worthy of mention are the Ross E.W.F. series, giving fully illuminated fields of diameters up to 66° (9°.4 with $\times 7$, 5°.5 with $\times 12$), and the Dollond 6×35 and Ross 6×43, which give fields of 8° or 9°. The latter—non-prismatic, inverting instruments with Kellner eyepieces—were made for naval use during the 1914–18 war, and with their relatively high transmission (light loss at prisms being avoided) give admirable low-power views of the sky. The Busch Olympic Spectacle Binocular ($D=1\cdot1$ in, $M=2\frac{1}{4}$, field about 20°, weight about 3 oz) is perhaps not much more than an amusing toy, though it provides very fine views of the Galaxy and might be of real value for such work as comet or nova hunting. The Zeiss Starmobi, on the other hand, is a very fine instrument: it is a binocular telescope, aperture 60 mm, with three magnifications ($\times 12$ giving a 4° field, $\times 24$ giving 2°, $\times 42$ giving 1°) provided by eyepieces mounted in revolving adaptors.

See also section 8.7.

22.2 Hand telescopes

The uses of a lightweight hand telescope of the coastguard variety are much the same as those of binoculars: for large and diffuse comets, 'sight-seeing', sweeping for novae, etc, and as a subsidiary finder. In addition it avoids the discontinuity between binocular and monocular vision in the observation of variables having very bright maxima.

Telescopes of this type typically have apertures of 2 to 3 ins, magnifications from about $\times 5$ to $\times 8$, and fields from about 5° to 8°. A good model by Cooke, Troughton, and Simms has $D=2\frac{3}{8}$, $F=9\frac{1}{2}$, $M=8$, field $6\frac{1}{2}$°. The Zeiss prismatic monocular ($D=6$ cms, $M=15$) is in effect one half of a powerful prismatic binocular.

An extremely useful army surplus instrument that has been available since the war is the predictor telescope. This is a remarkably compact instrument (overall length, 11 ins) which works at the phenomenally small focal ratio of $f/3$. The OG has an aperture of slightly less than 2 ins; the image is erected, and the convergent pencil turned through 90°, by means of a roof prism, to which is cemented the field lens of the positive achromatic ocular. It gives an 8° field with a magnification of about ×7.

22.3 Finders

There is little to say about finders except that they are usually much too small and are often mounted inconveniently far from the telescopic exit pupil. The finder of a Newtonian is apt to assume awkward positions, and it is a great convenience if more than one is fitted. As the image-reversal occasioned by a diagonal is always objectionable, the finder must be mounted well away from the telescope tube.

Low magnification, adequate light grasp and width of field are the characteristics to aim for. The method of mounting is such that the position of the finder's axis relative to that of the telescope can be adjusted, so that the two may be made parallel. The eyepiece usually contains some arrangement of cross-wires to indicate the centre of the field, though this can hardly be regarded as necessary.

22.4 Comet seekers and richest-field telescopes

The aim in designing a telescope for comet-seeking should be, primarily, width of field and light grasp. The relatively small focal ratios of comet seekers have, contrary to common belief, nothing to do with the fact that an instrument of small focal ratio is of greater photographic 'speed' than one of large. For further discussion of the qualifications of comet seekers see section 3.10.

The principles underlying the design of a richest-field telescope—i.e. one the dimensions of whose objective and ocular are adjusted so that it will show a maximum number of stars in any given field—are fully discussed by Walkden in B. 102.

22.5 Monochromator

The modern solar observer has at his disposal—if he is rich or an expert optician—an alternative to the spectrohelioscope for observing the limb in $H\alpha$ without suffering the restrictions of the open-slit method. The instrument in question is the narrow waveband filter known variously as the monochromator, the interference polarisation monochromator, the Lyot filter, the birefringent filter, monochromatic polarising filter, etc. It can be home-made (see especially B. 276), though the difficulties are considerable: in particular, the quartz raw material is expensive, and the optical work still more so; if, on the other hand, the amateur is to undertake the latter himself he must be a practical optician of the first order. The definition of the monochromator is superior to that of the spectrohelioscope, though the latter is in some respects the more versatile instrument (e.g. the ease with which radial velocities can be measured), and on the whole the spectrohelioscope is probably the more practical proposition for the amateur.

The principle of the filter, with certain differences of detail, was announced independently by Lyot (B. 278) in 1933 and Öhman (B. 280) in 1938. Since then, improvements and developments have been made, notably by Pettit, Evans and Billings (e.g. B. 281, 277). The filter, in one of its forms, consists of a series of polaroid screens separated by plates cut from a birefringent or doubly refracting crystal such as quartz (see section 21.3.4). Each quartz plate is double the thickness of its predecessor, and the wavelength of the final transmission is determined by the thickness of the first plate. The action of the first screen and plate is to produce two rays (the ordinary and the extraordinary) whose phase difference is, for a given thickness of quartz, a function of λ. The second polaroid screen eliminates those wavelengths whose phase difference is 180°. The spectrum at this stage in the filter is therefore continuous, with a series of broad absorptions whose separation is a function of the thickness of the first quartz plate. Each subsequent polaroid-quartz-polaroid combination introduces absorptions whose separation is half that of the absorptions in the preceding spectrum. Ultimately a spectrum can be produced which consists of a series of transmissions each from 1Å to 5Å wide, separated by absorptions several hundred Å wide. An ordinary red filter of glass or gelatin then eliminates all these maxima except that at the long-wave end of the series, whose wavelength is that of $H\alpha$. The total filter thickness required to transmit a range of about 5Å centred on the $H\alpha$ line is about 4·5 ins; Pettit's monochromator contained seven quartz units.

A complication is introduced by the fact that the difference between the refractive indices for the ordinary and the extraordinary rays in quartz is a function of temperature, the wavelength of the selected transmission shortening by about 0·8Å for every rise of 1° C. This necessitates the incorporation of a thermostat in the instrument.

In an ingenious development of the monochromator, due to Billings, the quartz units are replaced by artificially grown ammonium dihydrogen phosphate crystals, whose optical properties vary with the electric field in which they find themselves. By varying the field, the filter can in effect be tuned in to transmit any required wavelength.

22.6 Spectrohelioscope

The spectroheliograph was invented independently by Hale and Deslandres in 1890. The visual form of the instrument, the spectrohelioscope, though it was adumbrated by Janssen and Lockyer as long ago as 1868, had to await effective development by Hale during the years 1924–1929.

The essential principle involved in the spectrohelioscope is the employment of the persistence of vision in the observation of a rapidly (*c.* 25 per sec) oscillating monochromatic image of the Sun, thus allowing a considerably wider area than a single slit-width to be viewed.

Structurally it consists of: (*a*) An image-forming unit; usually an OG of about 18 ft *F* (forming a solar image about 2 ins in diameter), which need not be achromatic, since in the final image only a single wavelength is used; it is mounted permanently (apart for allowance for focusing adjustment) in a position where the solar image can be fed into it by a coelostat. (*b*) A high-dispersion spectroscope, either prismatic or diffraction, of about the same focal ratio as the objective; high dispersion is necessitated by the fact that the blaze of integrated photospheric light, when the slit is overlying the disc, makes it essential that all but a single wavelength be excluded; and this, without considerable dispersion, involves impracticably narrow slit openings. (*c*) A mechanism for synthesising the monochromatic images; this may be of several types:

(*i*) oscillating slits, yielding a final image whose orientation is a mirror reversal of the direct telescopic image (i.e. left and right reversed);

(*ii*) rotating prisms, one in front of each slit; the slits now being stationary, it is the solar image and the view-point that oscillate relative to them; the orientation of the image is here the same as that of the telescopic image of the Sun;

(*iii*) rotating glass disc, coated with black lacquer in which radial slits have been cut.

Hale's original model incorporated oscillating slits, which some observers still favour. However, it is nowadays the commoner practice—as, for example, at Cambridge and in the Newbegin instrument, now at Herstmonceux—to employ rotating Anderson prisms. These are $\frac{1}{2}$-in square prisms about 3 ins long, mounted on a single axis in front of the slits, which in the Hale model are in a vertical line.

Oscillating slits nevertheless possess important advantages:

(*i*) The brightness of the image is more easily variable during observation. Of numerous factors controlling image brightness (see B. 288) one of the most important is the field width; with oscillating slits this can be increased or decreased at will merely by increasing or decreasing the amplitude of oscillation, hence decreasing or increasing the brightness of the image. With Anderson prisms, on the other hand, image brightness can only be increased by changing to a smaller pair of prisms, stopping down the used aperture of the prisms having no effect upon image brightness or field width (which latter is equal to 0·66 × the width of the prism face, assuming $\mu=1·5$).

(*ii*) Oscillating slits give a slightly brighter image than prisms (other factors being equal), owing to light loss by partial reflection at the four prism faces.

(*iii*) The definition with oscillating slits is slightly better than that given by rotating prisms, which varies inversely with slit width; however, this does not in practice seem to have an important effect on performance.

The slits may be arranged either side by side or vertically above and below one another. The disadvantage of the latter arrangement is that the observer's head is only a few inches below the first slit, and convection results in deterioration of the definition. Hale's models have vertical slits, since this is the most convenient arrangement with Anderson prisms or the simple rocker bar mechanism he used for imparting unison motion to the slits. Slits mounted side by side require out-of-phase movement (like that of a tuning-fork, both of whose prongs move outward together and inward together, exactly in step and by precisely equal distances), which presents some problems in mechanical design. These difficulties have been overcome in Sellers' oscillating-slit mechanism, however; it is, moreover, ideally adapted to the needs of the amateur building his own instrument (B. 289, 290). The Sellers mechanism is employed in Ellison's Edinburgh spectrohelioscope with complete satisfaction.

The greater part of our knowledge of flares, of chromospheric eruptions, and of the physical basis of magnetic storms and short-wave radio fadeouts, has been provided by the spectrohelioscope. It is, in fact, the only satisfactory method of observing the transitory flares, and an enormous and virtually untouched field of work lies open before the amateur who builds himself a spectrohelioscope. Anyone contemplating building a Hale spectrohelioscope (though Ellison's autocollimating model, described below, is very much simpler) is fortunate in that Hale has himself described the construction and adjustment of the instrument in the greatest detail and the most practical manner, with the support of Russell Porter's incomparably revealing drawings (B. 284—a paper that every solar observer should read; B. 285-7 describe the uses to which the instrument can be put, rather than the instrument itself, though the whole series is of the first importance; B. 290 is also worth referring to). Very briefly, the Hale instrument consists of a 4-in non-achromatic planoconvex OG, of about 18 ft F, fed by a coelostat. The first, or scanning, slit lies at the focus of the OG, and slow-motion controls to the 'fixed' mirror of the coelostat allow the solar image to be adjusted on the slit jaws as required. From the scanning slit the light passes to a 3-in, 13 ft F, concave spherical collimating mirror situated at its focal length behind the slit. The parallel beam is reflected to a 15,000 rulings per inch reflection grating mounted just behind and below the scanning slit; it is tiltable with the necessary slow motion so as to reflect any required region of the spectrum to a second spherical mirror, identical with the first and mounted beside it. This brings the beam to a focus at the second, or viewing, slit, where the image is examined with a LP eyepiece. Just behind the viewing slit is the line-shifter, a plane parallel plate about $\frac{1}{16}$ in thick, which can be rotated about an axis parallel to the slit. By refraction it can displace the centre of the $H\alpha$ line to either side of the slit by an accurately known amount. Conversely, if the radial velocity of the source of the $H\alpha$ radiation has shifted the line away from the slit it can be brought back by the line-shifter, the amount of rotation required to effect this being a function of the radial velocity involved. Thus the wavelength of the light incident on the viewing slit can be varied at will during observation, without any complicated adjustments of the slit position, and the line-of-sight velocities of the prominences and other hydrogen forms can be immediately determined.

With the dimensions quoted, the width of the $H\alpha$ line (6563Å) is 0·008 ins, permitting slit widths up to about 0·004 ins to be used while retaining an effectively monochromatic transmission.

Ellison's autocollimating spectrohelioscope has a number of advantages over the Hale model and is probably the most suitable for amateur construction; it is described in B. 283, 290. Briefly, the layout is as follows. A 5·5-in achromatic OG of 18 ft focal length is fed by a siderostat and forms a solar image, 2 ins in diameter, on the scanning slit. Behind this lies a 4·5-in achromatic collimating lens of 16 ft focal length, and behind that again a 5-in Rowland plane grating (14,400 lines per inch). This reflects the first-order $H\alpha$ line back through the collimating lens, which brings it to a focus at the viewing slit mounted 5 ins to the side of the scanning slit, where it is examined with a LP ocular. The slits are operated by a Sellers linkage. Apart from the general advantages which it shares with all oscillating-slit over rotating-prism spectrohelioscopes, Ellison's instrument is notable for its reduction of the number of reflections (two, as against Hale's five) and its extreme simplicity.

The grating itself presents one of the main difficulties to the amateur who is building his own spectrohelioscope, since replicas (which are admirable for ordinary spectroscopic work) are rarely large enough or good enough; originals, on the other hand, are hard to come by and are always expensive. New replica techniques are being developed in America and at the National Physical Laboratory, and may eventually solve this problem for the amateur; but at the present time the grating is still his worst headache connected with spectrohelioscope construction.

22.7 Chronograph

An instrument which records on a continuous trace the instant at which an observation is made; either concurrently and automatically or at any required moment manually, and either on the same or a parallel trace, the instants of GMT or GST are recorded. A comparison of the two traces, or of the two sets of signals on the one trace, then allows the GMT or GST of the observation to be derived.

The arrangement of a typical chronograph is as follows. A screw carries an electromagnet slowly along the length of a drum which is rotated at a standard rate by clockwork or a synchronous motor; the uniformity of motion of the electromagnet is of little importance, but upon the unvarying speed of rotation of the drum depends the accuracy of the chronograph. A sheet of paper is wrapped round the drum and a pen is attached to the magnet in such a way that when the drum is driven the pen makes a continuous spiral trace. A make and break at the observatory clock is incorporated in the magnet's circuit, so that once per second the circuit is closed by the pendulum, the pen giving a kick which is recorded as a crotchet in the trace. The displacement of the magnet is adjusted so that, for the rate at which the drum turns, the crotchets in adjacent traces do not overlap. A signal key held in the observer's hand is also taken into the circuit; this he presses at the precise moment of transit, occultation, or whatever the observation may be. From the relative positions of the crotchets made by the observer and those made by the clock, the times of the observations can be determined with an accuracy which is, within limits, proportional to the speed of rotation of the drum. When the chronograph is started, the ST of a zero crotchet is determined from the clock, and all measurements made from this.

B. 292–3 describe a chronograph of such simple design that it can easily be made at home. A gram motor—in which the spring mechanism has been replaced by a weight drive—is used to rotate a wooden, rubber-covered roller; a second similar roller, with a hinged mounting, is kept in firm contact with the first by means of a small spring. A roll of paper strip is mounted on a freely rotating axle and the end of the paper led between the rollers. The action is to draw the strip steadily between the rollers when the motor is running. A fountain pen is mounted in a clip soldered to the clapper of a small electric-bell mechanism whose winding has been altered to produce a single beat instead of a continuous make and break. This is mounted so that the nib rests firmly on the paper strip where it passes over the idling roller. It is a good plan to mount the whole pen-and-armature unit on a pivot and provide it with an adjustable counterpoise, so that the pressure of the nib can be varied. When the rollers are turning, the nib will make a continuous linear trace on the paper strip. When the observer at the telescope closes the circuit by pressing the bell-push at the instant of, say, an occultation, the pen will give a kick and the trace will either be broken or distorted by a crotchet. It now remains to provide for the calibration of the trace. This can be done, speaking generally, by superimposing known signals on either the same trace or on a parallel trace, formed by a second pen mounted alongside the first. In the latter case there must be provision for producing a simultaneous crotchet in the two traces as a means of establishing temporally equivalent points. The trace can be calibrated by a variety of methods:

(a) Seconds impulses can be taken by brushing contact from a projection soldered to one of the gram spindles which has been timed and set to 1 r.p.s. by means of the gram governor.

(b) A separate synchronous motor can be used to give seconds impulses off the 1-r.p.s. spindle by brushing contact, if the required rate at which the paper is to be drawn over the roller makes it inconvenient to have any part of the motor revolving at precisely 1 r.p.s. In this case, as in (a), the seconds trace must be compared with the observatory clock.

(c) Impulses at the rate of one per swing can be taken from the pendulum of the observatory clock (or any pendulum beating seconds) by means of a mercury-trough contact.

(d) Calibration marks can be made manually by the press-button at instants determined visually from the observatory clock. It is wise to make two such zero marks—one before and one after the observations —and to measure from both, taking the mean.

(e) The seconds trace may be checked direct with radio time signals by linking a separate pen to a relay fed by a wireless speaker. This is also a convenient way of determining the clock error and clock rate, quite apart from the timing of observations. B. 291, 295 describe circuits of this sort.

Another type of homemade chronograph, described by Waterfield (B. 294), is probably rather more awkward to build and mount, and moreover can only be used with a clockdriven equatorial.

The precision with which times can be read from the chronograph trace depends on the speed at which the tape passes the pen. Measurements to $\pm 0^s.1$ are perfectly feasible.

22.8 Blink microscope

The blink microscope is a device for viewing pairs of photographs (of the same star field, taken at different times) in such a way that the attention is quickly attracted to any difference there may be between them. Such differences are, in particular: an image whose position on the two plates is different, relative to the general star field; an image whose brightness on the two plates is different; an image which occurs on one plate only.

The principle of the blink microscope is the alternating presentation of the two plates to the eye, the position of each within the visual field being identical. The vast majority of stellar images will then appear to be visible continuously and without change, since their positions and brightnesses on the two plates are identical (assuming equal exposure). But if an image occurs on plate A and not on plate B it will appear to 'blink', i.e. successively disappear and reappear as the field of vision switches back and forth between A and B. The image of an object which has moved during the interval between the exposure of the two plates will (assuming that both its images are contained in the same field of the microscope) appear to jump to and fro between its two positions, being seen alternately in each. Both these effects—blinking and jumping —are extremely eye-catching. The appearance under the blink microscope of an image which differs on the two plates only as regards its intensity is less obviously changing, and needs more careful scrutiny: its intensity, and to some extent its size, will simply appear to vary in successive jerks between the two limits.

The labour and time spent on any work involving the comparison of stellar photographs are cut to a minimum by the use of the blink microscope. For anyone undertaking the routine photographic search for novae it may be considered indispensable. In cases where plates have only occasionally to be compared, its function can be taken over by the superimposition of a glass positive and negative of each plate—a positive of A on a negative of B, and a negative of A on a positive of B. When viewed against a bright diffused light, any image which occurs on one plate only will immediately leap to the eye. The disadvantage of this method when a large number of comparisons have to be made is the additional darkroom work that it entails. The blink microscope is a great convenience in the location of minor planets, and may also be of limited usefulness to the variable-star observer. As an aid to cometary discovery it is only of value on the comparatively infrequent occasions when the cometary image is indistinguishable from that of a star. The photographic images of most comets, even at the time of discovery, are clearly non-stellar in appearance, and can be detected as easily, and more quickly, by the direct examination of the plate with a magnifying glass as by blinking it with a paired plate taken on an earlier occasion.

Summarily, then, the blink microscope, though a convenience in several fields of work, is only an essential part of the equipment of an observer undertaking regular search surveys—whether for novae, asteroids or variables. Such work is most economically performed by fast wide-angle cameras of the Schmidt and related families, whose plates are so crowded with images that any other method of comparison would be impracticable.

As regards design, blink microscopes differ among themselves chiefly in the manner whereby the two plates are alternately presented to the field of the viewing microscope or short-focus telescope. However this is arranged, there must be provision for (a) adjusting the positions of the two plates relatively to one another, so that corresponding images can be made to occupy precisely the same position in the field of vision; (b) moving the two plates as a single unit in two mutually perpendicular directions, so that their whole expanse can be scanned in a succession of overlapping strips.

The microscope designed by Will Hay and W. H. Steavenson, and constructed by Will Hay, now in the possession of the B.A.A., is probably of about as simple a design as can be achieved. The base is a wooden box in whose upper surface are two square windows covered by ground glass. Beneath each of these is an electric bulb. The bulbs are connected to the source of supply through a cam-operated switch, driven by a small electric motor which is also housed in the box. By this means the lamps are alternately switched on and off at any required speed; about one change-over per second from left to right, and right to left, is a comfortable running speed.

The two plates are mounted horizontally above the windows in a frame which can be moved both left-and-right and to-and-from the operator, allowing the whole extent of the plates to be scanned. The precise setting of the plates in this frame is adjustable as follows: the left-hand plate can be moved by a fine screw adjustment in the same two mutually perpendicular directions as the frame itself; the right-hand plate can be rotated in the horizontal plane about one corner through a small angle.

Mounted centrally above each window, and above the horizontal plane containing the plates, is a prism of the type illustrated in Figure 170, which reflects the vertical ray from the lamp into the horizontal

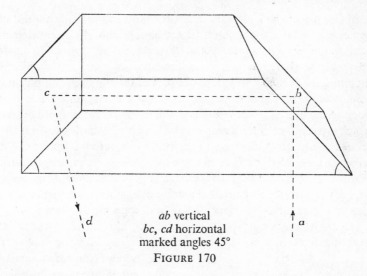

ab vertical
bc, cd horizontal
marked angles 45°
FIGURE 170

plane and then turns it through a right angle towards the central line of the apparatus. Here the rays from the two prisms encounter a half-silvered prism which reflects them into the OG of the view telescope, mounted horizontally in the median line of the instrument. The telescope and the three prisms are carried on a bridge, mounted on the box base, which spans the moving frame carrying the plates.

The initial adjustment of the plates to give image coincidence over their whole extent is carried out as follows: by means of the frame-moving screws the corner of the plates about which the right-hand plate rotates is brought into the field of vision, and the image of a star on the left-hand plate is brought into coincidence with the corresponding image on the right-hand plate by means of the former's rectangular adjustments. The frame is then slowly racked along so that one edge of the plates passes across the field. As this happens it will probably be found that corresponding images on the two plates diverge progressively from one another. This divergence is eliminated by means of the right-hand plate adjustment which, it will be remembered, rotates the plate about the original point of coincidence. When correspondence of the images has been achieved along the whole length of one edge of the plates it is automatically achieved over their whole extent. These initial adjustments are most easily made without the motor running, the cam being turned by hand to illuminate the right or left plate as required.

22.9 Ocular adaptors

In any work which involves frequent changes of magnification it becomes very tedious and time-wasting to be continuously unscrewing one ocular, screwing in another, and then refocusing. Revolving ocular adaptors, working on the same principle as microscope objective mountings, obviate all this; they are part of the standard equipment of some Continental and American telescope makers, and are a great boon to the user. They are usually made to hold three or more oculars, which are set in such a manner that little or no refocusing is necessary: all one has to do to change an eyepiece is to rotate the adaptor. With telescopes mounted in the open some trouble may be experienced from dewing up, but this would rarely occur under a dome. Lindemann (B. 135) describes a somewhat elaborate design incorporating electrical elements to prevent the condensation of dew on the eye lenses.

Failing a rotating adaptor, in which the alternative eyepieces are ready-mounted, a great deal of time can be saved by scrapping the screw fittings of the oculars, mounting them instead in sections of brass tube which fit into the drawtube. Refocusing can be avoided by fitting flanges on the eyepiece extensions in such a position as to give the correct focus, for a given setting of the drawtube adjustment, when they are pressed home against the end of the drawtube (Figure 171).

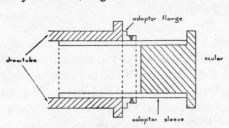

FIGURE 171

22.10 Verniers

A vernier is an index for the reading of a moving scale which does not consist of single pointer but of a zero pointer and a set of secondary graduations whose size is such that n graduations of the vernier are equal to $n-1$ graduations of the scale.

If $\theta=$ the required angle between zero on the scale and zero on the vernier,

$s=$ the angular value of 1 division of the scale,

$v=$ the angular value of 1 division of the vernier,

then
$$(n-1)s=nv$$
and
$$s-v=\frac{s}{n}$$

If, then, the vernier reads $\theta'+$, and the xth vernier graduation coincides with a scale graduation,
$$\theta=\theta'+[x(s-v)]$$

Suppose, for example, that the scale is graduated in degrees, 5 divisions of which are equal to 6 of the vernier. Then
$$s=1°$$
$$s-v=\tfrac{1}{6}°=10'$$

If, for example, the vernier zero lies between 30° and 31°, and the 4th vernier graduation from zero coincides with a scale graduation, then the reading of the scale is given by
$$\theta=30°+(4\times 10')$$
$$=30°\ 40'$$

22.11 Observing seats

A comfortable observing position is of importance only because its lack increases the difficulty and therefore the inaccuracy of the observations; and for this reason alone it is extremely important.

The requirements of any observing seat, whatever its type, are—apart from strength, lightness, and mobility—the ability to bring the eye comfortably to any position of the ocular, and to provide for small adjustments of the height of the seat being made without frequent interruptions to observation. The type and dimensions of a suitable seat are determined to a large extent by the focal length of the telescope, its type (refractor, Newtonian, Cassegrain, etc), and its mounting, and probably no one seat will suit all positions of the telescope.

With a small altazimuth refractor on a tripod mounting the most satisfactory arrangement is to use a low observing chair of fixed height

413

(such as a 'fireside' armchair) and to vary the height of the head from the ground by opening or closing the tripod legs; three metal or concrete strips in the hut floor or the ground, arranged radially at 120° from each other, and carrying sockets at 2-in intervals into which the spikes of the tripod can fit, allow this to be accomplished quickly and easily.

For an altazimuth mounted on a column, or an equatorial, some variety of observing-seat heights can be obtained by means of unequal-sided boxes—e.g. two boxes measuring $6 \times 9 \times 12$ ins and $16 \times 20 \times 24$ ins respectively. But though these may be a first approximation to a solution of the problem, they succeed in filling neither of the specifications of the ideal seat.

An improvement on them is a stepladder whose steps can be hinged upward. The disadvantage here is that either the steps must be uncomfortably narrow, or the vertical distance between steps must be undesirably great. The latter can be reduced by using a box, of height one-half the inter-step distance, as an auxiliary seat to be laid on any required step; but the arrangement is at best a makeshift. A similar arrangement, which has been strongly recommended, consists of a movable seat which can be hooked on to any desired tread of a pair of ordinary household steps.

Complete smoothness of height variation as well as ease of adjustment without moving off the ladder are provided by the Sellers observing chair (B. 297). A fixed step at the bottom of the ladder serves as a footrest during observation and a platform to stand on when adjusting the seat. The seat itself is constructed as in Figure 172, and the weight of the observer's body keeps it firmly jammed on the two uprights of the ladder. A tendency for the seat to slip if the weight is put too far back on it (also if, when the observer is standing and it is taking no load, the back of his coat happens to jog it) is overcome by notching the uprights at intervals of 2 ins or so (Figure 173); this also does away with the wood-screw grips, while retaining a fine enough height adjustment for practical purposes. In one development of this type of ladder (B. 296) spring plungers are used to lock the seat to the uprights.

In all ladder-type observing seats, the overall dimensions are dictated by the maximum and minimum heights that the ocular may assume. An angle of slope of from 60° to 70° will be found convenient; the ladder must not be of folding construction,* like a domestic stepladder, but solid; its back should be splayed to the base angles of a triangular base whose corners carry castors or domes of silence.

* Or must at least be fixed in the open position by means of a strut which should be attached not less than halfway up.

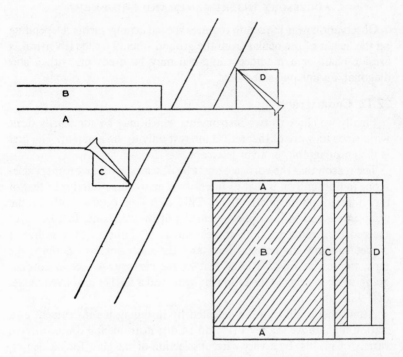

FIGURE 172

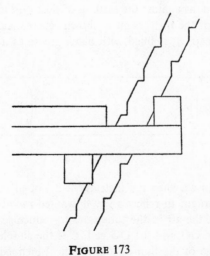

FIGURE 173

Observation near the zenith requires special arrangements. Depending on the height of the ocular from the ground when the tube is vertical, a fireside chair or a modified camp-bed may be used; or, with a star diagonal, an unequal-sided box.

22.12 Coronagraph

Finally we come to two instruments which may be summarily dealt with, since they are of theoretical interest only to the amateur. The first is the coronagraph, or Lyot telescope.

The reason that the corona cannot be observed with an ordinary telescope in full sunlight is that its intensity is many times inferior to that of the diffused photospheric light. This diffusion occurs both in the atmosphere and in the optical elements of the telescope. In 1930 Lyot succeeded in reducing both these sources of diffusion to a sufficient degree to show the coronal lines and the inner corona—perhaps the most impressive observational feat of the century. A second coronagraph was constructed the following year, and a further improved model of 7·9 ins aperture, in 1934.

Atmospheric diffusion he defeated by installing his instrument at a station high above sea level (Pic du Midi); instrumental diffusion by a variety of means: by a very careful selection of the glass for the lenses, so that it should be completely free from striae and bubbles, by high polishing of all glass surfaces, by using only simple lenses, by a system of diaphragms and occulting discs to eliminate scattered light, and by certain precautions against the settling of dust on the lens surfaces.

The layout of the instrument is shown schematically in Figure 174. The long dustcap, *A*, is lined with black grease as a dust trap, and is

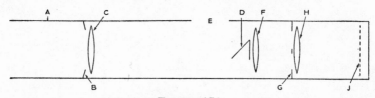

FIGURE 174

always kept closed when the instrument is not in use. *B* is a silvered concave diaphragm to reflect away unwanted radiation (which would otherwise heat the air in the tube) and to reduce scattered light from the edge of the OG and the OG cell. *C* is the simple objective, which forms an image of the Sun on the elliptical inclined silvered disc, *D*, which extends 15″ beyond the image of the photosphere; this unwanted light and heat is reflected from D out of the tube by the window, E. Behind D is the field lens, F, which forms an image of the objective at the diaphragm, G, whose function is to intercept stray light arising from diffraction and reflection at the lenses. Finally comes the camera objective, H, which images the corona at the focal plane, J.

Readers interested in fuller accounts of this fascinating instrument, and of the obstacles that Lyot had to overcome in perfecting it, should refer to B. 298–300.

22.13 Coronaviser

An interesting application of television technique to solar research. In its present still pioneering stage it does not offer definition and resolution comparable with that provided by the coronagraph, monochromator, or spectrohelioscope, but it is clearly capable of being further developed.

The first coronaviser was built by Skellett (B. 301–2) in 1940. It consisted of a T/V camera which scanned a radial path in the Sun's vicinity. The uniform photospheric glare gave rise to a direct current, while the lesser intensity variations due to the prominences and coronal streamers were translated as small alternating currents which, after amplification, activated the T/V image tube. Skellett's original model produced only two images per second, which resulted in too much flicker for visual observation. The image was therefore photographed.

CHOICE OF INSTRUMENT

23.1 Factors affecting choice

Both the reflector and the refractor possess specific advantages and weaknesses, the more important of which are listed in the two following sections. In addition, there are certain fields of work to which one is better adapted than the other, and if the observer is intending to specialise this also will affect his choice. In the case of most amateur installations, however, cost, and possibly the amount of space available, are likely to be the determining factors. The scales thus tip heavily on the side of the reflector: for even in cases where the refractor would be the superior instrument, aperture for aperture, the reflector is the superior instrument cost for cost—so important is the role of aperture in an instrument's performance.

Two words of warning should be given; the first against unfair comparisons: between telescopes of different apertures, between a good refractor and an admittedly inferior mirror, between a refractor and an improperly used reflector (e.g. unsuitable oculars). It is also wise to distinguish between those failings of the reflector that are intrinsic and those which are solely a function of focal ratio, therefore shared alike by a refractor and a reflector if their focal ratio is the same.

Secondly, it has been repeated almost *ad nauseam* in the literature of the past many decades that 'the reflector's definition is inferior to that of the refractor'. Sometimes the statement is unqualified, leaving the reader to suppose that some mysterious attribute of reflection makes it less perfect an image-former than refraction; sometimes the reflector's relative inferiority in this respect is attributed to the diffraction structure of its point image; sometimes, again, to the reflector's greater dependence on the atmospheric condition. But the supposed inferiority of the reflector—unsteadiness and poor definition of the image, which render high magnification impossible—is something that vanishes like an exorcised bogle, provided that the speculum is made of the right material, that steps are taken to eliminate tube currents, that the error of the flat is not greater than that of the parabolic figure of the speculum (and that both are small), and, finally, that suitable oculars, with or without a Barlow lens, are used. (See further, sections 8.8, 11.2, 13.4.)

The modification of the star image by the central obstruction of the reflector has also been built up into something of a nightmare by many writers; but provided its diameter does not exceed about $D/5$ its effect is negligible (see section 2.2).

Whilst it is admittedly true that the performance of many reflectors in amateur hands is markedly inferior to that of refractors with which they should be comparable, it is nevertheless very far from inevitable that this should be so.

23.2 Advantages of the reflector

(*a*) Perfect achromatism. The secondary spectrum of small refractors used for visual work may not indeed be of much importance, but in the photographic field the reflector is supreme—unless a photovisual can be afforded. (The $H\delta$ focus of the Lick refractor lies $81\cdot5$ mm beyond the D_2 focus.)

(*b*) Relative cheapness. Even for work in a field where the refractor is superior, aperture for aperture, one would hesitate to select a 4-in

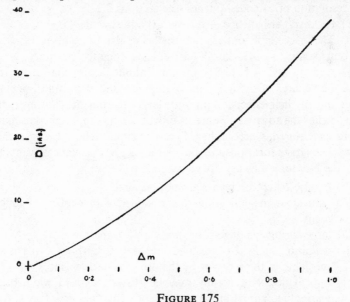

FIGURE 175

Difference between magnitude thresholds of refractors and reflectors

419

refractor in preference to the 8-in or 12-in Newtonian which the same money would buy.

(c) Superior light grasp (with fresh films). This is a negligible advantage with small instruments, but becomes significant when D reaches 15 ins or so (see Figure 175).

(d) Where large relative apertures are required, the reflector has the advantage. Furthermore, the possibility of employing relatively small focal ratios makes for compactness of design, with the attendant advantages of simpler and cheaper mountings, smaller observatory, and increased stability.

(e) Ease of construction: a single reflecting surface compared with a minimum of four transmitting surfaces, which have also to be very precisely centred and squared.

23.3 Advantages of the refractor

(a) Greater independence of temperature variations, with steadier images and higher possible magnifications than with a reflector of the same aperture. Much can be done, however, to bring the reflector nearer the level of the refractor in this respect (see section 11.2).

(b) Diffraction pattern which may, but does not necessarily, favour the resolution of extended images (see sections 2.2, 2.3).

(c) Relative permanence of the optical adjustments.

(d) The inaccessible optical surfaces being also those protected from the outside air, the OG seldom needs dismantling for cleaning.

(e) A small OG is easier to mount without flexure than a mirror, since the flexures of the individual components tend to cancel one another out in their effects. With very large instruments, however, the reflector has the advantage, since its mirror can be supported at as many points as required, whereas the OG can be supported at its edges only.

(f) The larger focal ratio of the refractor gives it some advantages over the Newtonian as usually constructed:

  (i) highly corrected oculars are not required;
  (ii) equivalent magnifications are obtained with oculars of longer focus;
  (iii) larger primary image-scale, aperture for aperture;
  (iv) definition at field edges tends to be better, though this is remediable in the reflector by the use of a Barlow lens;
  (v) optical defects generally may be cloaked at $f/15$ while being glaringly apparent at $f/5$.

(g) In one respect the refractor is easier to construct, for, the refraction being distributed over four surfaces, each need be figured to a tolerance four times less stringent than that to which the single parabolic surface of the speculum must be figured, in order to produce an image of the same quality.

23.4 Suitability for different types of work

As a rough guide, objects requiring light grasp rather than critical definition are best served by the reflector, those which make demands primarily on definition, by the refractor.

For work at the threshold of definition—e.g. lunar and planetary observation—a refractor is probably superior to a reflector of the same aperture, although, as already emphasised, the latter need by no means be as inferior as is often suggested; and, again, an 8-in reflector is preferable to a 3-in refractor. For precision measurements of double stars, as well as for transit instruments and the like, refractors are superior, reflectors being disqualified by the comparative mobility of their optical alignment.

For photography a reflector is to be preferred, unless a photovisual is available; this, however, is likely to be ruled out on the score of expense.

Solar work is best carried out with a small refractor unless the mirror (or mirrors, in the case of a specially built solar telescope, such as that described in *O.A.A.*, section 1.15) is constructed of Pyrex. The unimportance of light grasp should not make one forget that resolving power is also a function of aperture.

In all other fields covered by amateur observers, a moderate-sized reflector is to be preferred to a small (3-in or 4-in) refractor. The latter is perhaps chiefly of use to the beginner during his initial few months of 'star-gazing'; on the other hand, W. H. Pickering's remark that apertures smaller than 5 ins are useful for nothing but the observation of variables need not be taken too seriously.

The instrumental requirements in each particular field of observation are discussed more fully in *O.A.A.*, where more detailed information regarding apertures will also be found.

SECTION 24

VISION

24.1 Structure of the eye

The eye is a closed, approximately spherical structure. Its outer protective sheath (the *sclerotic*) is transparent at the front; this window is the *cornea*, which contributes largely to the lens action of the eye. A layer of pigment cells on the inner side of the sclerotic (the *choroid layer*) forms, at the front of the eye, the *iris*; this is a diaphragm lying within the cornea, whose diameter varies from about one-tenth to one-third of an inch according to the brightness with which it is illuminated. The central aperture of the iris is termed the *pupil*. The inner rear surface of the eye is the light-sensitive *retina*. The movements of the eye as a whole are controlled by six muscles attached to the sclerotic; by their means the whole eyeball can be moved and the eye's field of distinct vision (not more than a degree or two in diameter) may be displaced over the greater part of the whole field of vision (nearly 180°) which can be encompassed without moving the head.

Within the cornea lies the *crystalline lens*, an elastic biconvex structure whose hardness and refractive index increase towards the centre; when focused for distant vision, the radius of curvature of its outer surface is from 6 to 10 mm, that of its inner surface from 5 to 6 mm. The space between the cornea and the lens is filled with salt solution, the *aqueous humour*. The largest chamber of the eye, lying between the crystalline lens and the retina, is filled with a clear gelatinous substance known as the *vitreous humour*.

The crystalline lens is attached to the walls of the eye by the *ciliary muscles*, which are responsible for adjusting its curvature to accommodate the eye to the clearly focused vision of near and distant objects. When the ciliary muscles are completely relaxed, the eye is focused on its *far point*, which with normal eyes is infinity. To focus nearer objects, the ciliary muscles must be more or less contracted, increasing the curvature of the lens and at the same time moving it forward slightly; when fully contracted, the eye is focused on its *near point*. Between the near and far points, and closer to the former, is the *distance of distinct vision*,

422

normally about 10 ins. The distance from the near to the far point is eye's *range of accommodation*.

The function of the cornea, the aqueous humour, the crystalline lens, and the vitreous humour is to prepare the incident light for the retina: until the light falls upon the retina there is no vision. Their refractive indices lie between 1·33 and 1·42, those of the cornea and the aqueous humour being the same. The equivalent focal length of the combination, in the case of an average adult eye, is about 0·6 ins, giving a retinal image scale of rather less than 4° per mm.

The retina consists of a network of nerve endings, in close contact with one another, each of which is connected with the brain by the *optic nerve*. The retinal image is therefore granular, like a photographic image, though the grain is too fine ever to be perceptible.

The point at which the optic nerve enters the retina is insensitive to light: the *blind spot*. Almost on the optical axis lies a yellow spot, about 1 mm in diameter, the *macula lutea*, centrally placed within which is a minute depression (diameter about 0·25 mm) in the retinal surface, the *fovea centralis*, which is the point of most distinct vision.

The nerve endings of the retina are of two kinds. The *rods* are the more sensitive to faint illumination and occur in greatest numbers some 20° from the centre of the retina; at the periphery they are more common than in the central region, and they are entirely absent from the fovea itself. *Cones*, which are involved in the vision of colour and of brighter illuminations, predominate within the macula lutea, comparatively few occurring outside it, and at the centre of the macula the retinal surface consists of nothing else.

Within the macula the cones are about 0·003 mm in diameter, their centres being separated by about the same distance. Within the fovea the smallest cones are only about 0·0015 mm in diameter, though larger ones, up to about 0·005 mm diameter, also occur.

Both cones and rods are bathed in rhodopsin, a substance also known as the *visual purple*, which the eye manufactures and to which the rods owe their light-sensitivity. The visual purple disappears when the retina is exposed to light, and forms again as soon as it is placed in darkness or is only faintly illuminated. Over relatively long periods the aperture of the iris is related to the amount of visual purple present, but its sudden contractions and expansions are independent of this.

24.2 Clear vision, and vision of faint illuminations

Since the rods are increasingly numerous from the centre of the retina outwards, and the cones from the periphery of the retina inwards, it follows that clearest vision of a well-illuminated object is obtained when its image falls on the fovea. At 10° from the centre of the retina, visual acuity has already dropped to 25% that provided by the fovea, while at 40° it is only about 3%. Conversely, a faint object may be seen when its image falls on the outer region of the retina (indirect vision) although it is invisible when its image falls on the macula.

24.3 Defects of the eye and of vision

These are of various types: structural abnormalities producing defective sight which can be counteracted by suitable spectacles; defects of the normal eye, considered as an optical instrument; psychological defects of vision to which everyone is more or less prone. Some details of the commoner examples of each of these are given in the following sections.

The observer should be eye-tested as a matter of course, especially for colour blindness. Of the commoner functional defects of the eye, long and short sight can be corrected at the telescope without wearing glasses, by adjustment of the drawtube; astigmatism necessitates the wearing of glasses if clear vision is to be obtained with all magnifications, though with high powers the exit pupil is so small that they can often be dispensed with; slight or moderate colour insensitivity, providing its nature is known and can be allowed for, need not debar the observer from all work involving the estimation and comparison of stellar, lunar, or planetary colour. Both the acuity and the light-sensitivity of the two eyes are frequently different, and they should be reserved for different types of work accordingly.

24.3.1 Myopia: short sight: Due to abnormal length of the eyeball in the direction of the optical axis. The curvature of the crystalline lens cannot be reduced sufficiently to give it the long focal length necessary to throw the image of a distant object on to the retina. Hence the distance of distinct vision is less than that of a normal eye; it may be only a couple of inches or so. The condition is corrected by a diverging lens of suitable curvature.

Extrafocal images of point sources a degree or more in diameter are not uncommon with myopic subjects. One such sufferer put his disability to unexpected advantage, since he found he could quite easily estimate the brightness of the totally eclipsed Moon in stellar magnitudes by making step comparisons between it and the stars!

24.3.2 Hypermetropia: long sight: Due to abnormal shortness of the eyeball in the direction of the optical axis; curvature of the crystalline lens cannot be increased sufficiently to give the short focal length necessary to throw the image of a near object on to the retina. Hence the distance of distinct vision is greater than that of the normal eye.

It is corrected by a converging lens of suitable curvature. Thus a hypermetropic eye does not see distance objects more clearly than a normal eye, but only sees near objects less distinctly. The same is true, in the reverse sense, of a myopic eye.

The power of the correcting lens required to focus the relaxed eye on its near point is called the amplitude of accommodation.

24.3.3 Presbyopia: The eye tends to become long-sighted with age, owing to progressive changes in the elasticity of the lens, as shown below:

Age (years)	Average amplitude of accommodation (diopters)
10	14
40	4·5
70	0·0

This type of hypermetropia is known as presbyopia.

An eye which suffers from none of these defects is said to be emmetropic.

24.3.4 Astigmatism: Typically a condition of the cornea, less frequently of the crystalline lens: due to the radii of curvature being different in different planes. Thus a point source cannot be focused to a point image.

Correction is obtained by means of a cylindrical lens so oriented that it and the cornea together form a system whose surfaces are symmetrical about its axis.

An uncorrected astigmatic eye may produce lack of definition when low-powered oculars are used, but unless it is very marked its effect will not be noticeable with high magnifications, since the emergent pencil is then narrow, and only the central portions of the cornea and crystalline lens are involved in its transmission to the retina.

24.3.5 Spherical aberration: A condition of the normal eye. The aberration is overcorrected, and hence paraxial rays are brought to a shorter focus than peripheral (cf. biconvex lens).

24.3.6 Chromatic aberration: A condition of the normal eye. When the eye is focused for long sight, the violet rays are brought to a focus slightly nearer the crystalline lens (less than 0·5 mm) than the red.

The chromatic aberration of the eye contributes to the apparent increasing under-correction of an objective as lower and lower magnifications are used: increasingly abaxial zones of the crystalline lens then being involved in the formation of the image.

24.3.7 Other defects of cornea or crystalline lens: Local abnormalities of curvature, which cannot be compensated for by glasses, have been suggested as the cause of the instinctive preference shown by some observers for particular magnification ranges—those producing emergent pencils of diameters which render these faults least obtrusive.

The lens may also lack homogeneity, while lines of compression in the cornea or dust particles or droplets of different refractive index on its surface may be visible when very high magnifications are used.

24.3.8 Muscae volitantes: The so-called 'floating motes' which can be seen by staring at a bright sky. Their origin is uncertain, but it has been suggested that they are detached dead cells floating in the aqueous humour. Douglass estimates their diameter as of the order of 0·001 to 0·002 ins.

Telescopically, they become increasingly obtrusive as the magnification is increased, since their visibility is inversely proportional to the diameter of the emergent pencil. They also appear to be more noticeable when the head is tilted than when it is horizontal—providing an additional reason for using a star diagonal for zenith observation with a refractor or Cassegrain.

24.3.9 Irradiation: At the boundary between areas of markedly unequal brightness, the brighter appears to encroach upon the fainter to an extent that is proportional to their intensity difference. It can therefore be reduced by increasing the magnification.

Irradiation is a physiological effect, being caused by the spreading of the excitation from the retinal area which is actually stimulated by the light. It is thus somewhat analogous to photographic halation.

Examples are: the Martian caps, planetary and satellite discs, appear larger than photographically shown to be; dark spots, as on the Moon, appear conversely to be smaller.

24.3.10 After-images: The image of a conspicuous object which has been intently focused will survive for a few seconds after the object has been removed; it then fades gradually. After-images are of the complementary colour to that of the object. Thus, if a green ball is stared at for about half a minute, and the gaze then transferred to a sheet of white paper, a dim red disc of the same angular size as the ball will be seen.

24.3.11 Fatigue: Any operation involving concentrated visual attention fatigues the eye rapidly, so that its performance becomes increasingly unreliable. Such operations as bisecting a stellar image with a web, eliminating diffraction fringes, making visual estimates of the relative brightness of two stars, etc, should therefore be made as quickly as possible, and the eye given a chance to rest between observations.

For the same reason, movement of a faint object in the field (by tapping the telescope or even by movements of the eye itself, so that the image moves over the retina) often helps to render it visible.

A slight change of focus is also often very restful to the eye during prolonged sessions at the telescope.

(Colour fatigue is dealt with below, see section 24.4.)

24.3.12 Mid-point error: An innate and fairly constant inability in some observers to judge accurately the mid-point of a line (vertical or horizontal), or the point midway between two other points. The tolerably constant extent of the misjudgment is shown by the constant value of the personal equation throughout a long series of such observations by a single observer. In one experimental test, the average errors of 10 observers were:

horizontal: mid-point at 0·509: 9% error
vertical: mid-point at 0·486: 14% error

Estimating the position of the central meridian of a disc in order to obtain transit times of planetary markings is an instance of the type of observation susceptible to this error.

24.3.13 Leading error: The tendency, when a star, perceived to be in motion across the field (by comparison with fixed webs, for instance), is being kept bisected by a moving web, to adjust the web systematically p or f the star's image. This error is to be distinguished from mid-point or bisection error, which also occurs when the image is, or appears to be, stationary. It is unlikely to be of much practical concern to amateurs.

24.3.14 Scale error: In the estimation of the proportional distance of one 'event' between two others (e.g. the transit of a star at a micrometer web in relation to two beats of the clock, or the brightness of one star in terms of a brighter and a fainter comparison star), a systematic preference for certain proportions at the expense of others is sometimes encountered. This is termed scale error. Where present it is found that in a long series of observations, all the steps in the scale (e.g. the 10 steps of 0·1 between 0·0 and 1·0) would not occur with equal frequency.

24.3.15 Contrast error: A tendency to over-estimate the difference in brightness of two dissimilarly bright objects in the same field. Some observers suffer more than others from this source of error, but in each case it tends to be reasonably constant. If the luminosity-difference of the two points or areas is very great, decreased dark adaptation of the eye may be introduced as an additional factor tending to increase the error.

24.3.16 Other psychological pitfalls: It is vitally important to go to the telescope with a mind completely open and free from preconceptions or expectations. The insidiousness of the danger of preconceived ideas cannot be over-stressed. One should be prepared to see literally anything, and, ideally, each time an observation is made that does *not* surprise one, one should immediately be suspicious and on one's guard against a visual datum suggesting more than it in fact states.

The study of other observers' drawings, what one has read, unconscious conditioning by one's own previous observations—all these have a strong effect, of which one is usually unaware, upon what one 'sees'. The two schools of interpretation of Martian detail, and Hargreaves' theory of 'mentors' (B. 574) may be mentioned in this context.

The accuracy and objectivity of one's drawings can easily be gauged by making indoor tests on meaningless drawings prepared by a second person and observed with the same relative aperture and magnification as are employed for planetary work (see B. 310, 312–17).

24.4 Colour vision, and false colour effects

Wavelengths to which the normal eye is sensitive range from 4200Å to about 7000Å; exceptionally from 3200Å to 8350Å. The latter short-wave limit approaches the region of telluric absorption which obliterates the shorter wavelengths. The greater part of colour vision depends upon the action of the cones, the rods being comparatively colour-blind. The zone of maximum sensitivity occurs in the yellow at about 5600Å when the illumination is moderate or strong, but as the illumination is decreased towards the extinction threshold it shifts towards the green (Purkinje effect, see below).

The eye's absolute discrimination of colour is comparatively poor, and, furthermore, colour fatigue occurs rapidly, causing still greater insensitivity to uncompared tints. On the other hand, the eye is extremely sensitive to differences or changes of tint; this sensitivity is greatest in the yellow and green and lower in the red, blue, and violet, the latter being largely absorbed by the crystalline lens. An average eye can distinguish tints differing by from 10Å to 30Å between 5000Å and 6000Å, the minimal perceptible difference rising to 60Å at the ends of of the visible spectrum. A normal, untrained eye can thus distinguish a hundred or more tints between the blue and the orange.

The physiology of colour vision, involving the complex interaction of rods, cones, and visual purple, is still imperfectly understood. Colour blindness exists in many forms, from a slight relative insensitivity to a single colour, to completely monochromatic vision. Men appear to be about ten times more prone to it than women, 4% being afflicted with colour blindness of some sort or another. The colour vision of the two eyes is not infrequently different, and this also the observer should be aware of if it applies to him. If, over a long period, one eye has been continuously used for solar observation, it is likely that its colour vision will have been slightly affected.

From the macula to the periphery the retina is progressively more sensitive to short wavelengths and less sensitive to long. This is the physiological basis of the Purkinje effect, since the peripheral rods are the elements concerned in the vision of faint illuminations. This source of error may be described in various ways according to the form it takes in practice. Basically, it is that the wavelength of maximum sensitivity varies with the intensity of the illumination, shifting from the yellow (in daylight) into the green (dark-adapted night vision). Astronomically this is chiefly of interest as a source of uncertainty in the visual estimations of the brightness of faint stars when these are noticeably coloured. A source may be too dim to stimulate the colour-sensitive elements of the retina (the cones) while still being bright enough to stimulate the night-vision elements (rods), which are comparatively insensitive to colour.

Examples of the Purkinje effect are:

(a) The apparent brightness of a red star, estimated by comparing it with a white star, varies according as to whether it is relatively bright or faint. If it is faint, its estimated brightness will be too low; if it is relatively bright, or is observed against a faintly illuminated background, it will be judged too bright. More generally, if two stars of different colours are judged to be equally bright, and their brightness is then changed by an equal amount, they will no longer be judged to be equally bright.*

(b) In late twilight, red flowers lose their colour and appear black, though distinctly visible as shapes.

(c) If the brightness of a coloured light is decreased to extinction, it does not retain its colour till the last moment. Over a limited range of faint visibility immediately before extinction (the so-called achromatic interval) it appears grey; within this range the rods are alone functioning.

(d) The greyish-green colour which is the prevalent tone of objects in moonlight is an example of the Purkinje effect. Moonlight being reflected sunlight, its 'coldness' is dependent solely upon its lower intensity.

* The variable-star observer is never allowed to forget the Purkinje effect for long, and a further discussion of it in connexion with this type of work will be found in O.A.A., section 17.5.

Another effect of colour upon apparent brightness is shown by many observers who will habitually see one member of a pair of equally bright coloured stars as brighter or fainter than the other. Usually with a red and a blue pair, the blue appears to be brighter than the red. This is not, however, an example of the Purkinje effect, since it applies to stars far above the luminosity level at which the latter comes into operation.

Three well-known examples of defective colour vision which affect contiguous coloured areas are:

(a) The intensity of a patch of colour appears to be heightened if it is placed in contact with, or proximity to, its complementary colour.

(b) A white patch on a coloured ground appears to be faintly tinged with the colour complementary to that of the ground.

(c) When a bright and a faint object of different colours are observed together, the fainter will be tinged with the complementary colour of the brighter. For example, the white *comes* of a red primary appears bluish or greenish; dusky markings on the reddish Martian surface appear greener than they are.

24.5 Resolving and separating acuity of the eye

Depends upon the structure of the retina—i.e. upon the size and spacing of the nerve elements—and does not vary greatly from one individual to another, providing no pathological condition is present. The area of most distinct vision being the fovea, where the smallest cones are about 0·0015 mm in diameter, it follows that the centres of two cones will not be stimulated unless the diameter of the image is greater than about 20″. One function of the telescope is to spread the image over a larger area of the retina, increasing the number of cones that are stimulated, and hence the degree of resolution.

The resolution threshold of 20″ for 'point' images assumes, however, that the pupil is at its greatest aperture. Under these conditions the eye's abaxial aberrations will impair the definition of the image, and for clearly defined night vision the threshold is nearer 60″. In an extremely interesting series of experiments carried out by E. W. and A. S. D. Maunder (B. 315) a very consonant set of results was obtained, giving the minimum diameter of a visible black spot on a white ground (eye not dark-adapted) as 34″.

A linear object may stimulate a sufficient number of cones to produce sight even though its width is 20 or 30 times less than the threshold diameter of a spot. The Maunders' results (with independently obtained results by W. H. Pickering and Lowell for comparison) for a black thread viewed against a light background were:

	Maunders	Pickering	Lowell
Invisible	—	0″72	—
Glimpsed	1″2	0″82	0″73, 0″83
Steadily held but ill-defined	1″3	—	—
Held	—	1″15	1″13
Clearly defined	1″4	—	—

Again showing remarkable uniformity considering that the conditions of the experiments were not the same in each case.

Though 1″ may be taken as, roughly, the minimum angular width of a visible line, individual observers have considerably reduced it. Barnard, for example, was able to glimpse a wire suspended against the day sky when its diameter was only 0″44. In general, a lower threshold will be reached with a white line on a dark ground if viewed in daylight, probably on account of the assistance given by irradiation: with sufficient contrast, a bright line 1″ wide looks at least 1′ wide, and often much more.

Astronomical objects offer little opportunity for observing narrow lines with a dark-adapted eye (i.e. a bright line on a dark ground), and in the case of dark linear objects against a bright ground (lunar and Martian canali, for example) the threshold is comparable with that obtained experimentally from dark lines on a bright ground, rather than from bright lines on a dark ground where irradiation tends to widen rather than narrow the image. It was later discovered by Maunder that the threshold is raised if the length of the wire is considerably less than the diameter of the field of view, as is normally (but not necessarily) the case in astronomical application.

The threshold for the separation of two points (black on white) is much the same as that for the practical visibility of a single point, about 1′; for the centres of two points that are to be seen separately must be more than two cone diameters apart, i.e. more than about 40″. If, however, they are stellar points, the aberrations of the dark-adapted eye with which their observation will normally be made raise the threshold by a factor of about 3, though individual eyes vary rather widely. Naked-eye separation of ε Lyrae (3′5), for instance, requires at least average acuity. Pickering, however, obtained clear resolution of pinholes in black paper held against the light when their separation was only 86″. The threshold for the resolution of parallel lines is about that for the visibility of a 'point' image—namely, the region of 1′.

The acuity of naked eyesight usually decreases with age, though there is evidence that parallel decrease in the acuity of telescopic vision need not necessarily occur. This no doubt springs from the fact that the extreme threshold of visual acuity is not in practice called into play in telescopic work, atmospheric and optical causes having a greater effect in rendering fine detail invisible.

The figures quoted above assume optimal contrast conditions. These conditions do not apply in, for example, planetary work, and this account should be read in conjunction with section 24.8 below.

24.6 Visibility of extended detail

The complexities of planetary and lunar detail test the eye's powers of resolution and separation, as well as its power of discriminating between small differences and absolute values of colour and brightness, to the limit.

Although a spot may be glimpsed if its diameter is 20″ to 30″, and clearly seen when its diameter is about 60″, the shape of a visible spot does not begin to emerge until its diameter approaches 2′, or even 3′. Claims that, for example, the crescent of Venus (50″ to 60″) or the elliptical shape of Saturn (40″ × 18″) can be discerned without optical aid should therefore be treated with reserve. Differences of size between adjacent objects, on the other hand, may be detected even when as small as 10″.

There is experimental evidence that if the eye is abnormally acute in its perception of faint linear markings, it tends to be below average in its sensitivity both to faint spots and to tone gradients. Indeed, particular acuity in any one of these directions seems often to inhibit acuity in the other two (B. 314).

Among the commoner types of misperception that are liable to occur near the threshold of resolution or illumination are:

(a) The appearance of two adjacent spots, each individually below the threshold of visibility, as a single spot of approximately the threshold size.

(b) The appearance of a wavy line as a straight band.

(c) The appearance of a line of spots as a continuous, though faint and ill-defined, line. Spots no more than 8″ in diameter can produce this effect, and with a double row the diameters of the individual spots can be less than 4″. One of the Maunders' experiments, designed to throw light upon the possibly subjective nature of the Martian canali, consisted in exhibiting a 6-in disc, on which an irregular pattern of miscellaneous spots had been drawn, to a class of schoolboys sitting at varying distances from the disc; the boys then drew what they thought they saw. The results were:

15 ft from disc (angular diameter of disc=2°): correct delineation;
25 to 30 ft from disc (angular diameter≃1°): a number of canali drawn;
more than 40 ft from disc (angular diameter <43'): few details of any sort were seen or drawn.

24.7 Importance of focus

Failure to secure absolutely precise focus can produce deceptive appearances; similar conditions are provided by inferior seeing or intrinsic lack of definition. Owing to the nature of the focus there is, theoretically, no such thing as 'the' focus, the convergent and divergent cones overlapping by an amount equal to $2\dfrac{F}{D}.d$, where d is the diameter of the spurious disc. The focal plane, c, is in practice impossible to locate more precisely than somewhere between a and b:

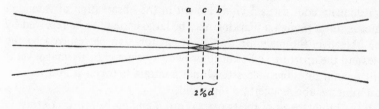

FIGURE 176

Nevertheless the importance of precise focus, especially in planetary and similar work, cannot be overrated. It is therefore unfortunate that inadequate data exist concerning the effect of fatigue upon the focus of the eye, though it appears normal for it to shorten during prolonged observation, necessitating a corresponding readjustment of the telescopic focus. There is always a tendency to focus too short when striving to make out the details or existence of an object near the threshold of visibility; this imposes a strain on the eye which rapidly becomes very tiring, and it is better to start with the focus if anything too long rather than too short.

It is certain that slight variations of focus are caused by the atmosphere, but the extent and effect of these—as of the eye's focus resulting from fatigue—are very difficult to estimate. At least it appears to be fairly well established that a star at low altitude requires a slightly shorter focus than one near the zenith.

When observing objects which themselves lack sharp definition, it is advisable to establish the best telescopic focus first by means of a star, or it may be very considerably in error. A triangular objective diaphragm allows a fine adjustment of the focus to be made. The diffraction pattern of a stellar image with such a diaphragm is 6-rayed; the extrafocal images are triangular, with the angles of the inside-focus image occupying the position of the sides of the outside-focus image. A precise focus is easily obtained by adjusting the drawtube so as to make the 6 rays of the focused image equal.

Some of the subjective effects of a slight displacement of the telescopic focus have been investigated by Antoniadi (see especially B. 307–308). He found, for instance, that a single band could be expanded into a pair of partially overlapping bands, of which only the darker core may be visible as an ill-defined dark line. The intersection of such bands may produce an illusion very similar to the appearance of Martian 'lakes' at the intersection of many of the line-like canali.

24.8 Importance of contrast

Adequate contrast is a vital element in the observation of extended images: even though a marking may be large enough to be resolved by the objective, the magnification sufficient for it to be resolved by the eye, and the aperture sufficient to provide a bright enough image, yet it will remain invisible unless the tonal contrast between it and its surroundings is above a certain threshold.

If the brightnesses of the two areas are B and b, then their contrast is measured by

$$\gamma = \frac{B-b}{B}$$

The threshold value of γ, below which B and b cannot be differentiated, may be designated γ', and is a function of B. Thus whilst a dark marking on a white ground viewed in a well day-lit room can be distinguished when $\gamma' = 0.005$, γ' increases to about 0.02 in the case of the Moon and the brighter planets, to 0.3 in the case of the fainter nebulae, and to about 0.5 in the case of the night sky itself.*

* This is only true if, as recommended by Danjon and Couder (B. 90), the magnification used on planets and the Moon is such that

$$0.1 > \left(\frac{M'}{m}\right)^2 > 0.007$$

(See section 1.3.)

These figures are 'average'. An experienced planetary observer may distinguish features two or three times less contrasted than quoted, providing the areas are not very small, and that they are contiguous. Furthermore, the contrast sensitivity of different eyes at different brightness levels varies considerably; hence one observer may be relatively more sensitive to faint variations of tone within a dark area, whilst to another they are most easily detected against a relatively bright ground.

The value of γ' also increases again if B is increased beyond a certain level. But this fact has little astronomical application except in the case of Venus, and possibly also of the Moon. Generally speaking, any reduction of brightness tends to increase the value of γ'. Thus lowering the magnification will make faintly contrasted detail more clearly visible, provided it covers a large enough area for the eye's resolving power not to enter the picture. Some astronomical objects, however, are so intrinsically faint that no practicable reduction of magnification will reveal their structure. A comparison of Lord Rosse's drawings of nebulae with later photographs brings this out clearly. In such circumstances there is, for visual observation, no solution but an increase of aperture.

Other agents of reduced contrast are mist, high and tenuous cloud, skylight from whatever cause, and dirty or dusty lenses.

The visibility of colour contrast, like that of tone, is also a function of image-brightness.

The artificial increase of contrast in certain cases by means of suitable filters, polarising devices, etc, is a widely practised device among planetary and lunar observers.

γ is not only a factor in the visibility or otherwise of a marking, but also in the threshold of size at which it emerges from invisibility. The subject is complex—as, indeed, is the whole field of the visibility of detail in extended images—and has never been thoroughly tackled.

The results of an investigation by Danjon and Chapeau (B. 309) are summarised below. Their test object was a grid of parallel equidistant lines, whose width was equal to that of the interspaces, drawn on a white ground. Though admittedly artificial, it nevertheless seems to uncover some of the factors controlling visual acuity.

Given adequate contrast between the lines and the interspaces, resolution is possible by an experienced observer when, as we have just seen, their angular separation is about 60″. By reducing either the brightness of the illumination of the test object, or γ, this value is increased. Furthermore, without varying either of these factors, the minimum separation for resolution may be increased merely by reducing the number of lines in the grid; one set of values derived empirically was:

11 lines in grid: minimum separation for resolution, 89"
6   ,,   ,,   ,,   92"
3   ,,   ,,   ,,   101"
2   ,,   ,,   ,,   115"

24.9 Illumination sensitivity of the eye

This varies (quite apart from real differences of sensitivity in different observers, which is partly innate and partly a function of age: night blindness tends to increase with age) with several factors:

1. degree of dark adaptation;
2. nature of the background;
3. position of image on retina;
4. nature of object;
5. colour of object.

1. Within a second or two of the eye being screened from light (or shielded by dense red goggles), the pupil dilates to a diameter of about 0·33 ins. The greater part of dark adaptation, however, concerns the production of the visual purple, since the area of the pupil only varies by a factor of about 16, whereas the total sensitivity range of the eye is of the order of a million. Most of the visual purple is produced during the first 20 minutes or half an hour of screening, though the final stage of adaptation may continue more slowly for as long as 2 hours. The dark adaptation of the cones, which is of little importance, is completed within the first 10 minutes.

Production of the visual purple is inhibited if the body suffers from vitamin-A deficiency. Carrots and codliver oil, whose efficacy was proved during the war years, might profitably be included regularly in the diet of the comet-seeker. Nicotine also retards dark adaptation, even one cigarette smoked half an hour previously having a noticeable effect.

Wide variations of night-vision sensitivity are shown by different individuals, and this should be taken into account before deciding in what branch of observation to specialise. The speed with which full adaptation is reached also varies considerably between one individual and another.

2. Whether the background is dark or more or less luminous may alter the threshold of perceptible illumination by as much as two magnitudes. On a clear, moonless night the normal adapted eye can just see

stars of mag 6·5. By observing the sky through a black screen which occludes all general starlight and skylight save that of the star under observation, whose light passes through a small hole, mag 8·5 can be reached. Under laboratory conditions, where all extraneous light can more easily be excluded, visibility of illuminations equivalent to mag 8·5 has also been recorded.

It will be noted that these optimum conditions approximate to those of telescopic observation with a high power ocular. For this reason, published tables relating aperture and minimum visible magnitude which are constructed by extrapolation from naked-eye observations (when the sky is considerably brighter than the field of an astronomical ocular) may be misleading by as much as 2 mags.

3. (a) The zone of the retina situated from about 20° to 50° from the centre is most profusely supplied with rods, and is therefore more sensitive to faint illuminations than either the macula or the extreme periphery. The focusing of an image on this outer region of the retina is termed averted, or indirect, vision. It may gain up to 2·5 mags, as compared with direct vision (the image falling on the macula), though the efficacy of indirect vision varies with individuals, and in some cases seems to offer no advantage at all; such differences are attributable to a different relative number of rods and cones in the inner and outer regions of the retinas.

(b) The upper part of the retina generally possesses greater sensitivity than the lower. Thus if two stars appear equally bright when the line joining them is parallel to the line joining the eyes, the lower star will appear brighter than the upper when the head is turned through 90°. It is on this fact that the variable-star observer's rule is based, always to incline the head so that the line joining the variable and the comparison star is parallel to the line of eyes. ·

(c) Less noticeably, the side of the retina nearest the temple tends to be more sensitive to faint illuminations than that nearer the nose.

4. Ill-defined objects respond better to indirect vision than well-defined objects (especially point sources), since the outer regions of the retina are incapable of providing sharp definition.

5. Since the outer regions of the retina are comparatively more sensitive to short- than to long-wave radiations, the observation of a blue star or a nebula (typically greenish) by indirect vision will yield a lower threshold than that of a red star. In other words, indirect vision will increase the visibility of a faint blue, green, or white object considerably, where little will be gained by observing a faint red star in this manner. It is because of this relative insensitivity of the dark-adapted

eye to the longer wavelengths that a red working light is used at the telescope, and that the boredom of dark-adapting in total darkness can be obviated by the use of red goggles.

Experiments performed by Waterfield suggest that the outer side of the retina is more sensitive to all colours than the nose side; and that the lower side is more sensitive to green, but less sensitive to red and yellow, than the upper side.

As with colour and size, the eye is extremely sensitive to differences of brightness, while being comparatively insensitive to absolute values. Variable-star observers habitually make their comparisons to 0·1 mag, and under laboratory conditions sensitivity to differences of 0·01 mag has been recorded. For very bright objects, as equally for those near the limit of perception, the eye's powers of discrimination are weakened.

24.10 Tests of naked-eye acuity and illumination thresholds

The observer should determine his own threshold of each type of acuity and illumination sensitivity. A wire (measured by calipers or microscope) stretched across a window may be used. It should be approached from a distance at which it is invisible, and the distances noted at which (*a*) it is glimpsed, (*b*) it is seen steadily. From these distances and the known linear diameter of the wire its angular subtention can be calculated. A white line drawn on a black ground and observed in daylight will yield a lower threshold.

Various natural test objects have been proposed from time to time in order to provide a standard of naked-eye acuity. Pickering gives the following graduated series of tests on lunar objects:

Scale Number	Test object
1	Bright region surrounding Copernicus, the most conspicuous lunar crater
2	Mare Nectaris
3	Mare Humorum
4	Kepler
5	The Gassendi region
6	The Plinius region
7	Mare Vaporum
8	The Lubiniezky region
9	Sinus Medii: dark line running E and W
10	Faintly shaded area in the region of Sacrobosco
11	Dark spot at the foot of Mt Huyghens, across the Apennines from Mare Vaporum
12	Ripheaen Mountains

1–10 should be visible to really good eyesight, 7 representing reasonably good sight if it can be steadily held; 11 represents exceptional acuity, and 12 is possibly just beyond the reach of the best. The objects can first be identified with binoculars and a map, and then looked for with the naked eyesight.

The Pleiades furnish a convenient test of the illumination-sensitivity of the naked eye. 9 or 10 stars are normally seen by perfect, but not exceptional, sight; 6 in full moonlight. Exceptional sight will show 12, though two or three of these will only be glimpsed. Denning frequently saw 14, and a note in *Nature* (1908) records the naked-eye visibility of 13 Pleiades stars, as well as Jupiter III at elongations.

Alternatively the Bowl of the Plough may be used to give some idea of one's sensitivity to faint illuminations. Normal eyesight will show about half a dozen stars in this area; up to a dozen may be visible to exceptional sight.

Stellar tests of this sort cannot, however, be regarded as pure tests of sensitivity, since the factor of acuity is also involved. Failure to pass them does not necessarily mean that the stars concerned would not be seen if sharply focused. In other words, naked-eye tests are not a safe guide to what can be accomplished by the same eye using a telescope. Those who fail to see 13 stars in the Pleiades need not therefore conclude that they will make inferior telescopic observers, or that their sensitivity to faint illuminations is below normal.

It should always be remembered that the sensitivities of the two eyes, both as regards resolution and the perception of faint illumination, are not necessarily the same. It is up to the individual to find out and make the best use of his natural powers.

SECTION 25

PERSONAL EQUATION

'Personal equation' is a term applied to an observer's systematic mis-observation of a particular phenomenon or situation; it also becomes, with a change of sign, the correction to be applied to his observations in order to obtain the true quantities. It is particularly applicable to estimations of the time at which an observed event occurs, but is to be encountered in most fields of observation. In some cases this systematic misobservation can be attributed to definite causes—either in the psychology or physiology of vision, or in the conditions under which the observations are made—while in others it can only be attributed to the 'personality' of the observer, which is equivalent to saying that its precise cause is unknown.

The discussion of personal equation here given will refer only to the observation of objects—stars, or the limbs of planets or of the Sun or the Moon—at a micrometer web, since it is in connexion with such work that most investigation of personal equation has been done; see also sections 24.3.12–24.3.16, 28.18, and *O.A.A.*, sections 2.7.5, 17.5, 17.7, 18.2. An observer's equation in the timing of stellar transits of webs, limb transits of webs, stellar occultations, etc, will not necessarily be the same.

The story began in 1796 when Kinnebrook, an assistant at Greenwich, was sacked by Maskelyne, the then Astronomer Royal, because his transit observations were systematically $0^s.5$ to $0^s.8$ later than his employer's. The method employed was eye-and-ear with a multiple-web eyepiece and a clock beating seconds, estimations being made to the nearest $0^s.1$. The problem of personal equation in transit work was dealt with, to the satisfaction of Maskelyne, by the dismissal of Kinnebrook. It was first seriously considered by Bessel, during the early years of last century. Since then it has been continuously in the front of the astronomer's consciousness, and the introduction first of the key-operated chronograph and then of the 'impersonal' micrometer has been his attempt to reduce it. The following conclusions have been reached:

1. Personal equation is normally a lag, i.e. observations are recorded too late, so that the correction to be applied to the recorded time in order to obtain the correct time carries a minus sign. This is not invariably the case, however, and to some extent it depends on the observational method used; thus the Greenwich observers (B. 325, 326) habitually anticipated the true time of transit when using the eye-and-ear method, but not when using the chronograph. Equations of the order of $0^{s}5$ are common with eye-and-ear.

2. For a given observer, working under uniform conditions, it may be regarded as sensibly constant for at least short periods. There is evidence of long-period variation, possibly due to the physiological effects of age. Bessel's relative eye-and-ear equation with Struve, for example, changed (probably to rather an abnormal extent in such a short period) from $0^{s}044$ (1814) to $0^{s}680$ (1820) to $0^{s}799$ (1821) to $1^{s}021$ (1823).

3. For a given observer, the personal equation depends upon the observational method employed. Three varieties of equation can therefore be distinguished: (*i*) absolute personal equation—between the observed times of a single observer and the actual times; (*ii*) relative equation between two observers using the same method; (*iii*) relative equation of two methods used by a single observer. As regards methods of observation, decreasing equation is experienced along the series:

eye-and-ear: 1-sec beats;
eye-and-ear: 0·5 sec beats;
key-operated chronograph;
hand-operated Repsold ('impersonal', or moving-web) micrometer;
mechanically operated Repsold micrometer.

The Repsold micrometer consists of a movable web which it is the observer's task to keep bisecting the stellar image; it is either hand-operated or mechanically driven with a hand-operated slow motion, and the conjunctions of the moving web with the fixed webs are automatically recorded, either by electrical contacts or by photography.

Examples of dependence of personal equation upon method are:

(*a*) The relative equation of Bessel and Argelander was reduced from $1^{s}22$ to $0^{s}5$ by changing from 1-sec to 0·5-sec beats.

(*b*) Average equation of the Greenwich observers over a period of 9 years was $0^{s}13$ to $0^{s}20$ greater by eye-and-ear than by chronograph, the former anticipating the latter.

(*c*) In another case, the average probable error of 3 observers was determined as $\pm 0^{s}081$ for eye-and-ear, and $\pm 0^{s}053$ with chronograph.

(*d*) Average equation of one set of 5 observers with a Repsold hand-operated micrometer was $+0\overset{s}{.}008$, and with a mechanically operated Repsold between $+0\overset{s}{.}001$ and $-0\overset{s}{.}004$, with a probable error of $\pm0\overset{s}{.}002$.

(*e*) At the Cape Observatory the relative equations of experienced observers were as large as $0\overset{s}{.}25$ when the eye-and-ear and chronograph methods were in use; with a hand-operated Repsold micrometer, the relative equations of 6 observers were reduced to $<0\overset{s}{.}06$; with a mechanically driven web they were further reduced to $<0\overset{s}{.}02$.

4. The brightness of the star, and also therefore the aperture of the instrument, will affect the personal equation. 'Magnitude equation', i.e. variation of personal equation with the brightness of the stellar image, affects observations by the chronograph method: the usual tendency is to press the key too early with a bright star, and too late with a faint one. General order of the magnitude equation is $0\overset{s}{.}05$ to $0\overset{s}{.}1$ over a range of several magnitudes; two specific investigations arriving at equations of $0\overset{s}{.}007$ and $0\overset{s}{.}009$ per mag. Both methods of using the key (see paragraphs 5(*a*) and (*b*) below) appear to be subject to magnitude equation. The moving-web micrometer is, on the other hand, free from it.

5. 'Declination equation' is a variation of the personal equation with the Dec of the observed star, where fixed webs are used with a chronograph. There are two methods of using the latter: (*a*) to start the muscular action of pressing the key slightly in advance of the bisection of the star image by the web, so that the electrical contact is completed at the precise instant of bisection; (*b*) to start pressing the key at the observed moment of bisection of the star by the web. The former method, involving the estimation of a minute angular separation, and the relating of this to the star's angular velocity towards the web, is subject to Declination equation, since a star's velocity across the field is a function of its Dec ($\propto \sec \delta$). (*b*), on the other hand, is immune from Declination equation. Observations made with the travelling-web micrometer are also free from it.

6. Personal equation is affected slightly by the direction of the star's motion across the field: (*a*) star south of zenith, observer's head north of the telescope, star's motion (in inverted field) right to left; (*b*) star north of the zenith, head south of the telescope, star's motion left to right. Affects both 5(*a*) and 5(*b*). It is eliminated by making half the observations with a reversing eyepiece.

7. Personal equation with both eye-and-ear and chronometer is partly due to scale error (see section 24.3.14).

8. Observations of limb transits between parallel webs are subject to the equation, differing from that of stars, which is partially caused by bisection, or mid-point error (see section 24.3.12). Eliminated by the reversing eyepiece.

9. A given observer has different personalities in respect of the p and the f limbs in case 8, above. This equation can be reduced by the reversing eyepiece.

10. A cause of accidental error in transit observations, which is not a contributory factor to personal equation, is the phenomenon known as 'jumping the wire': an illusion that the star pauses on the f side of the web and then, an instant later, has crossed to the other side—rather in the manner that electrons are supposed to move from one orbit to another. It is particularly noticeable when the seeing is poor. The moving-web micrometer is free from this source of error.

11. The main contribution to personal equation with the Repsold micrometer is from leading error (see section 24.3.13). Not eliminated by a reversing eyepiece, since the illusion of motion is still conveyed by the fingers operating the web-driving screw. Much reduced, however, by replacing manual by mechanical web drive.

12. Personal equation is probably subject to short-term variations of small amplitude, dependent upon such factors as the observer's health, freshness or fatigue, state of digestion, even his emotional life.

13. True 'personality' is a term that may be applied to the residual tendency for an observer's reaction to lag behind (occasionally to anticipate) his reception of the stimulus, when the other factors mentioned above have been eliminated or allowed for. Factors contributing mainly to this residual lag appear to be:

(*a*) Reaction times vary with the sense by means of which the stimulus is received. Reaction times, when the observer is immediately prewarned of the stimulus, are of the order of $0^s.12$ to $0^s.18$ for sound and touch, and $0^s.15$ to $0^s.225$ for sight.

(*b*) Reactions to multiple stimuli through more than one sense are subject to larger and less uniform errors than those to a single stimulus. Hence the superiority of the chronograph over the eye-and-ear method. In an extreme case, when simultaneous stimuli are received by two senses, the order in which they appear to occur may depend solely upon the relative concentration that the observer devotes to each.

(*c*) Degree of warning of the receipt of the stimulus that is given. Immediately prior warning results in shorter and more consistent reaction times.

(*d*) Predisposition of the observer's attention: either mainly on the

443

anticipated stimulus, or on the muscular response with which he will react when he receives it. Reaction times from the former are systematically longer than those from the latter by approximately 0ˢ.1.

The determination of the relative equation of two observers is a simple enough matter: it is the mean difference between the two sets of clock corrections indicated by a series of observations made under similar conditions by each of them. The determination of an observer's absolute equation involves the employment of one of the numerous types of 'personal equation machine' that have been evolved—instruments which record both the observer's reactions and, without any equation of their own, the occurrence of the stimuli to which he is reacting.

For a general survey of the subject from the historical standpoint, see B. 322, which also contains an extensive bibliography; B. 325, 326 discuss the personal equations of the Greenwich observers from 1885 to 1894; B. 324 contains an interesting discussion of the errors—personal and otherwise—which are encountered in the application of the micrometer web to the measurement of planetary discs.

SECTION 26

THE ATMOSPHERE AND 'SEEING'

In spite of the fact that astronomical observations may now be made from rockets and artificial satellites above the Earth's atmosphere, the great bulk of astronomical observations are still made through the atmosphere, which amounts to an additional element in the instrument's optical train; and its nature, behaviour, and optical effects concern the astronomer very closely.

26.1 Composition, density and extent of the atmosphere

The atmosphere consists, by weight, of

	%
Nitrogen	75·48
Oxygen	23·18
Carbon dioxide	0·045
Argon	1·29
Hydrogen ⎫	
Helium ⎪	
Neon ⎬	0·005
Krypton ⎪	
Xenon ⎪	
Ozone ⎭	

The average pressure of the atmosphere at Mean Sea Level is 1013·2 millibars, or 760·0 mm of mercury (where 1 mb$=0·7501$ mm of mercury$=0·02953$ ins of mercury$=10^3$ dynes/cm^2).

If $\rho_0=$atmospheric density at MSL,
$\rho_n=$atmospheric density at $3·33n$ miles above MSL,
$n=$any integer,

then
$$\rho_n=\frac{\rho_0}{2^n}$$

445

Hence:

Height above MSL (miles)	Density ($\times \rho_0$)
3·3	0·5
6·6	0·25
9·9	0·125
13·2	0·0625
16·5	0·0312
19·8	0·0156
23·1	0·0078
26·4	0·0039
29·7	0·0019
33·0	0·0009

Thus about nine-tenths of the atmosphere is to be found within 10 miles of the Earth's surface; the highest types of cloud (cirrus, composed of ice crystals) occur between 5 and $6\frac{1}{2}$ miles of MSL. The troposphere, or main mass of the atmosphere directly overlying the surface, is 10·5 miles deep at the equator, 7 miles in the latitudes of England, and rather less over the poles. The upper surface of the troposphere is known as the tropopause, and is characterised by a sharp change in the lapse-rate, or rate of fall of temperature with increasing altitude: throughout the troposphere the lapse-rate averages about 3·3°F per 1000 ft, or 17·5° per mile. The dry adiabatic lapse-rate, due solely to the expansion of a rising mass of air brought about by decreasing pressure, is about 5·4°F per 1000 ft. The relation between the actual and the dry adiabatic lapse-rates is of some interest to the astronomer.

26.2 The atmosphere's transmission characteristics

The effects of the atmosphere upon the appearance of an object lying outside it are direct functions of the length of the light path through it from the object to the observer. If h is the effective vertical thickness of the atmosphere above the observer, the length of the light path through the atmosphere of a star whose ZD is z is given approximately by

$$l = h \sec z \qquad \qquad (a)$$

the approximation being acceptable provided $z < 75°$.

If it is assumed that the atmosphere above a height of 10 miles exercises a negligible effect, we have:

Zenith distance	Length of light path
0°	10 miles
45	14
55	17
65	24
75	39
⎰80	58⎱
⎱85	115⎰

The importance of observing at minimum ZD (culmination) is made abundantly clear from this table: an external object is seen through about 8 times the thickness of atmosphere when it is close to the horizon as when its altitude is 45°.

If I_o and I are respectively the original and received intensities of the radiation from an external source (i.e. before and after passing through the atmosphere to the observer), and a is the transmission coefficient representing the proportion of I_o transmitted by unit atmospheric thickness l,* then

$$I = I_o a$$

From equation (a) above,

$$I = I_o a^{a \; \sec}$$

or $$\log I = \log I_o + \sec z \log a. \quad . \quad . \quad . \quad . \quad (b)$$

the precise value of a being a function of wavelength (see Figure 177). Figure 178 translates the relation expressed in equation (b) into stellar magnitudes: assuming that 78% of the light from a star at the zenith is transmitted by the atmosphere (the transmission at 5600Å approximately—see Figure 177), then the curve shows the light loss in magnitudes at ZDs from 0° to 85°.

The absorption of certain groups of wavelengths from mixed radiations passing through the atmosphere is responsible for the telluric lines in astronomical spectra. The most important of these lies in the ultraviolet below about 3000Å, and is due to atomic oxygen above about 60 miles, and to ozone above about 25 miles. Absorptions by carbon dioxide in the ultraviolet are relatively inconspicuous. The only significant tropospheric absorptions are those of water vapour, which are not conspicuous. The atmosphere's transmission curve through the

* Zenithal atmospheric extinction amounts to something like 0·3 or 0·4 mags on a good night.

visual and infrared regions to 20,000Å (with comparison curves for glass, rock salt, and fluorite) is shown in Figure 179.

Light loss in the terrestrial atmosphere is thus a function both of the

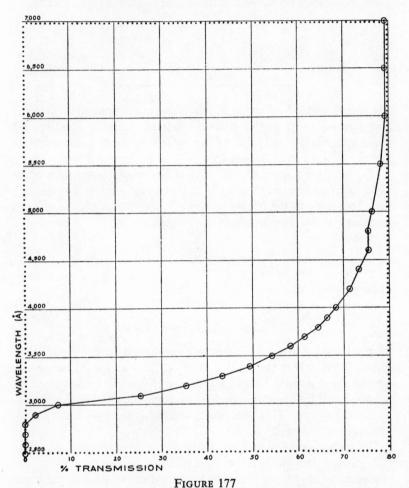

FIGURE 177

Percentage zenithal transmission of the atmosphere at Mean Sea Level

altitude above MSL at which the measurements are made and also the wavelength of the radiation with which they are made.

A proportion of the incident radiation that escapes absorption is scattered by molecules, solid particles, and the smallest water droplets, the diameters of all of which are smaller than the wavelength of light.

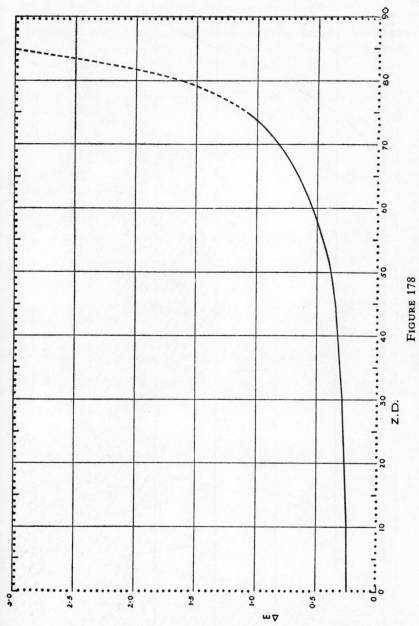

FIGURE 178

Atmospheric absorption in magnitudes, assuming $a=0.78$

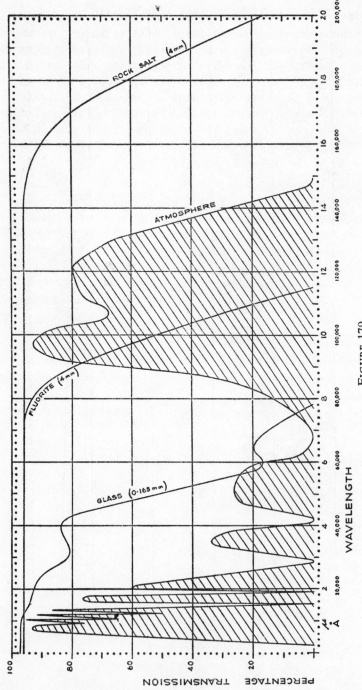

FIGURE 179

The shorter wavelengths are scattered more than the long, whence the terrestrial observer has a blue sky instead of a white one. Particles too large to scatter light cause diffuse reflection which, unlike scattering, is independent of wavelength. Thus clouds and wet fog are white, the diameters of their constituent particles ranging from 400,000Å to 1,000,000Å and 40,000Å to 300,000Å respectively. Town fog, on the other hand, contains a high proportion of small solid particles which cause selective scattering; hence the Sun seen through a country fog or mist is white, and through a pea-souper is red.

26.3 Atmospheric refraction

The refractive index of air at standard temperature and pressure is 1·00029; with increasing altitude, and consequently decreasing pressure, μ also decreases. A ray of light passing to the observer's eye through his zenith strikes the contours of equal atmospheric density normally, and is therefore not refracted; all other rays are refracted towards the Earth's centre, thus displacing the apparent source towards the zenith. Refraction therefore increases the altitude, decreases the ZD, and leaves unaffected the azimuth of a star.

Since the density of the atmosphere varies continously with altitude, atmospheric refraction—unlike that of a lens—is progressive throughout its extent. Hence the angle through which the ray is refracted depends upon the thickness of the atmosphere traversed, as also upon the wavelength of the light, and hence upon the ZD of the source.

Below are given two sets of values of the Mean Refraction, i.e. the refraction under standard conditions of temperature and pressure ($T=50°$F, $b=29·5$ ins). The rule for conversion to different conditions is: for an increase of 1 inch in b, add $3·5\%$; for a rise of $5°$F, subtract 1%.

The Pulkova Refraction Tables (e.g. B. 8) tabulate in full the factors necessary for the precise calculation of the correction to be applied to the apparent ZD. These are derived from the expression

$$R = \mu \tan z (B.T)^A \gamma^\lambda \sigma^i$$

where $z=$apparent ZD,

$\left.\begin{array}{l}\mu \\ A \\ \lambda \\ \sigma\end{array}\right\}$are functions of z,

B is a function of the barometric pressure,

$\left.\begin{array}{l}T \\ \gamma\end{array}\right\}$are functions of the temperature,

i is a function of the time of year.

Apparent ZD	Pulkova	Bessel	Apparent ZD	Pulkova	Bessel
0°	0' 0ʺ0	0' 0ʺ	65°	2' 1ʺ9	2' 3ʺ
5	0 5·0		66	2 7·6	
10	0 10·1	0 10	67	2 13·8	
15	0 15·3		68	2 20·5	
20	0 20·8		69	2 27·8	
25	0 26·7	0 27	70	2 35·7	2 37
30	0 33·0		71	2 44·4	
35	0 35·7		72	2 54·0	
40	0 47·9	0 48	73	3 4·7	
45	0 57·1	0 58	74	3 16·6	3 19
46	0 59·1		75	3 30·0	3 32
48	1 3·4		76	3 45·2	3 47
50	1 8·0	1 9	77	4 2·5	4 5
52	1 13·0		78	4 22·5	4 25
54	1 18·5		79	4 45·7	4 49
55	1 21·4	1 22	80	5 13·1	5 16
56	1 24·5		81	5 46	5 49
57	1 27·8		82	6 26	6 30
58	1 31·2		83	7 15	7 20
59	1 34·8		84	8 19	8 23
60	1 38·7	1 40	85	9 40	9 47
61	1 42·8		86	11 31	11 39
62	1 47·1		87	14 7	14 15
63	1 51·7		88	17 55	18 9
64	1 56·6		89	23 53	24 25
			89 30	28 11	29 3
			90 00	33 51	34 54

Another convenient tabulation of all the data necessary for the precise derivation of the refraction is to be found in B. 418:

(a) Normal refraction (standard conditions being taken as: air temperature, $t=0°$C; barometric pressure, $H=760$ mm; temperature of the mercury, $t'=0°$C; altitude (in metres), $h=$ MSL; latitude, $\phi=45°$ N) for altitudes—not ZDs—from $-1°$ to $+90°$ in steps of 10'.

(b) Values of correcting factor A as a function of temperature from $-30°$C to $+50°$C in steps of 1°.

(c) Values of correcting factor B as a function of barometric pressure from 490 mm to 790 mm in steps of 1 mm.

(d) Values of correcting factor α as a function of altitude from $-1°$ to $+45°$ in steps of 1°.

(e) Values of correcting factor β as a function of refraction corrected for temperature from 0' to 60' in steps of 2'.

(f) Values of correcting factor k as a function of altitude from $-1°$ to $+9°$ in steps of 1°.

Then if H' and t' are the observed barometric pressure and mercury temperature respectively, the barometric pressure corrected for air temperature, latitude, and height above MSL is given by

$$H = H'[1 - 0 \cdot 00264 \cos 2\phi - 0 \cdot 000000196h - 0 \cdot 000163(t' - t)]$$

The temperature correction is applied algebraically to the normal refraction R corresponding to the object's apparent altitude:

$$RA\alpha \frac{1 + 0 \cdot 00367t}{1 + kt}$$

and the barometric correction is applied to the normal refraction corrected for temperature, R', giving the fully corrected refraction:

$$R'B\beta$$

(The actual values of A and B are respectively $\dfrac{-0 \cdot 00383t}{1 + 0 \cdot 00367t}$ and $\dfrac{H}{760} - 1$.)

In the absence of tables, Comstock's formula gives a fair approximation:

$$R = \frac{983b}{460 + t} \tan z$$

where R is expressed in " arc,

z = apparent ZD,

t = temperature (°F),

b = barometric pressure (ins).

A still simpler approximation is given by

$$R = k \tan z$$

where $k = 57''$ at MSL (refraction at $z = 45°$ under standard temperature and pressure). The inaccuracy of R, so derived, increases with z, and becomes prohibitive when $z > 70°$ approximately.

The extent of the differential refraction, dependent on wavelength, is indicated below. The two right-hand columns refer respectively to light of the wavelength of the D lines of sodium, and to the difference between the refractions of wavelengths of the F and C lines:

ZD	Mean Refraction	
	R_D	$R_F - R_C$
25°	0′ 27″	0″3
45	0 57	0·6
65	2 02	1·2
72	2 54	1·8
76	3 45	2·4
79	4 46	3·0

To obtain the true difference of ZD of two stars (difference in Dec if they are on the meridian), the difference of their refractions $(R-R')$ must be added to their observed difference in ZD $(z-z')$. $R-R'$ is tabulated below for different values of $z-z'$ at selected ZDs:

$z-z'$	ZD					
	0°	10°	20°	25°	30°	35°
0′	0″00	0″00	0″00	0″00	0″00	0″00
2	0·04	0·04	0·04	0·04	0·04	0·04
4	0·06	0·06	0·08	0·08	0·08	0·10
6	0·10	0·10	0·12	0·12	0·14	0·16
8	0·14	0·14	0·16	0·16	0·18	0·20
10	0·16	0·16	0·20	0·20	0·22	0·26
12	0·20	0·20	0·22	0·24	0·26	0·30
14	0·24	0·24	0·26	0·28	0·30	0·36
16	0·26	0·23	0·30	0·32	0·36	0·42
18	0·30	0·32	0·34	0·36	0·40	0·46
20	0·34	0·36	0·38	0·42	0·46	0·52
22	0·36	0·38	0·42	0·46	0·50	0·56
24	0·40	0·42	0·46	0·50	0·54	0·62

26.4 Atmospheric humidity

The wetness of the atmosphere is given by the proportion

$$\frac{\text{Quantity of water vapour in unit volume of air}}{\text{Quantity of water vapour required to saturate it}}$$

and is usually expressed as a percentage. Air is said to be saturated if it will stand in equilibrium over a smooth surface of water at its own temperature.

The humidity may be determined directly by means of a hygrograph, or calculated from the readings of a wet-and-dry-bulb thermometer. This consists of a pair of ordinary mercury-in-glass thermometers; the bulb of one is surrounded by air, that of the other by a piece of flannel which is kept constantly damp by means of a small water reservoir into which its other end is dipped. The rate at which water evaporates from the wet bulb is, for a given temperature, inversely proportional to the relative humidity; also, the faster it is evaporating, the more thermal energy it will utilise, and the greater will be the discrepancy between the two thermometer readings. The humidity can be obtained from tables giving the relative humidity for different values of T_d-T_w at specified values of T_d.

The relative humidity is normally maximal during the night and minimal during the afternoon.

26.5 Dew

As the temperature of a volume of air is lowered, the quantity of water vapour required to saturate it falls likewise. Hence it can be saturated in either of two ways: by injecting more water vapour into it, or by cooling it. The temperature at which a given volume of air becomes saturated is known as its dewpoint: if it is cooled below its dewpoint, liquid water will condense out of it in the form of dew. A familiar example of this process—apart from the natural formation of dew, dependent upon the cooling of the ground by radiation after sunset— is the condensation of water droplets on the outside of a glass of cold water when taken into a warm room (the temperature of the glass being below the dewpoint of the air in the room). If the air is damp but unsaturated, its dewpoint temperature is lower than T_w; if it is saturated, its dewpoint temperature$=T_w=T_d$.

The condensation of dew is more marked on grass than on bare earth, since the latter's diurnal temperature change is smaller than that of grassed earth. For the same reason, certain types of soil favour the formation of dew more than others: increasing dew will be encountered along the series clay–loam–sand. Clear nights, especially if there is little or no wind, also provide the conditions most favourable for the deposit of dew.

26.6 Suspended matter

Astronomical 'seeing' does not mean, or include, clarity of atmosphere, which, indeed, is often exemplary when the seeing is at its worst. Atmospheric clarity becomes of paramount importance, however, in photometric work, the long-exposure photography of faint objects, and similar work.

Apart from such obvious drawbacks as a naturally cloudy climate, the worst enemy of atmospheric transparency is the proximity of large towns, which not only pollute the atmosphere but also illuminate the pollution. This was the cause of the removal of the Royal Observatory from Greenwich to Herstmonceux. Again, diffusion of the lights of Los Angeles by the layer of 'smog' that habitually hangs over the city reduces the maximum permissible exposure time (limited by fogging) of the Mt Wilson $f/5$ cameras from 80 minutes to about half an hour, as was proved during wartime dim-outs.

Matter in suspension in the atmosphere may be grouped under the following heads:

(a) *Cloud*:

Condensation occurs on small solid particles; the commonest are crystals of hygroscopic salts, such as marine brine, and oxides of nitrogen and sulphur, by-products of industrial and domestic combustion. Such particles have diameters as small as 1000Å or less.

The droplets will continue to grow until a dilution of the salt solution is reached at which no further attraction for water is exerted. Thereafter, increase in the size of the drops can only occur if the temperature falls, or more water is evaporated into the air containing the droplets, i.e. by an increase in the relative humidity of the air.

Inland, afternoons tend to be cloudier than other times of the day. There is a marked tendency for the sky to clear after sunset, and the times of minimum cloud are late evening and night. In SE England the sky is completely clear at midnight on about 122 nights per annum. During the period 1901–1930 there was a daily average of 4^h9 sunlight at Herstmonceux in Sussex, equivalent to 39·9% of the time the Sun was above the horizon. During the period 1927–1945 the corresponding figure at Greenwich was 30·55%. Measures of the clear night-hours at Greenwich, by means of a photographic trace of a circumpolar star, gave 36·82% of the total exposure time, over the same period. It therefore appears that there is, in SE England, little significant difference between daytime and night-time atmospheric clarity.

(b) *Fog and mist:*

Wet, or white, fog and mist consist of water drops only, and occur typically in the country. They not only absorb light from a distant source, but also form a screen of reflected white light between the source and the observer. It is this white veil that at times reduces visibility to a few feet. Sea fogs, relying for their formation on the presence of particles of extremely hygroscopic sodium chloride, can appear when the relative humidity is as low as 75%.

(c) *Smoke fog, or pea-soup fog:*

Consists of water drops whose condensation nuclei are solid carbonaceous by-products of combustion. Unlike a white fog, it derives its opacity solely from its action in cutting down the light from a distant source.

(d) *Solid particles in suspension, and corrosive gases:*

Most commonly dust and the by-products of combustion. Obscuration from this cause depends solely on the existence of fires or the nature of the soil, and is independent of the relative humidity.

The larger particles resulting from combustion settle near the source, depositing a layer of fine grime over the roofs of houses in urban districts. Gaseous by-products of combustion (particularly sulphur dioxide) are largely removed by solution in cloud droplets. The most troublesome element is the smaller particles comprising smoke, of which there may be 10^7 per cubic inch of town air, and are still about 10^3 in the cleanest country air. These may remain in suspension for long periods of time and be carried for astonishingly long distances. Smoke from the Ruhr valley, for example, is brought to England by a steady SE wind; again, smoke from the Midlands produces haze at Oxford when the wind is in the north. The ultraviolet is even more drastically eliminated by smoke particles than the visible wavelengths.

Artificial pollution, both smoke and acidity, is always worst in winter, and during the late afternoons at all times of the year. Results of two years' research on atmospheric pollution carried out in Leicester (B. 334) may be summarised as follows:

(*i*) Street-level pollution is minimal in the early morning, rising sharply to a maximum during the mid-morning, followed by a shallow minimum during the afternoon and a secondary maximum (which may be smaller or larger than the morning maximum) in the late afternoon.

(*ii*) It is comparatively little affected by wind: an increase of wind velocity by a factor of 3 reduces mid-city pollution by only about one-third.

(*iii*) Town air is to a large extent cleaned each night; thus a pea-soup fog of the evening is a white fog by next morning.

(*e*) *Condensation nuclei:*

The much smaller particles upon which (given the requisite relative humidity) cloud droplets form. In their dry state they scatter certain wavelengths of the light passing through them, causing, *inter alia*, the blueness of distances. If the relative humidity increases beyond the saturation point, droplets of increasing size form on the nuclei, as already described, and the blueness of the horizon changes to grey (haze —as often occurs on a hot day) when they become too large to scatter visible wavelengths, and finally to white (wet fog).

26.7 Astronomical seeing

The 'seeing' is the quality of the telescopic image, so far as this is dependent upon the condition of the atmosphere. Two separate factors in the state of the seeing can be distinguished, either or both of which may be present at a given time and place: deformation of the image, with or without bodily motion; and motion of the image without any appreciable degree of deformation.

Motion of the image can be eliminated—or, rather, converted into deformation—by a sufficient increase of aperture. Thus with average seeing a 3-in or 4-in will show star images which (if the objects are not too near the horizon) are clearly defined but wandering. The movements of each image will be contained in a circle which is typically of the order of 2″ diameter. With a 60-in, on the other hand, all the star images are enlarged, blurred, and comparatively or completely stationary, the diameter of the smallest again being about 2″. When seeing is bad, even with small apertures the image may lose all resemblance to the normal diffraction pattern, becoming an enlarged agitated, structureless blob, in which the rings are completely submerged.

This effect of atmospheric turbulence has its naked-eye counterpart in stellar scintillation or twinkling, and its extended-image counterpart in 'boiling', both of which can be reduced or eliminated by increasing the aperture (see also section 26.9). Generally, the effects of turbulence are loss of resolving power and impaired definition of the extended image.

Twinkling was at one time ascribed, erroneously, to physiological causes. It is in fact due to refractional inequalities in the atmosphere. If the surfaces of equal refractive index are resultantly convex to the eye, the intensity of a ray passing through them will be reduced; if concave, increased. Rapid deformations of the surfaces, or the horizontal passage of a deformed surface between the objective and the star, will thus produce a flickering in the perceived intensity of the star. Chromatic scintillation, visible in the neighbourhood of the horizon when the seeing is particularly bad (Sirius frequently exhibits it on frosty nights), is caused by the light being widely enough dispersed for different wavelengths to be independently affected by atmospheric turbulence. Telescopic scintillation not only jumbles the diffraction pattern but makes it 'shudder', the image's angular displacement in any direction from its mean position being, as we have seen, of the order of 1″ for average seeing, and exceptionally 10″ or, when the seeing is very bad indeed, 20″. There is scope for much more work on the correlation of seeing with local environmental conditions.

The deformations are rapid, normally between about 5 and 100 per second, only the low-frequency scintillations being perceptible to the naked eye, and may affect the components of even a close pair quite independently. Atmospheric waves of different refractive index have distinguishably different effects upon the image according to their angular size. This is particularly noticeable in the case of a planetary image: comparatively small irregularities will cause a blurring of the limb, without introducing any movement; with larger waves, on the other hand, the limb may be sharply defined though wavering or 'boiling'.

Some satisfactorily quantitative results concerning scintillation have been obtained at Edinburgh during the past three years by Ellison and Seddon (B. 335), using a photocell and the cine-camera recording of a cathode-ray tube.* As regards the frequency of scintillation they found that at low altitudes both low-frequency (about 5 cycles per second) and high-frequency (about 100 c/s) scintillation occur, but that as the ZD decreases the former dies out. The amplitude of the brightness scintillation invariably decreases as ZD decreases, ranging (with 36 ins aperture) from about ±20% to ±50% at ZD=80° down to a normal minimum of about ±5% at the zenith. A close correlation has been established between the amplitude of the scintillation and the quality of the visual telescopic image. The amplitude also varies approximately as $1/D$ for all ZDs.

Anderson (B. 327) has found that at moderate altitudes and with fairly good seeing, the frequency of scintillation is characteristically from a few to about 25 c/s (amplitude about 1 mag), increasing to as many as 150 c/s (amplitude 2 to 3 mags) when the seeing is bad—results in general agreement with those obtained at Edinburgh. Anderson describes a simple technique for estimating the frequency of the scintillation visually. A small telescope (such as those used for reading laboratory scales, $D \approx 1$ in, $F \approx 8$ ins) is held near its objective by the right hand; the eye end is supported at the observer's eye with the forked fingers of the left hand, the heel of which is pressed against the observer's chin for steadiness. If now the OG end is made to describe small and rapid circles, persistence of vision will draw out the image of a star into a continuous circle. The speed of movement of the OG should be just rapid enough (about 5 revs/sec) to close the circle, but not so fast that successive circles overlap. If the star is scintillating it will be found that the circle consists of alternate bright and less bright segments; the number of the former can with practice be estimated quickly and reasonably accurately. This number multiplied by the number of circles described by the OG per sec gives the frequency of the scintillations.

Some light has been thrown upon the probable size and location of

* See also the investigations by Butler at Dunsink (B. 329, 330).

the atmospheric disturbances which are responsible for scintillation. That movement of the image can be converted into confusion of the image by changing from an aperture of 4 ins to one of 60 ins suggests that the distance between adjacent convexities and concavities in the theoretically plane surfaces of equal refraction lies between these limits; observations of eclipse shadow bands suggest that it lies nearer the former, and Anderson accepts about 6 ins as the order of size. Ellison and Seddon found indications that the amplitude of the scintillation reaches a maximum at about $D=3$ ins, from which they deduced this to be the order of size of the refractional irregularities causing the scintillation. As regards the height of the distorting layer above the Earth's surface, it was found at Edinburgh that Jupiter and Saturn (angular diameters 49″ and 20″) showed no trace of stellar scintillation; the scintillation of Mars (4″4) was of about half the amplitude of that of a comparison star; and that of Jupiter III (1″5) was purely stellar. They concluded that the angular size of the refractional irregularities was of the order of 3″, the angular subtention of 3 ins at a distance of about 3 miles. Since the observations were made at an altitude of 30° this indicates a vertical height of about $1\frac{1}{2}$ miles.

The region of these refractional irregularities thus lies at a considerable height above the Earth's surface—probably between 1 and 3 miles. There is also evidence that they are confined to a comparatively thin stratum. Owing to their distance they affect fairly wide topographical areas similarly, i.e. the seeing at any moment will be equally good or bad over a wide area. Atmospheric currents themselves, if there were no temperature or pressure gradients through them, would have little or no effect upon the seeing. But owing to thermal differences, their densities and therefore refractive indices are various; constantly varying barometric pressure is another factor contributing to the optical instability of the atmosphere. Although the turbulence is directly proportional to the difference in the refractive indices of the currents and surfaces of fluctuating inclination to the ray, the number of such currents also has an important bearing on the visible effect in the telescope, there being evidence that two relatively small cross-currents have a more deleterious effect on the seeing than one conspicuous one.

Among the factors contributing to turbulence are: the settling of cold air masses during the night; the condensation of clouds; the mixing of masses of unequally heated air by convection and the wind; a lapse-rate in excess of the dry adiabatic lapse-rate.

Though turbulence increases rapidly with Zenith Distance (it is, approximately, a function of sec z), it may at any moment vary widely in different parts of the sky at the same ZD. The vicinity of clouds (where steep and erratic temperature gradients are liable to occur) in particular often provides a great variety of seeing. Also, it commonly varies by 1 or 2 divisions of the Standard Scale (see section 26.10) during the course of the night; usually, though by no means invariably, the seeing deteriorates during the latter half of the night. Very clear nights, again, are almost invariably bad; windless nights, in general, good.

A strange and rarely occurring type of stellar image deformation, first described by Dawes and subsequently noticed by independent observers, is a rhythmic oscillation of the disc from circular to triangular and back again. Dawes suspected that it most commonly occurred when the wind was in the E to SE. It is almost certainly due to tube currents.

Motion of the image without deformation is only visible telescopically, since the displacements are small, their total amplitude being usually of the order of 1″, exceptionally as large as 8″ or more. The movements are slower than those resulting from turbulence, the image wandering perhaps 0″.5 on either side of a mean position in a period of a minute or so. They are also shared by objects in the same vicinity—over, say, $\frac{1}{2}$° of sky: thus traces of Merope and Alcyone, trailing a photographic plate, are almost exactly parallel to one and another. This suggests that the cause is seated in the vicinity of the telescope, rather than high in the atmosphere—a conclusion which is confirmed by the fact that the movements are independent both of ZD and of the general turbulence.

These image wanderings are in all probability caused by currents in the telescope tube, or by the displacement of relatively large volumes of air in the dome slit or the immediate vicinity of the observatory. Another effect of tube currents (section 11.2) is a slow and continuous shifting of the focus. This is fatal to definition, especially of extended detail, the greatest precision in focusing being at all times a prerequisite of satisfactory observation.

The type of image motion known as 'jumping the wire', familiar to transit observers, is a psychological effect due to fatigue.

26.8 Direct visibility of atmospheric irregularities

Atmospheric thermal irregularities are visible to the naked eye as 'shadow bands' when the angular size of the source is reduced, as during a solar eclipse just short of the total phase.

Under certain circumstances they can also be observed telescopically. Given sufficient aperture—not less than about 12 ins—they can be seen as more or less straight, parallel bands rapidly crossing an objective flooded with light from a bright star or planet, the ocular having been removed and the eye placed at the focus. With small apertures the objective is too small compared with the thermal waves for them to be seen individually, though their integrated effect may be seen as a flickering of the total light received. Very similar, though artificially produced, effects can be seen with a small instrument by directing it above any source of heat.

If an ocular is used, and a bright star is thrown out of focus, they may nearly always be seen, though once again the individual waves are not distinguishable as such, owing to the rapidity of their motion. The general effect is that of straight, parallel linear streams moving in the direction of the wind prevailing at the height at which they are formed. They may also be seen crossing the disc and limb of a slightly out-of-focus planetary image; they come to a sharp focus just outside that for objects at infinity, whence their distance can be roughly measured in a telescope of considerable focal length. Rarely, they are seen crossing the projected solar disc of a Sun telescope (where they are not to be confused with the commoner phenomenon of 'boiling').

More slowly moving, wavy bands, often distorted by whorls, are caused by currents in the instrument itself.

26.9 Aperture and seeing

The waves visible at the focus are the 'shadows' of the surfaces of equal refractive index, moving relative to one another, locally distorted and mutually inclined. One can look upon them as a wave system of wavelength λ (Douglass concluded from his careful investigations—B. 331, 333—that the linear values lie usually between $\frac{2}{3}$ and 4 ins, in general agreement with the results of Anderson, and of Ellison and Seddon); or one can consider the waves, taken altogether, as producing a residual surface, the normal to which oscillates within a cone of angle $2t$, calling t the turbulence.

For twinkling to be observed, the aperture of the image-forming lens must be small compared with λ; hence twinkling is a characteristically naked-eye rather than a telescopic phenomenon. As D increases, so that the entry pencil includes more and more complete waves, the movement of the image decreases, but the breakdown of the diffraction pattern grows progressively worse. Thus what in a small instrument appears as a well-defined but much agitated stellar image is in a large instrument a practically motionless, structureless, circular blob. The following figures are necessarily very approximate:

$D < \frac{1}{2}\lambda$: twinkling;

$D = \frac{1}{2}\lambda$ to $D = 3\lambda$: image in motion, but rings and disc intact;

$D=3\lambda$ to $D=5\lambda$: decreased motion, disc destroyed, rings still intact;
$D=5\lambda$ to $D=8\lambda$: image static, rings still intact;
$D>8$: image static, neither disc nor rings visible.

The change-over from motion to confusion at about $D=5\lambda$ can be clearly witnessed by observing the limb of a planet with sufficient aperture to show it as fuzzily defined though free from 'boiling', and then progressively reducing the aperture. Conversely, these two types of poor seeing, as experienced with a given instrument, are determined by the size of the atmospheric waves at the time.

Generally speaking, image motion predominates over confusion of the diffraction pattern up to an aperture of about 4 ins; from 4 ins to 8 ins both factors are present, the former decreasing in importance; above about 8 ins deformation is always as great as is compatible with a given degree of turbulence, motion of the image being negligible. No doubt connected with this distinction is the observational fact that large apertures are at a greater disadvantage (compared with smaller) in the definition of extended images than in the resolution of stellar points. Thus under seeing conditions which prevent a 12-in from showing any planetary detail invisible with a 6-in it will nevertheless resolve closer doubles than the smaller instrument.

Since the quality of the image for a given turbulence, t, is a function of the aperture used, and the size of the diffraction disc is also (inversely) related to the aperture, it is possible to correlate t with r, the radius of the spurious disc. It is in this way that t, an unobservable, is derived and utilised in the third of the Standard Scales detailed below. Within the range of apertures which yield maximal image distortion for a given t-value—above 4 ins sometimes, and above 8 ins almost invariably—the extent of the deformation is a function of the relative angular magnitudes of t and r: for a given value of t, as r diminishes, so the deformation of the image increases; and conversely, for a given value of r, as t increases, so the deformation of the image increases. Hence the deformation of the image is a function of $\frac{t}{r}$.

When $\frac{t}{r}<\frac{1}{4}$ (i.e. if $D<5$ ins for average seeing, when $t=0''25$), the image will suffer little distortion, but will be displaced bodily through a total amplitude of about $2t$. With an aperture of 5 ins and the same t-value $\left(\frac{t}{r}=\frac{1}{4}\right)$, flickering and moving condensations begin to appear in the rings. With double this aperture, or double the t-value and the same aperture (in either case $\frac{t}{r}=\frac{1}{2}$), the rings begin to disintegrate, the fragments are in rapid motion, and the whole image suffers continuous distortion from the circular in random directions. Again doubling the aperture or the turbulence, giving $\frac{t}{r}=1$, turns the image into an agitated, structureless patch, the distinction between disc and rings being lost altogether. If $\frac{t}{r}>1\frac{1}{2}$, the image is not unlike a tiny, ill-defined, scintillating planetary disc of diameter $2t$ (i.e. equal in diameter to the amplitude of the image's displacements in a smaller telescope), 'boiling' but suffering no bodily displacement. The effect of this upon an extended image is seriously to impair the definition, with total suppression of the finer detail, since the image of every point on the object is a spot of radius t. Further increase of aperture will brighten the image, but will no longer reduce its angular dimensions, which remain at $d=2t$. These successive conditions can be observed with a given instrument at a given time by observing a sequence of stars of increasing ZD.

In an actual comparison of the performances of a 5-in and a 15-in, made by Pickering under varied atmospheric conditions, the following results appeared. They are in general agreement with the foregoing.

Poor seeing: moderately close, equally bright doubles: 5-in as good as the 15-in;
faint or unequal pairs: 15-in superior;

Seeing better than average: 15-in superior all round.

The conclusion to be drawn is that for work demanding fine definition of the image, the useful aperture is limited by the atmosphere, large apertures only coming into their own when the seeing is exceptionally good.

See further: section 27.1.

26.10 Scales of seeing

To be able to estimate the reliability of an observation, some indication of the condition of the seeing at the time must be given. Each period of observation should be preceded by an investigation of the degree of turbulence. There are numerous rough-and-ready methods of doing so —some are mentioned below—but the only satisfactory method involves the employment of a scale of atmospheric states, dependent upon the detailed appearance of a star image.

Such a scale should possess the following characteristics: (i) objectivity: it must be impersonal, independent of the observer's experience and local conditions, including the aperture of his instrument, so as to allow the direct comparison of his observations with others made by different observers at different times and places, with the confidence that the scale numbers, as used by all of them, refer to identical states of seeing; (ii) universal applicability; (iii) precision: involving numerical rather than verbal description; (iv) simplicity.

Many approximate criteria have been proposed at one time or another, none of which is ideal in all respects. Those based on the resolution of close doubles or the definition of extended images—a planetary limb, the fine detail of the lunar or a planetary surface, etc—allow rough qualitative estimates to be made; but in general they are less precise (being ill-adapted to numerical expression), and are also more dependent on instrumental factors, than scales based upon the stability or otherwise of a bright (mag 1 or 2) stellar image. Even as a criterion of the definition of planetary detail, the condition of the limb is an untrustworthy guide: it may be clearly defined, yet the definition of the detail be poor, or vice versa.

Five criteria, of the type described above, were devised by W. H. Pickering. In order of increasing sensitivity, they are:

(a) The resolution of double stars whose separation is near to the threshold R' of the instrument.

(b) The visibility of elliptical outline of Jupiter I or IV. It is difficult to see what reliability or validity this criterion can have, however, since the apparent ellipticity vanishes when the satellite is observed with a large enough telescope.

(c) Resolution of the double 'canal' in Aristillus. The 'canals' are two fine dark lines diverging from a point on the crater floor, whence they cross the NW wall, a few miles outside which they become invisible. Length, about 25 miles; width outside the crater varies with colongitude from about 100 yards ($0''05$) to 500 yards ($0''25$). Duplication visible between colongitudes 35° and 65°, and 110° and 140°, i.e. on only about five nights throughout the lunation. Resolution of the 'canals' inside the crater can just be glimpsed with $D=8$ ins, and is therefore a test of seeing with this and larger apertures; the external 'canals' can be resolved with $D=3$ ins, given the seeing.

(d) Visibility of a small group of 'canals' outside Aristillus, N and E of the double 'canal'.

(e) Visibility of a narrow bright line running parallel to the 'upper' reaches of the Mt Hadley 'river bed'; 5 miles long and probably not more than about 50 yards wide ($0''025$). Only visible between colongitudes about 43° and 57°.

The justification for the employment of such criteria of seeing as those described above is the claim that the visibility of extended objects of comparatively little contrast is more susceptible to turbulence than the minute images of stars, heavily contrasted with their background—at any rate, that a given degree of turbulence will affect the two classes of object differently. And that therefore the visibility of planetary and lunar detail does not run parallel with the perfection or otherwise of stellar images. And that therefore a scale of seeing based on the latter is not necessarily applicable to planetary and lunar observation.

In practice it is difficult to draw this distinction. Turbulence simultaneously affects the appearance of the stellar image, reduces the telescope's resolving power as adjudged by its performance on close doubles, and reduces the definition and resolution of extended images.

Many observers find it convenient to employ 'private' scales of seeing, which they have worked out for themselves to suit their own particular requirements. Such scales should be calibrated against one of the Standard Scales, so that the equivalent Standard grading can be appended to all observations.

Below are given details of three suggested Standard Scales of Seeing, based on the appearance of the stellar image. The first two, designed respectively for 6-in and for 5-in objectives, suffer from the following disadvantages: (a) The scale number that would be ascribed to a given state of seeing will vary with the aperture used. The calibration can be adjusted to another aperture on a basis of adding 2 to the scale number when the aperture is doubled: thus perfect seeing would be rated as 14 on Scale A with a 12-in, 12 on Scale B with a 10-in. But unfortunately even the descriptions of the several criteria vary to some extent with aperture: with small apertures an improvement of seeing causes the disc to separate out of the confused image before the rings; with larger apertures there is a progressive tendency for the rings to separate out first, the disc not becoming well defined till the rings, though they may still be in motion, are complete. (b) No means is provided for correcting the estimated scale number for the ZD of the star, thus converting it to standard conditions; for this reason the ZD of the star employed in the estimation of the scale number should always be given. As an example of the effect that differences of ZD can have, Douglass quotes the case of a difference of scale number of 5 when determined by a star at the zenith and one at a ZD of 80°; the amplitude of the motion of the image being 1″ in the former case and 5″7 in the latter.

SCALE A

Designed by W. H. Pickering for use with a 6-in objective employing a magnification of from $60D$ to $100D$.

0: disc and rings a confused mass, in violent motion; image larger than the true diffraction pattern, and varying continuously in size;

2: disc and rings a confused mass, in constant motion; image not enlarged;

4: disc well defined, rings blurred;

6: disc well defined, rings visible but broken;

8: disc well defined, rings complete but moving;

10: disc well defined, rings stationary; image moving in field;

12: disc well defined, rings stationary; image stationary in field.

SCALE B

A development of Scale A, compared with which simplicity is sacrificed for greater precision. Designed for use with a 5-in objective; recommended magnification, not less than $60D$; recommended test star, mag 1 or 2.

1: disc and rings undifferentiated; image usually about twice the size of the true diffraction pattern;

2: disc and rings undifferentiated; image occasionally twice the size of the true diffraction pattern; } Very poor

3: disc and rings undifferentiated; image still enlarged, but brighter at the centre;

4: disc often visible; also occasional short arcs of the rings;

5: disc always visible; short arcs of the rings visible for about half the time; } Poor

6: disc always visible, though not sharply defined; short arcs of the rings visible all the time;

7: disc sometimes sharply defined, and distinct from rings; } Good

8: disc always sharply defined; inner ring in constant motion;

9: disc always sharply defined; inner ring stationary; } Excellent

10: disc always sharply defined; all rings virtually stationary. } Perfect

SCALE C

The third Standard Scale to be described here is based upon the turbulence, t, and has the advantage over Pickering's scales that a correction is applied to remove the effect of ZD from the observations.

For several bright stars, differing in ZD, and observed with a magnification of not less than about $40D$, the quantity $\dfrac{t}{r}$ is estimated (see pp. 464–5). Let $\dfrac{t}{r} = x$.

Then, from equation (d) on p. 38:

$$t = \frac{5x}{D}$$

can be derived for each star. The value of t at the zenith (t_0) is given by

$$t_0 = \frac{t}{\sec z}$$

where z is the ZD. Hence for each value of t a corresponding value of t_0 is derived, the mean of which is the required criterion of the seeing.

Example: The observations of three stars and the deductions made from them are set out below, assuming $D=4$ ins:

ZD	$\sec z$	$x=t/r$ (estimated)	$t=\dfrac{5x}{D}$	$t_0=\dfrac{t}{\sec z}$
20°	1·064	0·75	0·94	0·88 ⎫
40	1·305	1·0	1·25	0·96 ⎬ 0·90
70	2·924	2·0	2·50	0·85 ⎭

Or t_0 can be derived in each case direct from (combining the two foregoing expressions for t and t_0),

$$t_0 = \frac{5x}{D \sec z}$$

26.11 Topographical and meteorological factors

Scattered references to the dependence of seeing upon local weather and topography are to be found in the literature, but probably very little in the way of universally applicable conclusions can be arrived at. The regular observer will sooner or later discover what meteorological conditions favour observation on the subsequent night or during the subsequent period; especially if he keeps a record of the wind direction and force, type of cloud prevalent, maximum and minimum temperatures, barometric reading and humidity. Annual and diurnal variations in the seeing should also be sought.

Below are collected a number of miscellaneous correlations.

Geographical:

1. The eastern States of America are much inferior to California, Arizona, and the SW generally.

2. England is superior to central France.

3. Conditions in France are better away from the Atlantic coastal zone.

4. Italy, and the Mediterranean seaboard generally, is superior to northern Europe.

5. Increasing atmospheric clarity: northern Europe (about 180 nights per annum), Italy, Greece, Egypt, Arizona (about 330 per annum). For atmospheric clarity at Greenwich, see section 26.6.

Topographical:

6. Avoid the vicinity of valleys, since cold air flows into them at night; if it overflows, usually during the latter part of the night, it may utterly ruin definition.

7. Avoid the foot or slope of a hill, since cold air flows down it at night, forming a lake at the foot which may be several hundred feet deep by the end of the night.

8. Avoid low-lying terrain (which usually contains a river or other water) owing to the high incidence of mists, especially during the winter (e.g. the Thames valley).

9. Avoid the crest of a hill: it is likely to be windy, and even on windless nights there will be constant currents owing to the sinking of cold air over the crest.

10. A high altitude is desirable, since the greater part of the dust and artificial pollution, as well as mists and fog, occur near sea level.

11. Surroundings covered by vegetation prevent the earth in the observatory's vicinity heating up badly during the day. The ever-present danger of forest fires, which may denude the mountain and impair definition over a period of years, is one of the few disadvantages of the site of the Mt Palomar Observatory.

12. The foregoing points would combine to suggest the ideal site as being a dry plateau of moderate elevation above sea level, covered with vegetation, located in a dry, clear climate, and far from valleys and the sea.

Latitude:

13. The advantages of low latitudes are: (*a*) small nightly variation of temperature, (*b*) calm nights, (*c*) they lie outside the anticyclone belt, with its constantly changing barometric pressure.

Barometer:

14. It has been noticed that in East Anglia the seeing is usually good during a period of rising barometer, following a depression.

Temperature:

15. Pickering noticed that in Jamaica there was a marked tendency for a temperature drop to occur within three days of bad seeing setting in.

16. Good seeing in NW Europe is favoured by moderate temperatures, i.e. spring and autumn.

Humidity:

17. Pickering noted that in Jamaica the best seeing occurred on damp nights. Herschel found the same at Slough.

18. In Melbourne the quality of the seeing often permits a correct forecast of rain or cloud on the following day.

19. In England winter fog conditions, clearing towards nightfall, are often followed by good seeing.

Wind:

20. Herschel noted that high winds usually bring poor seeing.

21. Dawes thought that the triangular deformation of star discs was favoured by an E or SE wind. No subsequent evidence as to the truth of this conjecture.

22. In East Anglia the seeing tends to be good when the wind is between W and NE.

23. Spain (1894): Landerer found that scintillation was maximum with NW winds and at weather extremes generally; minimum with damp off-sea winds.

24. When testing sites for the Radcliffe Observatory near Pretoria in 1930, W. H. Steavenson found: S wind, worst seeing; N or E wind, seeing almost invariably good. The same was found at Johannesburg.

The geographical conditions imposed by the twin considerations of maximum number of observing hours and quality of seeing often conflict—though common ground can be found—but this is not likely to be of more than theoretical interest to the amateur, who usually cannot choose an observing site apart from his home or place of work. He should, nevertheless, be conscious of how his own site fits into the general pattern.

DIAPHRAGMS AND FILTERS

27.1 Diaphragms

(a) Anti-glare or anti-reflection stops, exactly defining the truncated cone whose base, in the case of a refractor, is the OG and whose apex is the exit pupil; these diaphragms are normally incorporated in the tube as supplied by the telescope maker. (b) Field stop, limiting the visible field to the field of full illumination (see section 3.8). (c) Aperture stops, though found in cameras, are not normally incorporated in telescopes, whose objective itself is, functionally, the aperture stop for the ocular. (d) Additional diaphragms, at the entry or exit pupil, are justified when it is required (i) to reduce any of the characteristics of telescopic vision which are dependent on aperture: in particular, light grasp, or the effects of atmospheric turbulence on definition; (ii) to modify the diffraction pattern of the image by changing the size or shape of the entry pupil.

An entry-pupil diaphragm will reduce the instrument's light grasp, therefore the illumination of the image; will increase the size of the spurious disc, therefore reduce the resolving power; will reduce t/r and D/λ, therefore reduce the confusion of the image due to bad seeing (see section 26.9). Thus it alters the nature, not only the illumination, of the image.

An exit-pupil diaphragm will similarly reduce both the brightness and the resolution of the image. It would be natural to suppose—and is often stated—that the resolution of the image is unaffected by interposing a diaphragm at the exit pupil, since the characteristics of the light pencil that determine the resolution of the image are already established during its passage through the objective, and cannot be altered by its subsequent passage through a diaphragm.

This reasoning is fallacious, however, and might equally be used to prove that an exit-pupil diaphragm does not reduce the brightness of the image either, since light grasp is as dependent on D as resolution is. The exit-pupil diaphragm admittedly affects neither the light grasp nor the resolving power of the objective as such—the brightness and resolution of the image in the focal plane are those appropriate to the full aperture of the objective—but it does reduce the light grasp and resolving power of the telescope as a whole. The exit pupil is the image of the objective—rays from the boundary of the latter pass through the boundary of the pupil—and to reduce its aperture is *effectively* to reduce D.

Experiments on the resolution of double stars with exit-pupil diaphragms of graded sizes may in some cases appear to indicate an improvement of resolution when the pupil is restricted. Where this occurs, however, it means that the relative difficulty of resolving the pair without the diaphragm is due so preponderantly to excessive brightness that the apparent resolution of the pair is improved by reducing the brightness even at the expense of increasing the angular sizes of the images.

Hence the reduction of image glare (e.g. of the Moon or Venus) can be effected equally by an objective diaphragm or an exit-pupil diaphragm, with the latter having the advantage of extreme convenience in use. No expensive iris diaphragm, with controls leading back to the observing position, is required. Each ocular is fitted with a slide to take a strip of aluminium, down the central line of which a series of graded holes have been drilled. Care must be taken that each hole comes centrally into position in front of the exit pupil, and that the slide does not force the eye away from the exit pupil; this will probably mean cutting away the flange or rim surrounding the eye end of the ocular.

The use of objective screens—such as perforated zinc, or closely reticulated material like some types of mosquito netting, or even ¼-in mesh wire for an aperture of 12 ins—was first suggested by J. F. W. Herschel and Dawes, and has been advocated since by W. H. Pickering and others, both for planetary and for stellar observation. If the apertures in the screen are not very small compared with the total aperture of the objective, most undesirable modifications of the diffraction pattern of the image could result. In any case numbers of diffraction spectra are produced radially round the object. The latter, however, is seen in its natural colours with the full effective resolution, though not the full light grasp, of the whole aperture. The remarks below—on diaphragming to reduce the effects of poor seeing—apply equally to the use of perforated screens; though ultimately the proof of the pudding must always be its eating.

The position regarding the use of an objective diaphragm when the seeing is poor is an interesting one. Planetary observers in general advocate the reduction of aperture if the full aperture reveals too grossly the distortion of the image or the confusion of definition which springs from a high value of t, the turbulence; and, indeed, since the deformation of the image is greater the larger the aperture, this would seem to be reasonable. Douglass, for example, recommended a fairly small aperture for planetary observation unless the seeing is excellent (though full aperture always for such work as micrometer measures). Jarry-Desloges, working with a 20-in refractor under abnormally favourable conditions in Algeria, found that on 3 nights out of 4 a reduction of aperture to $8\frac{1}{2}$–14 ins gave better planetary images than the full aperture; and that only on about 2–5 nights in 100 did the performance of the full aperture surpass that of the 14-in diaphragm. He considered that 20 ins is the effective limit for an instrument intended primarily for planetary work, a conclusion with which Pickering—also working under exceptionally favourable atmospheric conditions in Jamaica—agreed. With seeing 2–3 on the Standard Scale he found 5 ins to be about the maximum useful aperture, increasing to 8 ins with seeing 5–6, and from 11 to 20 ins only when the seeing was better than 6. Pickering has further stated that although a 20-in will, when the seeing is poor, divide binaries which cannot be separated with a 10-in, the 20-in will nevertheless show no more planetary detail than the 10-in; and that when t is high a 5-in will show all the planetary detail visible under the same conditions even with the Mt Wilson 100-in.

All this is no doubt perfectly true, but the word 'better' in the statement, 'Stopping down, when the seeing is bad, gives a better image', conceals a confusion. The image, it is true, will appear superficially better, but only because it is neither bright enough nor highly enough resolved to reveal fully the effects of the atmospheric turbulence*: it is not more highly resolved—it is indeed less so, of course. Stopping down suppresses the more glaring effects of turbulence, but suppresses also the finer detail that might otherwise have been glimpsed during the fleeting moments of improved seeing that so often occur. The full-aperture image looks horrible, but contains the resolved detail; the reduced-aperture image looks a great deal smarter, but only because its threshold of resolution is higher—the detail is actually not there, in the focal plane image.

In short, the use of an objective diaphragm reduces the distortion of the image, but at the same time reduces the resolving power (by increasing the angular size of the spurious discs) and the light grasp of the instrument. A purely superficial improvement has been gained, no real

* For the same reason it is seldom necessary to employ an objective stop with apertures smaller than about 4 ins.

improvement. If the seeing really is so bad that observation with full aperture is 'impossible', it is better (unless the telescope is of such size that even reduced aperture is useful) to persevere or pack up, rather than to stop down. The employment of diaphragms is something of an obsession with many observers; used at random, and without a proper realisation of their action, they will more often do harm than good.

It is generally agreed that in stellar observation reduction of aperture is seldom if ever a help, such authorities as Schiaparelli, Burnham and Aitken all speaking against the practice. A possible exception to this general embargo is the case of very unequal pairs, or possibly very bright pairs. A polygonal objective diaphragm, capable of being rotated about the optical axis by means of controls from the observing position, is useful in the case of unequal doubles, not because of the resultant reduction of light grasp (amounting to a little over 17% in the case of a

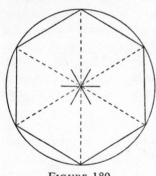

FIGURE 180

hexagon), but of the modification of the diffraction pattern that it produces. Figure 180 shows a hexagonal diaphragm and the corresponding disposition of the rays of the diffraction pattern. If the diaphragm is rotated so that the line joining the stars is parallel to one of its diagonals the *comes* will lie in one of the artificially darkened areas between the arms of the primary image, and its visibility will be slightly augmented. Another device, suggested by Sir J. Herschel, is

a triangular diaphragm, producing an illumination reduction in the region of 40% and a 6-rayed diffraction pattern; these rays, however, are brighter than those of the hexagon, which is therefore to be preferred.

A similar improvement can be effected by rotating the flat and ocular ring through 45°–60° if the *comes* lies on one of the arms of the diffraction cross.* Given sufficient aperture, the diffraction effects of the central obstruction can, of course, be avoided by using an eccentric diaphragm. Thus C. J. Tenukest's 20-in Cassegrain at Sydney can still give 7 ins unobstructed aperture with an eccentric stop, and W. H. Steavenson reports that his 30-in reflector gives excellent refractor-type images with a 11½-in eccentric stop.

* The modifications of the image structure due to a central obstruction are described in sections 2.2, 2.3.

27.2 Filters

There are two distinct factors involved in the improvement of the image by means of a filter:

(a) reduction of glare, expecially with low magnifications and large aperture, without either reducing and resolving power or increasing magnification beyond the point dictated by the state of the atmosphere;

(b) improvement of definition and augmentation of tonal and colour contrasts.

As regards (a) almost any colour filter will fill the bill, a neutral one being the obvious choice, though various tints have been used successfully.

The second factor, (b), operates partly through the filter's selective transmission (eliminating some of the secondary spectrum in refractors), and partly, it is likely, by an actual improvement of the diffraction pattern of the image (the spurious discs of bright stars can be reduced in size by suitable filters). As examples of the use of filters in planetary work may be cited: increased contrast between Jovian white spots and the belts in which they are situated, by means of a Wratten K.3; increased tonal contrast in the Martian image with the use of a very pale-red filter; wide use of filters in lunar observation.* Green and yellow filters have proved to be particularly useful in increasing contrast between extended detail.

Graded neutral filters may easily be made by exposing film for different periods—according to the degree of absorption required, and ultimately therefore on the type of observation for which they are required. Six or eight, mounted between glass slips in a circular brass disc, rotatable about a central pivot which is carried on a clip fixing it eccentrically to the ocular, provide an accessory whereby the intensity of the image can be quickly varied over as wide a range as required (Figure 181). Neutral filters (granular silver or colloidal carbon) are also supplied by the makers of colour filters,

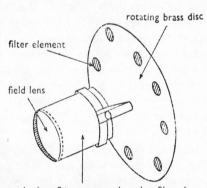

rotating brass disc

filter element

field lens

ocular (eye flange removed so that filter does not force the eye back from the exit pupil)

FIGURE 181

* See *O.A.A.*, section 2.5.

to specified densities accurate to about 5%.* Complete neutrality is, however, impossible, a typical (Ilford) 'neutral' filter transmitting about 5% at 4000Å, 20% at 7400Å, and 10% throughout the range 4600Å–6400Å.

Such firms as Eastman Kodak, Ilford and Dufay-Chromex supply a wide range of colour filters (Kodak's Wratten series alone contains over one hundred), and will supply on request full details of the transmissions in different wavelengths, as well as the photographic exposure factors, of their products. The Ilford 'Spectrum' series is particularly to be noted, since these filters are virtually monochromatic, their transmissions being limited to as little as 450Å.

The transmission of a neutral filter, or of a colour filter in a particular wavelength, is the intensity of the transmitted light expressed as a percentage of the intensity of the incident light, the incident beam being parallel and normal to the surface of the filter. Hence if I_0 is the intensity of the incident radiation, and I that of the transmitted radiation, we have for the transmission of the filter in wavelength λ,

$$T_\lambda = 100\, \frac{I}{I_0}$$

The density, D, of the filter is given by

$$D = \log \frac{I_0}{I}$$

or

$$D = \log \frac{1}{T}$$

The combined density of two or more filters is equal to the numerical sum of their individual densities, thus making D a more convenient quantity to manipulate than T. D is tabulated against T below.

Colour filters consist of organic dyes held in suspension in gelatin or cellulose acetate film of uniform thickness. This film, as employed for astronomical purposes, is cemented between two plane parallel (or, for the utmost accuracy, optically flat) glass slips. It should be remembered that unless the glass is replaced by quartz, filters of whatever transmission characteristics will remove the ultraviolet beyond about 3300Å. All gelatin filters, on the other hand, transmit freely in the infrared.

Gelatin swells and becomes distorted when exposed to moisture. Hence even glass-cased filters should always be kept in a dry place when not in use, and particular attention paid to the edges of the filter where, if anywhere, moisture may get at the film. Provided they are stored in a

* See also sections 21.2.1, 21.3.1.

Density	Transmission (%)	Density	Transmission (%)
0·00	100	0·84	14
0·02	96	0·86	14
0·04	91	0·88	13
0·06	87		
0·08	83	0·90	13
		0·92	12
0·10	79	0·94	11
0·12	76	0·96	11
0·14	72	0·98	10
0·16	69		
0·18	66	1·00	10·0
		1·02	9·6
0·20	63	1·04	9·1
0·22	60	1·06	8·7
0·24	58	1·08	8·3
0·26	55		
0·28	52	1·10	7·9
		1·12	7·6
0·30	50	1·14	7·2
0·32	48	1·16	6·9
0·34	46	1·18	6·6
0·36	44		
0·38	42	1·20	6·3
		1·22	6·0
0·40	40	1·24	5·8
0·42	38	1·26	5·5
0·44	36	1·28	5·2
0·46	35		
0·48	33	1·30	5·0
		1·32	4·8
0·50	32	1·34	4·6
0·52	30	1·36	4·4
0·54	29	1·38	4·2
0·56	28		
0·58	26	1·40	4·0
		1·42	3·8
0·60	25	1·44	3·6
0·62	24	1·46	3·5
0·64	23	1·48	3·3
0·66	22		
0·68	21	1·50	3·2
		1·52	3·0
0·70	20	1·54	2·9
0·72	19	1·56	2·8
0·74	18	1·58	2·6
0·76	17		
0·78	17	1·60	2·5
		1·62	2·4
0·80	16	1·64	2·3
0·82	15	1·66	2·2

Density	Transmission (%)	Density	Transmission (%)
1·68	2·1	2·38	0·42
1·70	2·0	2·40	0·40
1·72	1·9	2·42	0·38
1·74	1·8	2·44	0·36
1·76	1·7	2·46	0·35
1·78	1·7	2·48	0·33
1·80	1·6	2·50	0·32
1·82	1·5	2·52	0·30
1·84	1·4	2·54	0·29
1·86	1·4	2·56	0·28
1·88	1·3	2·58	0·26
1·90	1·3	2·60	0·25
1·92	1·2	2·62	0·24
1·94	1·1	2·64	0·23
1·96	1·1	2·66	0·22
1·98	1·0	2·68	0·21
2·00	1·00	2·70	0·20
2·02	0·96	2·72	0·19
2·04	0·91	2·74	0·18
2·06	0·87	2·76	0·17
2·08	0·83	2·78	0·17
2·10	0·79	2·80	0·16
2·12	0·76	2·82	0·15
2·14	0·72	2·84	0·14
2·16	0·69	2·86	0·14
2·18	0·66	2·88	0·13
2·20	0·63	2·90	0·13
2·22	0·60	2·92	0·12
2·24	0·58	2·94	0·11
2·26	0·55	2·96	0·11
2·28	0·52	2·98	0·10
2·30	0·50	3·00	0·10
2·32	0·48	4·00	0·01
2·34	0·46	5·00	0·001
2·36	0·44		

dustproof container, and their surfaces never touched with the fingers, cemented filters should require cleaning only at very long intervals. When it does become necessary, they should first be brushed with a sable brush to remove dust, and then gently polished with a piece of soft dry material such as silk or high-grade flannel; then, if this is not enough, with a screw of tissue paper damped, but not saturated, with absolute alcohol. Undue heating will soften the balsam cement, hence filters of this type cannot be used for solar work.

Cellulose acetate filters (manufactured by Dufay-Chromex) are more resistant to rough handling than gelatin, being little affected by fingering, damp, or grease; in particular, they are not distorted by water. Cellulose acetate filters, cemented between glass, are unaffected even by immersion in water, and may therefore be used without fear of the consequences in the heaviest dew. Within the range −40°C to 70°C they are unaffected by temperature.

Most filters are affected in their transmission characteristics by exposure to light over long periods (from one to twelve months' total exposure) and should therefore be kept in a lightproof as well as dust- and damp-proof container when not in use.

SECTION 28

TIME AND CLOCKS

Two systems of time reckoning are based upon the position of the Sun: Apparent (Solar) Time and Mean (Solar) Time.

28.1 Apparent Time (AT)

The Hour Angle of the true Sun, measured from lower meridian transit (i.e. midnight). This is the time told by a sundial.

AT is useless for civil time reckoning, since, for two reasons, it is non-uniform: (a) the Sun's motion in the ecliptic is not uniform, (b) Hour Angle is measured in the equator, which is inclined to the ecliptic.

28.2 Mean Time (MT)

To avoid these sources of non-uniformity of the time units a fictitious *mean sun* is invoked: this moves in the equator with uniform angular velocity.

Mean Time is then defined as the Hour Angle of the mean sun measured from lower meridian transit (i.e. midnight). Thus the MT at a given moment is the Hour Angle of the mean sun $\pm 12^{\mathrm{h}}$.

28.3 Equation of Time (E/T)

The E/T is the difference in RA between the mean sun and the true Sun, or $\Upsilon M - \Upsilon T'$ in Figure 182. It is thus the correction to be applied to MT in order to derive AT, being positive when the true Sun is W of the mean sun, negative when E of the mean sun. The value of the E/T varies continuously throughout the year, the limiting and zero values being as follows:

December	0	minutes
February	−14·25	
April	0	
May	+ 3·75	

480

June	0
July	− 6·5
September	0
November	+16·25

For Table of the daily values of the E/T, see the *N.A.*

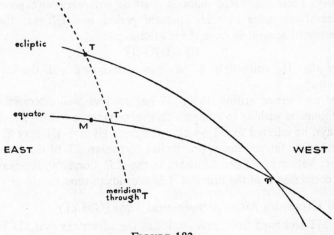

FIGURE 182

T : true Sun
M : mean Sun
♈ : First Point of Aries

28.4 Greenwich Mean Time (GMT)

Mean Time for the meridian of Greenwich, reckoned from midnight. Before 1925 the term 'GMT' was used for MT reckoned from midday, and to avoid confusion it is recommended by the I.A.U. that the term *Universal Time* (UT) now be used for GMT reckoned from midnight. *Greenwich Civil Time* is synonymous with GMT.

28.5 Ephemeris Time (ET)

Following upon recommendations of the International Astronomical Union, various changes have been made to the content of the *N.A.*, and the tables for 1960 *et seq.* will be published in London and Washington under the respective titles *The Astronomical Ephemeris* and *The American Ephemeris* (*A.E.*).

The changes which have been made have arisen because of the need for greater accuracy in the predictions of astronomical events. Predictions tabulated for years prior to 1960 were based on the rotation of the

481

Earth. However, the Earth's axial angular velocity is not quite uniform and hence equal intervals of UT, based on the Earth's rotation, are not equal by absolute standards. In an attempt to overcome this difficulty, accurate predictions will in future be referred to a new time scale, the time on this scale being called Ephemeris Time (ET), based on the orbital motions of the Earth and Moon calculated from gravitational theory. These calculated motions must be compared with positional observations made over an extended period, and ET may then be determined accurately from the relationship

$$ET=UT+\varDelta T$$

only after the correction $\varDelta T$ has been determined with the requisite accuracy.

At the time of writing (1960), $\varDelta T$ has not been well-determined and astronomers wishing to compare their observations—which should, as always, be referred to UT—with predictions given in ET may add the appropriate (at present, approximate) correction $\varDelta T$ to the Universal Time. Values of $\varDelta T$ are tabulated in the *A.E.* For 1960, for example, the correction is of the order of $+35$ seconds of time.

28.6 Greenwich Mean Astronomical Time (GMAT)

GMT reckoned from midday. It has the advantage over UT of not involving a change of date at midnight, in the middle of the astronomer's working period; consequently employed in some fields of observation, though UT is the generally employed system, and should always be used in any rewriting of the observations. In practice it is convenient to ignore the change of date when recording observations in UT at the time they are made (thus carrying on the reckoning to 25^h, 26^h, etc), and only to make the necessary and easy adjustment when reporting or publishing relevant observations.

Note that the astronomical date does not change at 24^h GMT but at 12^h. Hence the date of the period midnight-to-midday on, say, February 10 (civil date) is still February 9.

$$UT=GMAT+12^h$$
$$GMAT=UT-12^h$$

28.7 Standard Time, or Zone Time

The Local Mean Time (LMT) in any of the recognised Time Zones into which the Earth's surface is divided. E.g. the 1-hour-separated Eastern Standard Time, Central Standard Time, Mountain Standard Time, and Pacific Standard Time of the North American continent. Similarly GMT is the Standard Time for all places lying between longitudes $7\frac{1}{2}°$ East and West. Standard Times normally differ from GMT by a whole number of hours.

28.8 British Summer Time (BST)

In operation between April and October; it is 1^h ahead of GMT:

$$GMT = BST - 1^h$$
$$BST = GMT + 1^h$$

British Double Summer Time: a wartime measure, now discontinued, which was operative between May and August:

$$GMT = BDST - 2^h$$
$$BDST = GMT + 2^h$$

28.9 Julian Day (JD)

Calendar consisting of a continuous series of numbered days, and decimals of a day, unbroken by subdivisions into months and years. The Julian Day begins at midday, that beginning on 1951 January 1·5 being 2,433,648. Tables in the *N.A.* and *A.E.* give the JD Number for Jan 0 of every leap year from A.D. 0 to A.D. 2296.

O.A.A., section 17.14 contains a Table giving the JD of the zero day of each month from 1950 to 1975.

28.10 Solar days and years

The mean solar day is the interval between successive transits of the mean sun over the same meridian:

$$24^h MT = 24^h 3^m 56^s\!.555 \text{ mean Sidereal Time}$$

The solar year, or Besselian fictitious year, begins when the sun's mean longitude is $280°$, which instant is written 1950·0, 1951·0, etc.

28.11 Recording the time of observations

(*a*) Employ UT rather than GMT.

(*b*) Always specify whether using UT or GMAT.

(*c*) Never use BST, BDST, or a LMT other than that of Zone 1 (i.e. GMT) for recording observations made in Great Britain.

28.12 Sidereal Time (ST)

Owing to the orbital motion of the Earth, causing the Sun to advance eastward along the ecliptic, the star sphere completes one revolution in a slightly shorter period than the Sun. ST is based upon the rotation of the star sphere, and a ST clock and a MT clock will tell the same time only once a year—at the autumnal equinox (about September 23), when the Sun and the First Point of Aries (FPA) are 12^h apart in RA.

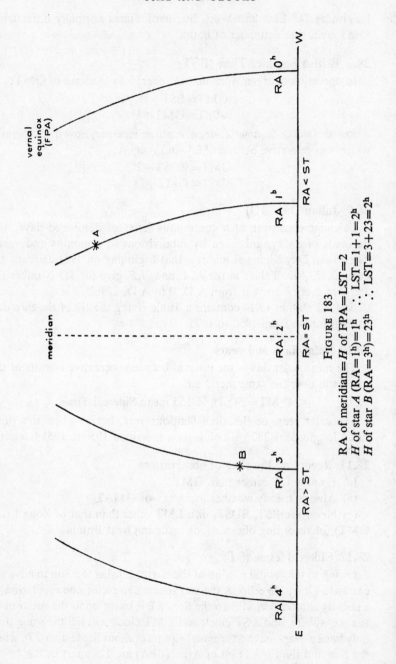

FIGURE 183

RA of meridian = H of FPA = LST = 2

H of star A (RA = 1^h) = 1^h ∴ LST = 1 + 1 = 2^h

H of star B (RA = 3^h) = 23^h ∴ LST = 3 + 23 = 2^h

The Local Sidereal Time (LST) at any moment and place is the RA of the meridian at that place. The difference between the RA of a star (α) and that of the meridian is known as the star's Hour Angle (H); it is measured westward from the meridian (in practice, for greater convenience, it is often reckoned in both eastward and westward directions to 12^h). Since RA is measured eastward from the FPA, it follows that the RA of the meridian is the Hour Angle of the FPA; also that the Hour Angle of the FPA is the ST at any moment (0^h occurring at upper meridian transit—cf. AT and MT). Hence:

LST=Hour Angle of the FPA=RA of the meridian=$\alpha + H$

These relations are illustrated in Figure 183, which represents the view of the southern sky.

It is now easy to see the relation of the Sun to ST. ST=RA+Hour Angle whether the object concerned is a star, the true Sun, the mean sun, or anything else; and since MT=Hour Angle of the mean sun,

ST=MT+RA of mean sun

In the special case of Mean Noon, when the mean sun is on the meridian,

ST=RA of mean sun=RA of true Sun+E/T

Owing to nutation, ST is not uniform, which leads to a distinction between *Mean* or *Uniform Sidereal Time* and *Apparent* or *True Sidereal Time*, analogous to MT and AT; the relation analogous to E/T is termed the *Nutation in RA*:

Nutation in RA=Apparent ST−Mean ST

Sidereal Day: the interval between successive transits of a star across the same meridian—

24^h Mean ST=23^h 56^m $4^s.091$ MT

This is, on average, $0^s.009$ shorter than the Earth's rotation period, owing to the Precession of the Equinoxes.

28.13 Greenwich and Local Times

Greenwich Sidereal Time (GST) and Local Sidereal Time (LST) are related as follows. If H_G=Hour Angle of a star for an observer in the Greenwich meridian, and H_λ=its Hour Angle for an observer in longitude λ, then

$$\left.\begin{array}{l} H_G = H + \lambda \\ H_\lambda = H_G - \lambda \end{array}\right\} \text{when } \lambda \text{ is West}$$

the signs of λ being reversed if it is East, and λ being converted to time by means of $1° \equiv 4^m$.

Thus $\qquad$ LST=GST$-\lambda$ (West of Greenwich)

$\qquad\qquad$ LST=GST$+\lambda$ (East of Greenwich)

A similar relationship holds between GMT and LMT, the Hour Angle of the mean sun replacing that of the star:

$$LMT=GMT-\lambda \quad \text{(West of Greenwich)}$$
$$LMT=GMT+\lambda \quad \text{(East of Greenwich)}$$

28.14 Conversion of ST/MT intervals

The relation between ST and MT can be expressed in a variety of ways, of which two of the more important are:

(a) The Tropical Year contains 365·2421 mean solar days,

$\qquad$,, $\qquad$,, $\qquad$,, $\qquad$,, $\qquad$ 366·2421 sidereal days.

(b) Since 24^{h} ST$=23^{h}$ 56^{m} $4^{s}09$ MT,

$\qquad$ 24^{h} MT$=24^{h}$ 3^{m} $56^{s}56$ ST,

a sidereal clock gains on a MT clock at the rate of

$$9^{s}8565 \text{ per MT hour,}$$
$$9^{s}8296 \text{ per ST hour.}$$

The conversion of a ST interval to the equivalent MT interval, and vice versa, may be carried out in various ways, of which the most practical and convenient is (a) below:

(a) *N.A.* and *A.E.* Tables give the MT equivalent of every ST hour from 1 to 23 and of every ST minute and second from 1 to 60; also the ST equivalents of the same MT intervals. Summarised, these read:

ST interval	Equivalent MT interval			MT interval	Equivalent ST interval		
	h	*m*	*s*		*h*	*m*	*s*
1^{h}	0	59	50·17	1^{h}	1	0	9·86
1^{m}	0	0	59·84	1^{m}	0	1	0·16
1^{s}	0	0	0·99(7)	1^{s}	0	0	1·00(3)

(b) A convenient approximation, involving errors not exceeding a few secs:

$$T=t+\frac{t}{360}$$

(where T is the number of sidereal h.m.s. in an interval, and t the number of mean h.m.s. in the same interval). To divide h.m.s. by 360, divide by 6 and enter the h. in the m. column, the m. in the s. column.

(c) A less convenient approximation is
$$T = 1 \cdot 003t$$
(d) Table showing the gain of ST on MT for every h., m., s. A summarised Table of this sort will be found in section 28.16.

28.15 Equivalent ST and MT instants

The simplest and most direct method of deriving the ST at a given MT and longitude is to make use of the two *N.A.* or *A.E.* tables: GST of midnight, and conversion of MT to ST intervals.

Suppose we want to know the Local ST at $14^h 30^m$ MT on 1950 Jan 1 at a point 20° W of Greenwich. The *N.A.* gives the GST of the preceding midnight as $6^h 40^m 17\overset{s}{\cdot}9$. From the *N.A.* conversion table we find that $14^h 30^m$ MT$\equiv 14^h 32^m 22\overset{s}{\cdot}9$. Therefore the GST at 2.30 p.m. is $21^h 12^m 40\overset{s}{\cdot}8$. But we want the LST at 20° W longitude, not GST. $20° \equiv 1^h 20^m$, and since the longitude is W we subtract. Therefore the required LST$= 19^h 52^m 40\overset{s}{\cdot}8$.

The derivation is the work of a moment, and slips are less likely to occur if it is laid out as below:

MT	14. 30. 0
Equivalent ST (from *N.A.*)	14. 32. 22·9
GST of midnight (*N.A.*)	6. 40. 17·9
∴ GST at $14^h 30^m$	21. 12. 40·8
20° W (subtract)	1. 20. 0
∴ required LST	19. 52. 40·8

The procedure for calculating the LMT at a given ST is analogous and equally simple, the required MT being the MT of the last transit of the FPA added to the MT equivalent of the given ST. For example: what is the LMT at $19^h 52^m 36^s$ ST on 1950 Jan 1 at a place 20° W of Greenwich?

ST	19. 52. 36
Equivalent MT (from *N.A.*)	19. 49. 20·6
MT of FPA transit (*N.A.*)	17. 16. 51·7
∴ GMT at given ST	37. 6. 12·3
20° W (subtract)	1. 20. 0
∴ required LMT	35. 46. 12·3
	=11. 46. 12·3

28.16 Precision timekeepers

Precision timekeeping received an impetus with the construction of the Shortt free-pendulum clock in 1921, in which the pendulum is freed from both escapement and gearing. The master-and-slave arrangement, developed from this, probably represents the limit of precision attainable by the oscillation of a pendulum. It consists of a master pendulum swinging in a vacuum, and a slave pendulum synchronised with it. The master pendulum has nothing to do but to continue swinging uniformly. The slave operates the chronograph signalling mechanism and is also responsible for operating the catches of both master and slave remontoirs (the mechanism which, at $\frac{1}{2}$-sec intervals, transmits an impulse to the pendulum when it is at mid-swing). Thus the master pendulum is relieved of all work, and stability of rate to $0^s.03$ per day is obtainable.

(See further, B. 341–2.)

Even more recently, various types of oscillator have been employed for frequency control and, through synchronous motors, for precision timekeeping. Most commonly a quartz crystal, cut as a plate or ring, is used, though metal rods and tuning forks have also been tried. In each case the principle is the same, the natural period of vibration of the oscillator being employed to provide the invariable time unit. The quartz crystal is made to vibrate in its natural period by an alternating current (usually 100 kc/s); the frequency in the circuit is stabilised by the crystal to within about 0·000002% of its natural period, and the output—after being stepped down by a frequency divider and then amplified—is used to operate a synchronous motor to which the seconds contacts or other recording device are geared. Owing to their superior uniformity of rate, extrapolation from quartz oscillators is much more reliable than with free-pendulum clocks. (See further, B. 344, 346.)

28.17 The observatory clock

Clocks of the order of precision of those described above are, however, of no more than theoretical interest to the average amateur. Even the standard Synchronome Master Clock (whose variations of rate, though its pendulum is not completely free, are smaller than those introduced by barometric variations) is an expensive instrument.

Any clock whose rate (i.e. rate of gaining or losing) is reasonably uniform can be used as an observatory clock. It is of little importance that the clock is running fast or slow, provided that the rate is both uniform and known, and that checking of its uniformity can be carried out at regular intervals by means of time signals. The behaviour of the clock should be kept under constant supervision, the most convenient arrangement being to plot clock error against GMT at frequent intervals. The following terms are used:

Clock running fast: gaining rate; clock error+; clock correction−.
Clock running slow: losing rate; clock error−; clock correction+.

The clock error can be determined from meridian transit observations, from other stellar observations, or by means of radio or TIM signals (see section 28.17). The transit instrument is not described here, since its function can be more conveniently and cheaply performed by time signals. The procedure for setting up the instrument and determining the azimuth, level and collimation errors, etc, is, moreover, fully described in such works as B. 8, 90, 343, 345.

The clock error can be determined by an equatorial with an accuracy depending on the fineness of division of the hour circle, without making any observation of meridian transit:

If α is the RA of a star near the zenith, T the clock time as it transits the central web of a micrometer eyepiece, and t the reading of the hour circle (corrected, if necessary, for index error), then the clock error, E, is given by

$$E = T - \alpha - t$$

If, for example, a star whose RA is $10^h\ 0^m\ 0^s$ is observed to transit the central web when the clock reads $10^h\ 59^m\ 56^s$ and the Hour Angle as given by the circle is $1^h\ 0^m\ 0^s$, then

$$E = 10^h\ 59^m\ 56^s - 10^h - 1^h$$
$$= -4^s,$$

∴ correction = $+4^s$, the error being slow.

In the case of meridian transits,

if observed transit time $>$ star's RA: error is fast, correction−,
if observed transit time $<$ star's RA: error is slow, correction+.

The determination of the clock error will yield the clock rate, which may be defined as the rate of change of clock error, or the change of clock error per unit time:

$$C/R = \frac{\Delta T' - \Delta T}{T' - T}$$

where $\Delta T'$ = clock error at time T',
ΔT = clock error at time T.

If, for example, the observed times of meridian transit of a star whose RA is $10^h\ 42^m\ 20^s9$ are on consecutive days

$10^h\ 42^m\ \ 9^s4$ (whence clock error = -11^s5)
$10^h\ 42^m\ 10^s3$ (whence clock error = -10^s6)

the clock has a gaining rate of 0ˢ9 per sidereal day, which, if the clock is a good one, will remain fairly constant.

A clock rated to gain 9ˢ8565 per MT hour will keep sidereal time. But even a clock running at the MT rate can be used for supplying ST with sufficient accuracy to find objects by means of the circles. The following Table shows the discrepancy between MT and ST for various intervals of the former, together with the equivalent angular discrepancy at the equator:

MT interval	Gain of ST on MT	Angular equatorial equivalent
10ᵐ	1ˢ7	25″5
30	5·0	1′ 15
1ʰ	9·8	2 25
2	19·7	4 55
6	59	14 45
12	1ᵐ 58	29 30
24	3 56	59 0

Hence, using a LP ocular—of diameter, say, 3°—for finding, it will only be necessary to reset the clock to ST every alternate night for an equatorial object to be found within the field. Objects in higher Decs will be nearer the centre of the field in proportion to the cosine of their Dec.

If, however, it is decided that the bi-nightly resetting is undesirable, it will be necessary to accelerate the clock's rate so as to gain approximately 10ˢ per hour.

Periodic checking and resetting of the sidereal clock may be carried out by one of the following methods:

1. *By transit observation:* Stop the clock. Hold the pendulum to one side by means of a small electromagnet controlled by bell-push from the telescope. Take from the *N.A.* the RA of the next clock star* which is due to culminate. Set the clock's hands to this RA, if possible to the nearest second. If the observation is to be made with an equatorial, not with a transit instrument, set the telescope to the star's Dec, with the hour circle reading zero. Clamp the polar axis. Observe the transit with a micrometer eyepiece, and at the instant the star's image is bisected by the central web start the clock by releasing the pendulum.

2. *Without transit observation:* Set an accurately rated MT clock by means of a time signal. Select the MT instant at which the sidereal clock is to be started. Determine the ST of this instant by means of the *N.A.*

* Clock stars are stars whose RAs are given in the Almanac with a precision equivalent to 0·001 seconds of time.

conversion tables (see section 28.14). Set the sidereal clock hands to this reading and release the pendulum when the MT clock shows the pre-selected reading.

3. *Without transit observation:* A quick method which is accurate enough for most practical purposes is simply to read the Hour Angle of any convenient star and set the clock accordingly.

28.18 Time signals*

Since the Greenwich time service was opened in 1833 with the midday dropping of the time ball, more and more accurate time has been purveyed to the world from the Royal Observatory. Today GMT may be had correct to a few milliseconds.

(*a*) *The 'six pips':* This service, started in 1924, is transmitted 96 times a day to some part of the globe on the long, medium, and short wave-bands. The schedules are frequently changed, and up-to-date information regarding transmissions must be obtained from the latest BBC publications. At the time of writing (April 1952) time signals are transmitted on the Home Programme frequencies (692, 809, 881, 908, 1052, 1088, 1151, 1457, 1484 kc/s) at

0700, 0800, 1100, 1300, 1800, 2300 GMT on weekdays
0800, 1300, 1800, 2300 GMT on Sundays

and on the Light Programme frequencies (200, 1214 kc/s) at

1000, 1400,† 1500,† 1900, 2200 GMT on weekdays
0900, 1900, 2200 GMT on Sundays.

The contacts are controlled by the MT master clock (pendulum) at Greenwich, and the pips (whose claimed accuracy is 0^s1) are normally accurate to within 0^s05, which is sufficient for most amateur and all civil purposes. The pips (1000-cycle notes) mark 1^s intervals, the sixth falling on the exact hour or quarter-hour.

(*b*) *The rhythmic (or vernier) time signals, and MT signals:* Transmitted daily from $9^h 54^m 00^s$ to $10^h 06^m 15^s$ and from $17^h 54^m 00^s$ to $18^h 06^m 15^s$ from Rugby by the long-wave transmitter (GBR, 16 kc/s) and by a number of short-wave transmitters. The latter, which are normally altered on April 1 and October 1 each year (on account of seasonal propagation conditions and other factors), are usually selected from the following: GIA (19,640 kc/s), GIC (8640 kc/s), GID (13,555 kc/s), GKU2 (17,685 kc/s) GKU3 (12,455 kc/s). The long-wave GBZ (19·4 kc/s) at Criggion is a standby transmitter in case of failure of GBR.

* See page 559 for a list of 1980 time signals.
† Except Saturdays.

The detailed schedule of the transmissions is as follows (the evening transmission following exactly the same course as the morning one, here given):

$09^h\ 54^m\ 00^s$ to $09^h\ 54^m\ 45^s$: 'GBR GBR TIME' repeated four times in Morse, followed by a tuning dash.

09 54 55 to 10 00 00 : MT signal: $0^s.1$ dots at 1^s intervals, $0^s.5$ dash at the minute (the beginning of each signal being the timing reference point).

10 01 00 to 10 06 00 : Rhythmic signal: 60 dots and 1 dash per minute, the dash marking the minute.

10 06 05 to 10 06 15 : Tuning dash.

The dots are thus separated from one another (beginning to beginning, or end to end) by $\dfrac{60^s}{61}$. To use the 'time vernier' the procedure is to count the number of inter-dot intervals from a dash to the next coincidence of clock beat and signal (there being 1 such coincidence per minute in the case of 1^s beats, 2 per minute for $\frac{1}{2}^s$ beats). This number is then the number of sixty-firsts of a second that the clock is fast of the signal. In other words, if t is the time of the clock beat immediately preceding a dash, t' the exact minute given by this dash, and n the number of dot-intervals between the dash and the coincidence, then

$$t' - t = \frac{n^s}{61}$$

The following table gives the decimal of a second of gain corresponding to the numbers of rhythmic intervals from 1 to 60 (61 intervals being equal to 60^s):

1	0·016	11	0·180	21	0·344	31	0·508	41	0·672	51	0·836
2	0·033	12	0·197	22	0·361	32	0·525	42	0·689	52	0·852
3	0·049	13	0·213	23	0·377	33	0·541	43	0·705	53	0·869
4	0·066	14	0·230	24	0·393	34	0·557	44	0·721	54	0·885
5	0·082	15	0·246	25	0·410	35	0·574	45	0·738	55	0·902
6	0·098	16	0·262	26	0·426	36	0·590	46	0·754	56	0·918
7	0·115	17	0·279	27	0·443	37	0·607	47	0·770	57	0·934
8	0·131	18	0·295	28	0·459	38	0·623	48	0·787	58	0·951
9	0·148	19	0·311	29	0·475	39	0·639	49	0·803	59	0·967
10	0·164	20	0·328	30	0·492	40	0·656	50	0·820	60	0·984

The rhythmic time signals, initiated in 1927, are derived from a quartz clock.

492

(c) *TIM:* The speaking clock service—in operation in London since 1936, and now available in many provincial towns and cities—is synchronised by a signal from the Royal Greenwich Observatory at each hour. It is normally accurate to $0\overset{s}{.}5$.

For details of international and foreign systems of radio time signal, reference should be made to B. 347, 348.

28.19 Timing observations

The instant at which an observed event occurs may be determined in a variety of ways.

28.19.1 Stopwatch and clock: Check the rate of the watch to see that it is uniform (variation $<0\overset{s}{.}5$) over periods of at least several minutes. Over longer periods, even good watches cannot be relied on.

Check the clock with TIM immediately before and after the observation; or if TIM is not available, ascertain that its rate does not vary by more than $0\overset{s}{.}5$ during the interval from the observation to the nearest radio signal. In the case of a clock whose rate is regularly determined and recorded, this step will not be necessary.

Start the watch from zero at the instant of the observed event. Carry it to the clock. Pick up the rhythm of the clock beat, and as its second hand comes up to a complete minute, stop the watch. Then if

T_0 = the required UT of the event,

T_1 = watch reading (secs),

T_2 = clock time when watch stopped (UT),

Δt = clock correction (+ if slow, − if fast),

$\Delta t'$ = correction for personal equation (error +, correction −, assuming normal lag),

$$T_0 = T_2 - T_1 + \Delta t + \Delta t'$$

For example: Clock rate = $+0\overset{s}{.}8$ per hour,

Clock reading = $17^h\ 1^m\ 5\overset{s}{.}0$ at $17^h\ 0^m\ 0^s$ (TIM),

$T_2 = 17^h\ 31^m\ 0^s$,

$T_1 = 46\overset{s}{.}6$,

Clock error = $+(1^m\ 5\overset{s}{.}0 + 0\overset{s}{.}4) = +1^m\ 5\overset{s}{.}4$,

$\Delta t = -1^m\ 5\overset{s}{.}4$,

Personal equation = $+0\overset{s}{.}3$,

$\Delta t' = -0\overset{s}{.}3$,

$T_0 = 17^h\ 31^m\ 0^s - 46\overset{s}{.}6 - 1^m\ 5\overset{s}{.}4 - 0\overset{s}{.}3$

$= 17^h\ 29^m\ 7\overset{s}{.}7$

Alternatively, the watch can be taken to the phone and T_2 obtained from TIM; or from a radio signal, if there happens to be one within a few minutes of the observation.

28.19.2 Eye and ear: (*i*) *Clock beat audible at telescope:* Estimate the fraction of the interval ($\frac{1}{2}/\frac{1}{2}$ or even $\frac{1}{4}/\frac{3}{4}$ is quite feasible with 1^s beats) between beats at which the observed event occurs, and start counting the ticks. 'Carrying the beat' in the head, note this down, go to the clock and identify the beat. Suppose, for example, that the event is judged to occur $\frac{3}{4}$ of the inter-beat interval after one tick and $\frac{1}{4}$ of the interval before the next. Counting starts with the latter tick (nought). On going to the clock, still mentally counting, it is found that the 31st tick occurs at $7^h 4^m 0^s$. Then the time of the observation was

$$7^h 4^m 0^s - 31^s.0 - 0^s.25$$
$$= 7^h 3^m 28^s.75$$

An alternative method, which is safer but needs rather more practice, is to look at the clock and start counting shortly before the expected event, estimate the inter-second fraction of the instant of observation, record it, carry the beat to the clock and verify the count.

(*ii*) *Clock beat inaudible at telescope:* An assistant watching the clock is notified of the instant at which the observation is made, by means of an audible signal from the observer. He will be spared strain, and therefore make his estimation more accurately, if a system of warning bells is pre-arranged, so that he is expecting the observation signal when it occurs.

28.19.3 Chronograph: The advantages of the chronograph are that fractions of a second do not have to be estimated by ear, that neither a clock nor an assistant is required, that it is less distracting to the observer than the eye-and-ear method, and that it involves a more systematic personal equation than the latter.

See section 22.7.

28.19.4 Impersonal micrometer: The advantages and method of employment of this micrometer in transit observations are described in section 25. By its means personal equation can be reduced to the order of $0^s.02$.

28.19.5 Other methods: Ingenuity might suggest, e.g. a method of evolving an impersonal time-recording system employing a photo-electric cell linked to a chronograph; or one employing a movie film moving at the known rate of frames per second, time calibration being achieved by light flashes from a bulb at the edge of the field operated by the clock or by radio or other time signals.

All observations are now recorded in MT, sidereal clocks being used only for finding purposes. If, therefore, the sidereal clock is used in the determination of the time of an observation, the derived time must be converted to MT for the record.

MATHEMATICAL TREATMENT OF OBSERVATIONS

Any set of measures of a physical quantity will exhibit a greater or smaller degree of dispersion. That is to say, the individual observations are to some extent inaccurate, and are scattered round the bull's-eye represented by the true value of the quantity.

The errors responsible for the scatter in the observational material are of two kinds, random and systematic. Random errors are the inescapable result of the humanity of the observer and the imperfection of his instrument, and occur, as the name implies, entirely fortuitously; they alone are susceptible of treatment by statistical methods based on probability theory. Systematic errors are not susceptible to such treatment, since they spring from some tendency to scatter the observed values which is not reduced by increasing the number of the observations. This tendency may be intermittent or periodic in action; when the latter, and its period happens to coincide with that of the quantity under measurement, it is often extremely difficult to disentangle.

Freedom from unknown sources of error can, as a general rule, be increased by varying the conditions under which the observations are made. This is exemplified by the old laboratory maxim 'Reverse everything reversible, and repeat.'

Three main stumbling blocks lie in the way of the application of probability considerations to observational material: the normal law is inapplicable to systematic errors, and therefore to observational data which are still tainted by such error; the number of observations must be adequate for the operation of probability among them; and the relative values of the individual observations must be correctly estimated, i.e. the system of weighting (*v. inf.*) must be carefully considered.

If a large number of readings of a certain physical quantity—micrometer readings of the separation of two stars, for example—are taken, the type of their distribution can be represented conveniently by means of a histogram (Figure 184) in which divisions of the range of the observations—e.g. 0″1, where the measures themselves are to 0″01—are represented on the x-axis, and the y-axis level of the histogram corresponding to each such interval is determined by the number of measures

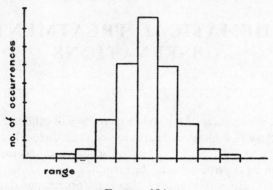

FIGURE 184

that fall within it. The outline of the histogram is in fact an approximation to a curve of the form shown in Figure 185, the so-called Normal Curve, which closely represents the distribution of values obtained

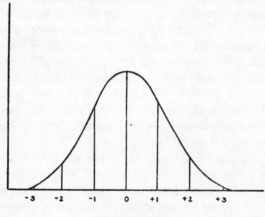

FIGURE 185

under conditions exemplified by, for example, those of astronomical observation.

The aim of any theory of error and probability is at least twofold: to provide a means of estimating the most probable value of a quantity from a set of more or less discordant measures of it, and to provide a means of estimating the accuracy of the derived result and the degree of precision of the observations on which it rests. The normal, or Gaussian, law of error, which is embodied in the Normal Probability Curve, has the additional advantages of simplicity, the yielding of unique solutions, and convenience and uniformity in the application of the rules derived from it. In practice these advantages are as important as the achievement of the very highest degree of probability. The Normal Curve is to be regarded as the appropriate mathematical model to use in cases of the type we have been discussing, and its value resides to a large extent in the fact that its characteristics and properties are both definite and well known. This point will be returned to later.

Faced, then, with a set of observational data, discordant measures of a given quantity, the problem resolves itself thus: what are the errors of the individual readings? what is the best value to take? what is the error of the best value? The normal law of error gives as the most probable value of a quantity q the arithmetical mean of the measures $q_1, q_2 \ldots q_n$. Thus

$$\bar{q} = \frac{\Sigma q}{n}$$

where $\bar{q}$ represents the arithmetical mean, and Σq the sum of n equally reliable observations of q. The most probable value of a quantity, understood in this sense, is applicable only to the observations themselves, and not to functions of the observed quantity. Thus, for example, the most probable value of the distance of a star is that distance corresponding to the arithmetical mean of the parallax measures, and not the mean of the derived distances, which are a function of the observable.

In more complex problems, where it is required to derive the most probable values of several different quantities, each a function of q, resort must be had to the method of least squares; this is an application of the simple proposition, deducible from the normal law, that the sum of the differences between the observed and computed values of the variables, when raised to a common power, should be minimal.* An example of such a problem is the derivation of the elements of an orbit from a set of observations of position. Providing the sum of the squared residuals is kept to a minimum, the errors of solution are confined within tolerable and known limits.

If certain members of a series of measures are considered on good grounds to be more reliable than the others, they must be given accord-

---

* For detailed treatments of the operation of the method, the reader should refer to the textbooks (see Bibliography).

ingly greater weight in the calculation of the most probable value. Thus if the observations of several observers are being combined, and one observation made by A is considered to be worth w observations by B—i.e. that A's observations are w times as reliable as B's—then the weight of A's observations is w when that of B's is unity.

If the weights allotted to $q_1, q_2 \ldots q_n$ are respectively $w_1, w_2 \ldots w_n$, then the weighted mean is given by

$$\bar{q} = \Sigma\frac{(wq)}{\Sigma w}*$$

(cf. preceding expression for the corresponding unweighted mean), and the weight of $\bar{q}$ is Σw. For example, if q_1 is considered to be three times as reliable as q_2 and q_4, q_3 twice as reliable as q_2 and q_4, and q_5 and q_6 twice as reliable as q_1, then they will be weighted as follows:

q_1	3
q_2	1
q_3	2
q_4	1
q_5	6
q_6	6
$\Sigma q=$	19

If the values of $q_1 \ldots q_6$ are respectively 50·3, 49·8, 50·2, 50·1, 49·7 and 49·9, then the weighted mean is

$$\bar{q} = \frac{(3 \times 50\cdot3) + (49\cdot8) + (2 \times 50\cdot2) + (50\cdot1) + (6 \times 49\cdot7) + (6 \times 49\cdot9)}{19}$$
$$= 49\cdot94$$

To answer the remaining questions we must derive some expression for the scatter of the observations, i.e. discover a parameter which will convey significant and sufficient information regarding the manner in which the individual measures are dispersed about the most probable value. One such parameter of dispersion is the *range*. The range of the six quantities used in the example above is 50·3−49·7=0·6, i.e. the difference between the extreme values. This is not a very satisfactory criterion, for the very reason that it is based upon exceptional rather than on normal or typical values. Moreover it tells us nothing about the way in which the measures cluster about the best value.

A more satisfactory description of the dispersion of a set of observations is provided by the *mean deviation* (or mean error). This is defined as the average deviation of a set of measures of a quantity from the

* Numerator expanded as $w_1q_1 + w_2q_2 + \ldots + w_nq_n$.

most probable value of that quantity. If $q_1, q_2 \ldots q_n$ are the measured values of a quantity q, then the most probable value of q is given by

$$\bar{q} = \frac{\Sigma q}{n}$$

and the residuals are $q_1 - \bar{q}, q_2 - \bar{q} \ldots q_n - \bar{q}$. Call these respectively $d_1, d_2 \ldots d_n$, giving each a positive sign. Then the mean deviation is

$$\frac{\Sigma d}{n}$$

Taking the same six unweighted values as before, the most probable value of q is

$$\frac{\Sigma q}{n} = \frac{300}{6} = 50 \cdot 0$$

and the residuals are 0·3, 0·2, 0·2, 0·1, 0·3, 0·1. Hence

$$\Sigma d = 1 \cdot 2$$
$$\frac{\Sigma d}{n} = 0 \cdot 2$$

The mean deviation of the six measures is thus 0·2, and the derived value of q would be given as

$$\bar{q} = 50 \cdot 0, \quad \text{M.D.} = 0 \cdot 2$$

Mean deviation gives a truer and more significant picture of the dispersion of $q_1 \ldots q_n$ than does the range, since it does not depend solely upon two exceptional values, but upon all the values in the set.

A word may be said at this point concerning residuals. The term has somewhat different connotations according to its context, but in general the residuals of a set of observations are the deviations of the individuals from some mean, most probable, computed, independently derived, or true value of the quantity observed. If O is the observed value of a quantity, and C is the approximately computed or arbitrarily chosen value (its true value being unknown), then $O - C$ is the *residual* (or apparent error) in the strict sense. It provides a check on the accuracy of the computed value, and thence on the computation or assumptions upon which C rests—e.g. observations of cometary positions to improve an orbit. If, on the other hand, C is the known true value of the quantity, then $O - C$ is not a residual but a *true error*, and provides a check on the accuracy of O. $C - O$ is then a *correction* to be applied to the observed value in order to obtain the true value—e.g. testing an instrument or observer for defects or personal equation.

Even better than mean deviation, as an expression of dispersion, is

499

the so-called *standard deviation*. This is the square root of the sum of the squared residuals (deviations from the most probable value) divided by the number of measures in the set. Thus if, as before,

$$d_n = q_n - \bar{q}$$

and
$$\Sigma d^2 = d_1^2 + d_2^2 + \ldots d_n^2$$

then the standard deviation, σ, is given by

$$\sigma = \sqrt{\frac{\Sigma d^2}{n}}$$

Using the same set of figures as before,

$$\sigma = \sqrt{\frac{0 \cdot 28}{6}}$$
$$= 0 \cdot 216$$

The importance of standard deviation as an interpretation of the scatter of a set of values lies in the fact that, if the latter is assumed to be represented by a Normal Curve, then we at once know that approximately $\frac{2}{3}$ of the distribution lies within 1σ of the mean, 95% within 2σ of the mean, and $<1\%$ more than 3σ from the mean. Whatever the shape of a particular Normal Curve (depending upon the parameter σ) its form is such that the area beneath it is divided as follows:

within 4σ	(+ *and* −) of the mean:			99·994%
„ 3σ	„	„	„	99·73
„ 2σ	„	„	„	95·4
„ 1σ	„	„	„	68·1
„ $0 \cdot 6745\sigma$	„	„	„	50·0
„ $0 \cdot 25\sigma$	„	„	„	20·8

The name *probable error* is given to the deviation whose value of $0 \cdot 6745\sigma$ is equally likely to be exceeded as not. The probable error in the arithmetical mean of a set of observations being inversely proportional to $\sqrt{n}$, there is little to gain in multiplying the number of the observations beyond a certain point. On the contrary, if a reduction of 50% in the number of observations is accompanied by a similar reduction of the random errors, then the value of the result is doubled. Probable error has rather fallen into disfavour with the statisticians, who today tend to regard standard deviation as a more useful and significant guide.

REDUCTION TO EQUINOX OF DATE

The derived position of any object, determined by reference to the stars, will depend upon the epoch of the Catalogue from which the positions of these stars are taken. Similarly, the current position of a star will differ from that given in a Catalogue constructed for a past epoch.

The system of geocentric coordinates is not static, but is subject to progressive and non-uniform change. In addition, the actual position of the object, relative to the Earth, may change. The factors concerned in both these types of change are:

Precession: a secular change due to motion of the ecliptic round the equator; it affects both the RA and Dec of an object. Precession in RA is a function of both the RA and Dec of the object; precession in Dec is a function of its RA only. General Precession is the term applied to the total effect of the two components, Planetary Precession and Luni-solar Precession.

Nutation: a periodic change which can be broken down into long- and short-period terms, the latter being negligible in most cases.

Aberration: a periodic change, depending on the finite velocity of light and the changing direction of the Earth's motion.

Parallax: a periodic change due to the Earth's changing position during the course of the year.

Proper motion: a secular variation of the object's position relative to the coordinate system itself.

Over a long period, the secular variations tend to become increasingly important, as compared with the periodic ones; of them, precession is in the general run of cases much more important than proper motion. Successive refinements would therefore involve consideration and correction of the following:

(a) precession only;
(b) precession, aberration, and long-period terms of nutation;

(c) precession, aberration, long- and short-period terms of nutation;

(d) precession, aberration, nutation, proper motion and parallax.

The terms Mean Place, True Place and Apparent Place must be defined:

Mean Place: the position of an object at a given epoch, referred to the Mean Equinox. MP's of selected objects are quoted in Almanacs for the beginning of the current year, and in Catalogues for a stated epoch. MP's are affected by precession and proper motion.

True Place: the position of an object, referred to the True Equinox. The TP is equivalent to the MP corrected for nutation.

Apparent Place: the position of an object as seen from the Earth's centre. Equivalent to the TP corrected for aberration.

The quantity that it is required to derive is the AP of the object, referred to the equinox of date, i.e. at the time of observation. If its AP is quoted at intervals in the Almanac, it may be obtained for the required date by simple interpolation; if its MP on Jan 0 of the current year is given in the Almanac, its AP at date may be reduced as described below; if its MP at some past epoch is given in a Catalogue, the reduction must be performed in two stages: from MP at past epoch to MP at Jan 0 of the current year (see below), to AP at date (as before).

30.1 Derivation of MP at Jan 0 of current year (α, δ) from MP at past epoch (α_0, δ_0)

Two corrections are to be applied: for proper motion and for precession:

(i) *Proper motion:*

Given that $\Delta\alpha, \Delta\delta =$ the proper motion,

$\qquad t =$ mean epoch of the observations on which the Catalogue is based (not the epoch of the Catalogue itself),

$\qquad t' =$ current year,

then
$$\alpha' = \alpha_0 + \Delta\alpha(t' - t)$$
$$\delta' = \delta_0 + \Delta\delta(t' - t)$$

where α', $\delta' = \alpha_0$, δ_0 corrected for proper motion only. They are then corrected for precession, as described below.

(ii) *Precession:*

(a) Approximate corrections may be derived by graphical methods;

see Figures 186, 187; similar diagrams are to be found in *B.A.A.H.*, 1923, 1949.

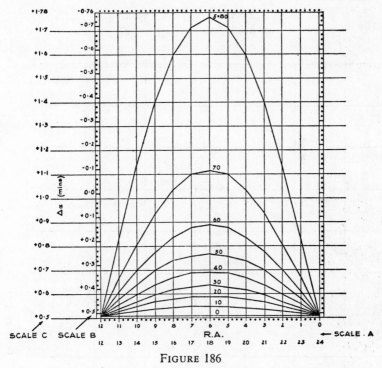

FIGURE 186

Precession corrections (RA) for intervals of 10 years.

Enter the diagram with the object's RA on Scale A ; take its Dec from the appropriate curve (interpolate if necessary) and read the value of $\varDelta\alpha$ from Scale B or C. These are used according to the object's RA and Dec, as follows :

RA	Dec	Scale
$<12^h$	$+$	C
$>12^h$	$+$	B
$<12^h$	$-$	B
$>12^h$	$-$	C

Examples :

$$\text{Position of object, } 7^h, +70° : \varDelta\alpha = +1^m10$$
$$20^h, +70° : \varDelta\alpha = -0^m02$$
$$3^h, -70° : \varDelta\alpha = +0^m08$$
$$13^h, -70° : \varDelta\alpha = +0^m67$$

503

(*b*) Calculation: Given that the position at epoch was α_0, δ_0, and that the interval between the epoch and the current year is t years, then:

$$\alpha = \alpha_0 + t(m^s + n^s \sin \alpha_0 \tan \delta_0)$$
$$\delta = \delta_0 + t(n'' \cos \alpha_0)$$

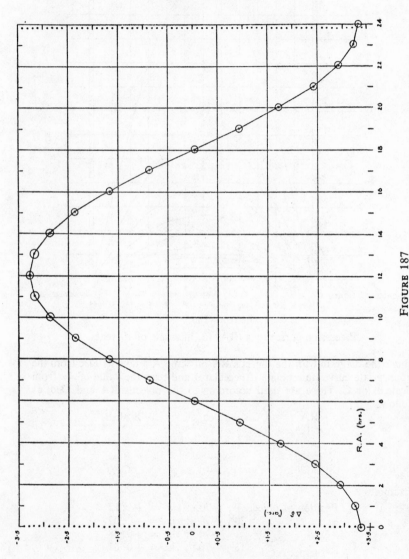

FIGURE 187

Precession corrections (Dec) for intervals of 10 years. The sign of $\Delta\delta$ is the opposite of that given if the object is in S Dec

where m^s, n^s and n'' (m being precession in RA, n precession in Dec) are tabulated below:

	m^s	n^s	n''
1950	3.07327	1.33617	20.0426
1952	3.07331	1.33616	20.0424
1954	3.07334	1.33615	20.0422
1956	3.07338	1.33614	20.0421
1958	3.07342	1.33613	20.0419
1960	3.07346	1.33612	20.0417
1970	3.07364	1.33606	20.0409
1980	3.07383	1.33600	20.0400
1990	3.07401	1.33595	20.0392
2000	3.07420	1.33589	20.0383

30.2 Derivation of AP at date (α, δ) from MP at Jan 0 of current year (α_0, δ_0)

The complete procedure would consist of the following stages (using p, μ, N and A to denote the corrections for precession, proper motion, nutation, and annual aberration respectively, during the interval of τ tropical years (<1) between Jan 0 and the current date):

$$\underset{\text{Jan 0}}{MP} + p + \mu = \underset{\text{date}}{MP}$$

$$\underset{\text{date}}{MP} + N = \underset{\text{date}}{TP}$$

$$\underset{\text{date}}{TP} + A = \underset{\text{date}}{AP}$$

This procedure would be exceedingly laborious, and in practice one or other of two methods initiated by Bessel is employed, which combine the various corrections into a single expression:

(i) employing Besselian day numbers and Besselian star constants: the former are taken from the *A.E.*, and the latter computed or, in practice, taken from a star catalogue; since the values of the constants vary slowly, they will not be accurately given by a catalogue of distant epoch;

(ii) employing independent day numbers: this method is used when the star constants are not known with sufficient accuracy.

(i) *Besselian day numbers:*

$$\alpha = \alpha_0 + \tau\mu_\alpha + Aa + Bb + Cc + Dd + E$$
$$\delta = \delta_0 + \tau\mu_\delta + Aa' + Bb' + Cc' + Dd'$$

where α, δ = the required AP at date,

      α_0, δ_0 = the MP at Jan 0,

         τ = the fraction of the tropical year between Jan 0 and date

         μ_α = proper motion in RA,

         μ_δ = proper motion in Dec,

$A, B, C, D, E =$ the Besselian day numbers (derived from the constants of precession, nutation, and aberration; E, a function of nutation and precession, can usually be neglected),

$\left.\begin{matrix} a, b, c, d \\ a', b', c', d' \end{matrix}\right\} =$ the Besselian star constants, quoted in the catalogue under each star. They are in fact derived as follows:

$$\left.\begin{aligned} a &= \frac{m}{15} + \frac{n}{15} \sin\alpha \tan\delta \\[1mm] b &= \frac{\cos\alpha \tan\delta}{15} \\[1mm] c &= \frac{\cos\alpha \sec\delta}{15} \\[1mm] d &= \frac{\sin\alpha \sec\delta}{15} \\[1mm] a' &= n\cos\alpha \\ b' &= -\sin\alpha \\ c' &= \tan\epsilon\cos\delta - \sin\alpha\sin\delta \\ d' &= \cos\alpha\sin\delta \end{aligned}\right\}$$

where

$\alpha, \delta =$ apparent RA, Dec (not sensibly different from α_0, δ_0)

$\epsilon =$ obliquity of ecliptic

$\left.\begin{matrix} m = \text{precession in RA} \\ n = \text{precession in Dec} \end{matrix}\right\}\,''\text{arc}$

Of the terms on the RHS of the above expressions for α and δ, the 2nd concerns the star's proper motion during the interval τ, the 3rd precession, the 4th the long-period terms of nutation, the 5th and 6th the annual aberration, and the 7th nutation and precession.

If required, the additional correction for (a) parallax, and (b) the short-period terms of nutation, are:

(a) $\left\{\begin{aligned} \Delta\delta &= \pi(cY - dX) \\ \Delta\delta &= \pi(c'Y - d'X) \end{aligned}\right.$ $\left\{\begin{array}{l} \text{where} \\ X, Y \text{ are solar coordinates } (A.E.), \\ c, d, c', d' \text{ are the star constants.} \end{array}\right.$

(b) $\left\{\begin{aligned} \Delta\alpha &= A'a + B'b \\ \Delta\delta &= A'a' + B'b' \end{aligned}\right.$ $\left\{\begin{array}{l} \text{where } A', B' \text{ are tabulated in } N.A. \text{ The short-} \\ \text{period terms of nutation in longitude and} \\ \text{obliquity are tabulated in } N.A. \text{ and } A.E. \end{array}\right.$

The data, required for reducing α_0, δ_0 to α, δ by this method are given in the $N.A.$ and $A.E.$ for every day of the year.

(ii) *Independent day numbers*:

$$\alpha = \alpha_0 + \tau\mu_\alpha + f + \frac{g\sin(G-\alpha)\tan\delta}{15} + \frac{h\sin(H-\alpha)\sec\delta}{15}$$

$$\delta = \delta_0 + \tau\mu_\delta + g\cos(G-\alpha) + h\cos(H-\alpha)\sin\delta + i\cos\delta$$

where $\left.\begin{array}{c} f \\ g \\ G \\ h \\ H \\ i \end{array}\right\}$ are the independent day numbers (functions of the Besselian day numbers), the first three being concerned with precession and nutation (long-period terms), the last three with aberration.

The additional correction for parallax, if required, is the same as before. That for the short-period terms of nutation is:

$$\Delta\alpha = f' + \frac{g' \sin (G' - \alpha) \tan \delta}{15}$$

$$\Delta\delta = g' \cos (G' - \alpha)$$

The following data, required in the operation of this method, are tabulated in the *A.E.* for every day of the year:

f

g

G

h

H

i

Sid. T. at 0^h

τ

f', g', G'.

The *A.E.* also contains useful supplementary tables, brief descriptions of which are given below, for reference:

(*a*) Approximate reduction from standard equinox of 1950·0 to true equinox of date:

$$\alpha_{\text{date}} = \alpha_{1950} + f + 4g \sin (G + \alpha_{1950}) \tan \delta_{1950}$$

$$\delta_{\text{date}} = \delta_{1950} + g \cos (G + \alpha_{1950})$$

values of $4 \tan \delta$, f, g and G being tabulated.

It will be seen that these approximate expressions are derived from the full equations involving the independent day numbers by the omission of the expressions for proper motion and aberration.

(*b*) Reduction of star positions from the mean equinox of the current year to the standard equinox of 1950: Two methods, with the relevant tables, are given; one accurate for all values of δ, the other inapplicable to high Decs.

(c) Differential aberration, precession and nutation: Useful in the reduction of observations of, e.g. a comet or asteroid, whose positions are obtained in the form $\varDelta\alpha$, $\varDelta\delta$ relative to a fixed star. If the star's position (α', δ') is reduced to the equinox of Jan 0, then the position of the object, referred also to Jan 0 of the current year, is given by

$$\alpha=\alpha'+\varDelta\alpha+[A]_\alpha+[P]_\alpha+[N]_\alpha$$
$$\delta=\delta'+\varDelta\delta+[A]_\delta+[P]_\delta+[N]_\delta$$

where $[A]=$ differential aberration,

         $[P]=$ differential precession,

         $[N]=$ differential nutation,

   $\varDelta\alpha$, $\varDelta\delta=$ RA/Dec of object *minus* RA/Dec of star, in units of 1^m and $1'$ respectively.

$$[A]_\alpha=a\varDelta\alpha+b\frac{\varDelta\delta}{10} \text{ in units of } 0^s\!.001$$

$$[A]_\delta=c\varDelta\delta+d\frac{\varDelta\delta}{10} \text{ in units of } 0''\!.01$$

where a, b, c, d, are derived from the tables given.

$$[P]_\alpha+[N]_\alpha=e\varDelta\alpha\,\frac{10\tan\delta}{15}-f\varDelta\delta\,\frac{10\sec^2\delta}{225},$$
$$\text{in units of } 0^s\!.001$$

$$[P]_\delta+[N]_\delta=f\varDelta\alpha \text{ in units of } 0''\!.01$$

where e, f, $\dfrac{10\tan\delta}{15}$, $\dfrac{10\sec^2\delta}{225}$ are tabulated.

SECTION 31

CONVERSION OF COORDINATES

Equations for some of the more commonly required transformations are given below. The following symbols are employed:

α	RA
δ	Dec
h	Altitude
Az	Azimuth
H	Hour Angle
ϕ	Latitude (terrestrial)
ϵ	Obliquity of the ecliptic
λ	Longitude (celestial)
β	Latitude (celestial)
z	Zenith Distance
θ	Sidereal Time

$$\left.\begin{array}{l} \tan H = \dfrac{\tan Az \sin X}{\cos(\phi-X)} \\[2ex] \tan \delta = \tan(\phi-X)\cos H \end{array}\right\} \tan X = \tan z \cos Az$$

$$\sin Az = \frac{\cos \delta \sin(\alpha-\theta)}{\sin z}$$

$$\cos Az = \frac{\sin \delta - \cos z \sin \phi}{\sin z \cos \phi}$$

$$\left.\begin{array}{l} \tan Az = \dfrac{\tan H \cos Y}{\sin(\phi-Y)} \\[2ex] \tan z = \dfrac{\tan(\phi-Y)}{\cos Az} \end{array}\right\} \tan Y = \frac{\tan \delta}{\cos H}$$

$$\left.\begin{array}{l} \cos z = \sin \delta \sin \phi + \cos \delta \cos \phi \cos(\alpha-\theta) \\ \sin h = \sin \delta \sin \phi - \cos \delta \cos \phi \cos H \\ \sin \delta = \sin h \sin \phi - \cos h \cos \phi \cos Az \end{array}\right\} \quad \cdots \cdots \quad (a)$$

509

$$\left.\begin{array}{l} \tan \lambda = \dfrac{\cos (Z-\epsilon) \tan \alpha}{\cos Z} \\[2mm] \tan \beta = \tan (Z-\epsilon) \sin \lambda \end{array}\right\} \tan Z = \dfrac{\tan \delta}{\sin \alpha}$$

$$\sin \frac{H}{2} = \pm \sqrt{\frac{\sin \frac{1}{2}[Z+(\phi-\delta)] \sin \frac{1}{2}[z-(\phi-\delta)]}{\cos \phi \cos \delta}}$$

Alternatively, specially constructed diagrams may be used for approximate solutions; e.g. that of W. Heath (B. 403), from which any two of δ, h, H, Az can be read, given the other pair. B. 401–2 also refer to the graphical solution of equations (a) above. Nomograms given in B. 20 permit the same conversions to be made with an accuracy of about $\pm 2°$. See also B. 576, Appendix 2.

For conversion between equatorial and galactic coordinates, see B. 404, 432.

SECTION 32

OBSERVATIONAL RECORDS

It is generally true to say that an unrecorded observation is an observation wasted. Communication of results is an essential condition of the progress of any science.

When planning the form that the observational record is to take, and subsequently while making the actual notes at the telescope, the characteristics to keep constantly in mind are: clarity and unambiguity, conciseness, objectivity and comparability, orderliness, legibility. The record should be immediately intelligible to anyone familiar with the subject, without any supplementary explanation.

Whenever the nature of the observations allows it, arrange the record in tabular form: this enforces orderliness in the arrangement of the data, makes for ease and speed of reference, and is some safeguard against the omission of relevant facts. In the interests of objectivity, precision, and comparability, numerical observational data are superior to both drawings and verbal descriptions.

The particular requirements of satisfactory records in the various branches of observation are dealt with under the several sections of *O.A.A.* The miscellaneous points mentioned here are more generally applicable.

It is often convenient to record the actual observations at the telescope in one book, and subsequently to transfer the entries to a permanent record book, or possibly card or looseleaf index, in which it may be desirable to lay out the material somewhat differently. If this is so, the record made at the telescope must nevertheless be legible, ordered and sufficiently comprehensive. The entering up of the permanent record of one night's observations should not be delayed beyond the following day. Nothing must be added to the permanent record from memory: as regards observational data, the sole source for the permanent record is the observer's notebook; any amplifications or explanatory notes that have to be added to either record after the observations are ended must be distinguished from observational data recorded at the telescope— by the use, e.g. of a different-coloured ink.

511

The two mental processes to be most feared and shunned are expectation and memory: expectation when making the observation, and memory when recording it. The time to record the observation is whilst still at the telescope: if the nature of the object permits, it should be reobserved and compared with what has just been recorded.

It is advisable to include with the record of each observation a note of the degree of reliance placed upon it. This may take the form of a numerical weight or of a code letter (e.g. A, clearly and definitely seen; B, glimpsed at intervals, or visible only to indirect vision; C, doubtful); in either case a separate column should be set aside in the record for this note.

Never make an erasure in the record. Anything to be altered should be deleted by a single line, which allows the original as well as the amendment to be read.

In a tabular layout, draw a line through any compartment in which no entry is made.

Always quote full instrumental details and an index, as impersonal as possible, of the seeing and sky condition.

In the case of drawings, record the time at which the main outlines were laid down. The subsequent redrawing of rough sketches made at the telescope is not to be recommended: human memory is not that infallible.

Employ UT or GMAT throughout.

The indexing of photographic material falls into a rather different category. A serial number (plus a distinguishing letter for each camera if more than one is used) should be scratched on each plate immediately before it is removed from the plateholder for development. The record book should devote a separate page to each night's exposures, and be headed with the instrumental data and date. Against each serial number should be recorded in tabular form:

> Name or catalogue number of object
> RA and Dec of object or of plate centre
> Start of exposure: UT
> > Hour angle
> > Approximate ZD
> > Sky condition
> End of exposure: UT
> Length of exposure
> UT of any interruptions to the exposure
> Focus and focal ratio
> Type of plate and of any filter used
> Remarks

In the history of astronomy can be found numerous cautionary tales which illustrate the fatal consequences of messy and muddled observational records, as well as of preconceived ideas regarding what is likely or possible, and of emotional bias—expectation, disappointment, surprise, hope. No fewer than 19 pre-discovery observations of Uranus have been identified, from 1690 (by Flamsteed) onward. It is true that many of these in no way reflect upon the technique of the observers, since a single observation would quite possibly not reveal its planetary character.* But that *none* of these observations should have led to the discovery of Uranus is incredible. The case that is most relevant to the matter of observational records concerns the French astronomer Lemonnier, who in January 1796 observed Uranus six times over a period of nine days, including observations on four consecutive nights. His records of observations were kept in a particularly untidy and unsystematic fashion (one of the Uranus observations was noted down on a paper bag that had contained hair powder), and this certainly contributed to his failure: in a well-kept record the anomalies between these nine observations could not have failed to strike him—and he would have anticipated Herschel in the first planetary discovery of historical times by twelve years.

Finally, having made your record, have confidence in it and be prepared to stand by it. On 1795 May 10, Lalande observed a star-like object and noted that its position had apparently changed since he observed it two days previously. But he 'assumed' that his earlier observation must have been inaccurate, or his record of it mistaken, and did not investigate further. The object was Neptune.

The lesson is clear: avoid preconceptions, avoid bias, record systematically and unambiguously, and always investigate apparent anomalies. Never accept an anomalous observation as being the result of error until investigation, not assumption, has proved it to be so. In the history of science small anomalies and unexpected discrepancies have led to more discoveries than, perhaps, any other single factor.

* On the other hand Herschel detected its non-stellar character before its motion had been established.

SECTION 33

OBSERVATIONAL AIDS

Both the *Astronomical Ephemeris* (B. 417)* and the *Handbook of the British Astronomical Association* (B. 421) contain vast quantities of information invaluable to the observer, whatever his particular field of work may be. Under sections 1, 2, 4–11, 16 of *O.A.A.* are listed the relevant data to be found in these two sources. In addition to these the *B.A.A.H.* annually gives tables of Astronomical and Physical Constants, Miscellaneous and Terrestrial Data, and notes on Time Reckoning, including a table of Julian dates. General articles dealing with some branch of observational technique, asteroid ephemerides, lists of variables especially requiring observation, etc, are included from time to time.

The B.A.A. *Circulars* (B. 423) are also invaluable to the practical observer. Distributed at irregular intervals to subscribers, they contain up-to-the-minute news regarding new discoveries and objects urgently requiring observation—comets, novae, planetary phenomena such as notable and ephemeral surface markings, variable stars requiring observation, outstanding solar activity, meteor showers, unusual asteroid news and discoveries, etc. The primary aim of the *Circulars* is to notify observers of new discoveries quickly enough for the objects to be adequately and widely observed before their disappearance.

As regards star charts and atlases, the reader is referred to B. 426–447 for details.

The British Astronomical Association is a world-wide association of amateur astronomers. Every observer should not only join such an Association but should also become an active member of the Observing Section devoted to his particular field of work. These Sections—each under the directorship of an experienced observer who collates and periodically publishes the observations of the Section, as well as giving information and advice whenever required—are as follows: Sun, Moon, Mercury and Venus, Mars, Jupiter, Saturn, Comet,

* The French, American, and German equivalents are respectively B. 418, 419, 420.

514

Meteor, Aurora, Variable Star, Instruments and Observing Methods, Computing, Historical, Radio and Electronics, Artificial Satellites.*

The following information regarding the nature and aims of the Association is reprinted by permission :

The British Astronomical Association was originally formed in 1890 and now numbers over 4000 Members. Its leading features are as follows:

Membership—Open to all persons interested in astronomy. Nomination Forms may be had on application to the Assistant Secretary, and must be signed by one Member from personal knowledge, or by two resident householders (giving addresses) from personal knowledge.

Objects—(1) The association of observers, especially the possessors of small telescopes, for mutual help, and their organisation in the work of astronomical observation.

(2) The circulation of current astronomical information.

(3) The encouragement of a popular interest in astronomy.

Methods—(1) The arrangement of Members, for the work of observation, in Sections or departments of observation, under experienced Directors.

(2) The publication, at regular intervals, of a *Journal* containing reports of the Association's Meetings, and of its Observing Sections, papers by Members, and notes on current astronomy.

(3) The formation of Branches of the Association in the provinces and in the Commonwealth and Empire.

(4) The holding of Meetings in London, and at the seats of the Branches.

(5) The formation of a library and of collections of astronomical instruments and lantern slides, for loan to Members of the Association.

(6) The affiliation of schools and societies at a minimum annual subscription of £3·25.

Entrance Fee—£1·00 (50p under 21).

Annual Subscriptions (due on August 1 each year)—£3·25 (£2·50 under 21). The *Circulars*, giving early news of discoveries etc, are a separate publication—annual subscription 50p.

All Cheques and Postal or Money Orders are to be made payable to The British Astronomical Association, crossed 'National Provincial

* The work of the Sections is described in B. 424.

Bank', and sent to the Assistant Secretary at the Registered Office of the Association: Burlington House, Piccadilly, London, WIV ONL.

Meetings—The Meetings of the Association are held on the last Wednesday in each month (unless the Council shall decide otherwise), from October to June inclusive, at 17.00 hrs at the Scientific Societies' Lecture Theatre, 23 Savile Row, London W.1.

Another national association of amateur astronomers is the Junior Astronomical Society (J.A.S.), which holds meetings four times a year at the Alliance Hall, Westminster, London. The J.A.S. has a variety of observing sections, publishes a journal ("Hermes", quarterly) which is distributed free of charge to members, and also news circulars. Membership is open to all at £1·25 p.a., and enquiries should be directed to the J.A.S. Secretary, 58 Vaughan Gardens, Ilford, Essex.

Of the many foreign societies with amateur or partly amateur membership, the following may be mentioned particularly:

La Société Astronomique de France: 28 rue Saint-Dominique, Paris 7ᵉ.

Amateur Astronomers Association: American Museum of Natural History, New York.

American Association of Variable Star Observers: Harvard College Observatory.

Astronomical Society of the Pacific: San Francisco, California.

Royal Astronomical Society of Canada: Toronto, Canada.

B. 425 summarises a great deal of information regarding the amateur societies and publications of the world.

FIELD ORIENTATION

The normal orientation of the telescopic field—i.e. a field lying between the zenith and the S point of the horizon, as seen in an astronomical ocular—is

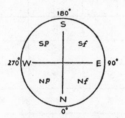

FIGURE 188

It is, compared with the naked-eye field, inverted (top and bottom reversed) and reversed (left and right reversed). Position angles, whatever the orientation of the field may be, are always measured from the N point eastwards.

Field orientation becomes complicated with the introduction of diagonals and, in the case of the Sun, of projection. An odd number of reflections causes inversion or reversal of the image; an even number of reflections either produces inversion and reversal or else leaves the image as it was presented to the first reflecting surface; projection is equivalent to one reflection.

In the following Table, the image orientation specified in the left-hand column is obtained by a combination of those succeeding columns which are, on any one horizontal level, marked with a ×. Thus (second horizontal line) an erect image is obtained by projection with an astronomical ocular and a diagonal turned vertically downward, the screen being viewed from above by an observer standing behind the telescope and facing the OG.

517

| | Astronomical ocular | Terrestrial ocular | Diagonal | | Opaque projection screen | Translucent projection screen |
			Turned vertically (Observer facing OG)	Turned horizontally (Observer shoulder to OG)		
Erect		×				
	×		×		×	
	×			×	×	
	×					×
Inverted	×			×		
	×		×			×
		×			×	
		×	×			
		×		×		×
Reversed	×		×			
	×			×		×
	×				×	
		×	×			×
		×		×		
Inverted and Reversed	×					
		×	×		×	
		×		×	×	
		×				×

The four possible orientations of a field on the meridian between the zenith and the S horizon are:

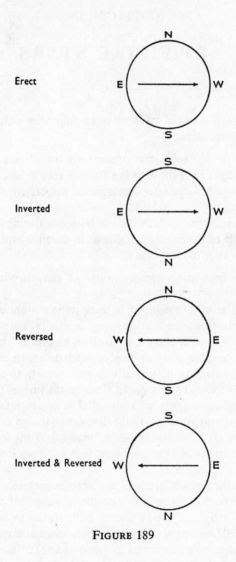

FIGURE 189

ARTIFICIAL STARS

As test objects, artificial stars have the following advantages compared with natural stars:

(a) they are far less subject to atmospheric turbulence, owing to the reduced light path through the atmosphere;

(b) they are not subject to atmospheric dispersion, for the same reason;

(c) they are stationary relative to the telescope during the test;

(d) the effects of temperature gradients in the tube and in the objective are easier to eliminate.

They are also interesting demonstrators of the phenomena of diffraction.

Whatever form of artificial star is used, its minimum distance from the telescope must be such that its image is smaller than the spurious disc of a point source. It should, if possible, be greater than this minimum permissible value, since as its rays approximate to parallelism the plane in which its image is formed will approximate to the true focal plane of the objective (i.e. the plane in which the images of 'infinitely' distant astronomical objects are formed). The separation of these two planes is inversely proportional to the distance of the artificial star; thus in the case of a $f/15$ 3-in the image is formed 0·57 ins from the focal plane when the star is 100 yards distant, 0·28 ins from it when the star is 200 yards distant, etc.

Two convenient types of artificial star may be mentioned:

(i) Circular aperture made with a needle in a sheet of tinfoil. If this is to be illuminated simply by an electric bulb, its diameter will have to be as large as 0·02 ins approximately. A more efficient arrangement is a very much smaller hole (diameters as small as 0·0005 ins can be managed) illuminated by focusing the image of the filament of a gas-filled, unfrosted electric lamp on to it by means of a condensing lens; without this arrangement, insufficient light will pass through the hole. Holes of this calibre can only be made with a needle that has been very carefully honed on glass (its progress checked with a microscope); the punctures are made with the tinfoil glued along its edges to a sheet of glass, the needle being twisted through at least one turn while resting, with very little pressure, on the foil.

(*ii*) Reflection of the Sun in any polished convex surface of short radius, such as the field lens of a Huyghenian eyepiece, a spherical thermometer bulb, a globule of mercury on a dark non-reflecting background, or a black glass bead. The diameter of the solar image will be about 0·0025 that of the bulb or globule, which makes the calculation of the minimum permissible distance a simple matter.

SECTION 36

ANGULAR ESTIMATES

In some branches of work (e.g. Zodiacal Light, meteors, etc) it is essential to be able to make accurate estimates of angular distances by eye; and in any case it is a useful ability to have.

An object x units long subtends an angle of $1°$ when its distance from the eye is $57\cdot3x$. Thus:

1 in at 4 ft 9 ins subtends:	$1°$
1 in at average arm's length (23 ins) subtends:	$2°5$
6-in rule at arm's length subtends:	$15°$

Other approximate measurements are:

width of index nail at arm's length:	$1°$
1st–2nd knuckles of clenched fist at arm's length:	$3°$
2nd–4th knuckles of clenched fist at arm's length:	$5°$
1st–4th knuckles of clenched fist at arm's length:	$8°$
span of open hand at arm's length:	$18°$

Among the stars, the following distances are useful for training the eye:

α–η UMa	25°	α–β Aql	$2\frac{1}{2}°$
α–δ UMa	10°	α–γ Aql	2°
α–β UMa	5°	δ–ε Ori	$1\frac{1}{4}°$
α–β CMi	4°	Solar=lunar diameter	$\frac{1}{2}°$
δ–ζ Ori	3°		

SECTION 37

THE NORTH POLAR SEQUENCE, AND OTHER STANDARD REGIONS

The NPS is a series of stars in the vicinity of the N Pole (chosen for visibility at all times and seasons) whose positions and other data—particularly including stellar magnitudes—have been determined with precision.

B. 509 contains:

(a) the NPS: 46 stars of mags 4·47–20·96 (photographic); positions, catalogue numbers, spectroscopic types, photometric and photographic magnitudes;

(b) the NPS of Red Stars: 12 stars of mags 6·69–13·34 (photographic); same data;

(c) Supplementary Standard Stars near the NCP: 38 stars of mags 2·71–21·00 (photographic); same data;

(d) photographic chart of the region of the NCP on scale $5'' = 1$ mm (Mt Wilson 60-in reflector, 4^h exp.).

B. 514 contains:

(a) all the B. 509 stars with the addition of colour indices, derived both from magnitudes and exposure ratios;

(b) Seares' additional list of Supplementary Stars: 56 stars of mags 11·00–16·28 (photographic); same data.

B. 507 is useful for users of small instruments, being a selection from B. 509, 514 in which the lower magnitudes are omitted:

(a) 68 stars from B. 509 mags 2·55–18·16 (photographic); photographic and photovisual mags, c/i from mags and exposure ratios;

(b) the 56 stars from B. 514 (b); same data;

(c) 2 photographic charts of the NCP region: $4' = 1$ mm, and $23'' = 1$ mm, approximately.

B. 506 gives a general account of the whole NPS work at Harvard, and contains:

(*a*) the 3 lists from B. 509, slightly amended;

(*b*) 36 Pleiades stars of mags 3·01–14·82 (photographic); photometric and photographic mags;

(*c*) 18 Praesepe stars of mags 6·31–10·55 (photographic); same data;

(*d*) 2 charts of the NCP region: 20″=1 mm (see also B. 504), and 5″=1 mm (see also B. 509).

B. 500 contains:

(*a*) photographic magnitudes of 262 stars in the NCP vicinity, mags 8·83–15·80 (photographic);

(*b*) chart of the area 50′ in diameter, centred on the NCP: 30″=1 mm.

B. 505, 502 contain data relating to stars in the Harvard Standard Regions, which are widely distributed in both RA and Dec so that one at least may always be available near the meridian.

B. 504 gives no photometric information, but lists the positions of 589 stars in the vicinity of the NCP; also includes a photographic chart of the region on a scale of 20″=1 mm.

SECTION 38

I.A.U. CONSTELLATION
ABBREVIATIONS

The method of designating constellations throughout this book is the 3-letter abbreviation system recommended by the I.A.U. This is concise, convenient, and unambiguous, and amateurs are urged to employ it in their observation records and elsewhere.

The constellation names fall into four groups, according to the method of their abbreviation:

(*a*) Sixty-five constellation names are abbreviated to the first three letters (e.g. Andromeda, And). The twenty-four exceptions to this rule are:

(*b*) Nine contractions to the first two letters of the name (or of its genitive) and a later one:

Apus	Aps	Hydrus	Hyi
Aquila	Aql	Phoenix	Phe
Aquarius	Aqr	Sculptor	Scl
Crater	Crt	Scutum	Sct
Hydra	Hya		

(*c*) Ten two-word names are contracted as follows:

Canis Major	CMa	Leo Minor	LMi
Canis Minor	CMi	Piscis Austrinus	PsA
Corona Australis	CrA	Triangulum Australe	TrA
Corona Borealis	CrB	Ursa Major	UMa
Canes Venatici	CVn	Ursa Minor	UMi

Note: Coma (Berenices) is contracted to Com

(*d*) Five contractions are formed irregularly:

Cancer	Cnc	Sagitta	Sga
Corvus	Crv	Sagittarius	Sgr
Pisces	Psc		

Note: Caelum is the modern spelling of Coelum, hence Cae. Malus (once part of Argo) is replaced by Pyxis. Puppis, Vela, Carina (once parts of Argo) are now autonomous.

525

SECTION 39

WAVELENGTHS AND FREQUENCIES

Wavelengths are commonly expressed in one of the following units:

micron	μ	10^{-4} cm
millimicron	$\mu\mu$	10^{-7} cm
micromillimetre	mμ	
Ångström unit	Å	10^{-8} cm
X-ray unit	XU	10^{-11} cm

If λ=the wavelength of a radiation (cms),
  ν'=its wave number (number of waves per cm),
  ν=its frequency number (number of waves per sec),
  T=its period (interval between successive occurrences of the same phase at a given point),
  c=its velocity,

then
$$\nu=\frac{1}{T}=\frac{c}{\lambda}$$

$$\lambda=\frac{1}{\nu'}=\frac{c}{\nu}=cT$$

Electromagnetic wavebands of most interest are (approximately):

30,000	..	1000 m	long-wave radio
1000	..	100 m	medium-wave radio
100	..	10 m	short-wave radio
10	..	0·1 m	T/V
0·1	..	0·001 m	radar
20,000	..	7700Å	infrared
7700	..	6500Å	red
6500	..	6000Å	orange
6000	..	5500Å	yellow
5500	..	5000Å	green
5000	..	4500Å	blue
4500	..	3600Å	violet

red, orange, yellow, green, blue, violet } visible light

3600	..	200Å	ultraviolet
32	..	1·0Å	X-radiation
1·0	..	0·1Å	γ-radiation
0·001	..	0·00001Å	cosmic radiation

By international convention (International Solar Union, 1907) the red line of cadmium at 6438·3696Å is taken as the fundamental reference point in the solar spectrum.

Spectral lines are denoted by their wavelengths in Å. Some of the more important solar lines are also specified by letters assigned to them by Fraunhofer. The lines in the Balmer series of the hydrogen spectrum are also denoted by the letter H followed by letters from the Greek alphabet (e.g. $H\alpha$).

O.A.A., section 1.11, contains a table and map of the more important lines of Fraunhofer's spectrum.

BIBLIOGRAPHY

[See pages 557–558 for a Supplemental Bibliography listing books published since the original bibliography was compiled.]

Abbreviations used in this Bibliography are listed below; wherever possible the style of the *Astronomischer Jahresbericht* or the *World List of Scientific Periodicals* has been followed. Bibliographical data are enclosed in round brackets ( ); my own comments, if any, in square brackets [ ]. Each division of the Bibliography is arranged alphabetically by authors unless some other arrangement (e.g. chronological) is specified.

A.J. Astronomical Journal.
A.N. Astronomische Nachrichten.
Ann. Cape Obs. Annals of the Cape Observatory.
Ann. d'Astrophys. Annales d'Astrophysique.
Ann. Lowell Obs. Lowell Observatory Annals.
Ann. Obs. Strasbourg. Annales de l'Observatoire de Strasbourg.
Ap. J. Astrophysical Journal [previously 'Sidereal Messenger', 1883–92; 'Astronomy and Astro-Physics', 1892–95].
Astr. Mitt. Göttingen. Astronomische Mittheilungen der Königlichen Sternwarte zu Göttingen.
Astr. Pap., Wash. Astronomical Papers of the American Ephemeris & Nautical Almanac.
B.A.A.H. Handbook of the British Astronomical Association.
Brit. J. Physiol. Optics. British Journal of Physiological Optics.
B.S.A.F. Bulletin de la Société Astronomique de France ['L'Astronomie'].
Connais. Temps. Connaissance des Temps.
Contr. Dunsink Obs. Dunsink Observatory Contributions.
Contr. Lick Obs. Contributions from the Lick Observatory.
Contr. Perkins Obs. Perkins Observatory Contributions.
Contr. Princeton Obs. Princeton University Observatory Contributions.
C.R. Comptes Rendus des Séances de l'Académie des Sciences.
Engl. Mech. English Mechanics [formerly 'The English Mechanic'].
Harv. Ann. Annals of Harvard College Observatory.
Harv. Bull. Harvard Bulletins.
Harv. Circ. Harvard College Observatory Circulars.
Harv. Monogr. Harvard Observatory Monographs.
Harv. Repr. Harvard Reprints.
H.d.A. Handbuch der Astrophysik (Julius Springer, Berlin).
H.M.S.O. Her Majesty's Stationery Office.
I.A.U. International Astronomical Union.
Internat. Ass. Acad. International Association of Academies.

Irish Astr. J. Irish Astronomical Journal.
J. Appl. Phys. Journal of Applied Physics.
J.B.A.A. Journal of the British Astronomical Association.
J. Can. R.A.S. Journal of the Royal Astronomical Society of Canada.
J. Opt. Soc. Amer. Journal of the Optical Society of America.
J. Sci. Instr. Journal of Scientific Instruments.
L.O.B. Lick Observatory Bulletins.
Lund. Medd. Meddelande från Lunds Astronomiska Observatorium.
Mem. Amer. Acad. Arts Sci. Memoirs of the American Academy of Arts and Sciences.
Mem. B.A.A. Memoirs of the British Astronomical Association.
Mem. R.A.S. Memoirs of the Royal Astronomical Society and Memoirs of the Astronomical Society of London [consecutive volume numbering].
M.N. Monthly Notices of the Royal Astronomical Society.
M.W.C. Mount Wilson Contributions.
N.A. The Nautical Almanac and Astronomical Ephemeris (H.M.S.O., annually).
Nat. Nature.
Obs. The Observatory.
Perkins Obs. Repr. Perkins Observatory Reprints.
Phil. Trans. Philosophical Transactions of the Royal Society.
Photogr. J. Photographic Journal.
Phys. Rev. Physical Review [U.S.A.].
Pop. Astr. Popular Astronomy.
Proc. Amer. Acad. Arts Sci. Proceedings of the American Academy of Arts and Sciences.
Proc. Amer. Phil. Soc. Proceedings of the American Philosophical Society.
Proc. Phys. Soc. Lond. Proceedings of the Physical Society.
Proc. R. Irish Acad. Proceedings of the Royal Irish Academy.
Proc. Roy. Soc. Proceedings of the Royal Society.
Publ. A.A.S. Publications of the American Astronomical Society.
Publ. A.S.P. Publications of the Astronomical Society of the Pacific.
Publ. Tartu. Publications de l'Observatoire Astronomique de l'Université de Tartu (Dorpat).
Publ. Astrophys. Obs. Potsdam. Publikationen des Astrophysikalischen Observatoriums zu Potsdam.
Publ. Carneg. Instn. Carnegie Institution Publications, Washington.
Publ. Cincinn. Obs. Cincinnati Observatory Publications.
Publ. Dom. Astrophys. Obs. Dominion Astrophysical Observatory Publications.
Publ. Lick Obs. Lick Observatory Publications.
Publ. L. McC. Obs. Publications of the Leander McCormick Observatory of the University of Virginia.
Publ. Michigan Obs. Michigan Observatory Publications.
Publ. Univ. Pa. Pennsylvania University Publications (Astronomical Series).
Publ. Yale Obs. Yale University Observatory Publications.
Result. Obs. Nac. Argent. Resultados del Observatorio Nacional Argentino.
Rev. d'Opt. Revue d'Optique Théorique et Instrumentale.
Rev. Mod. Phys. Review of Modern Physics.

BIBLIOGRAPHY

Rev. Sci. Instr. Review of Scientific Instruments.
Sci. Amer. Scientific American.
Sci. Progr. Science Progress.
Sky & Telesc. Sky and Telescope [amalgamation of 'Sky' and 'Telescope'].
Smithson. Contr. Smithsonian Contributions to Knowledge.
Trans. I.A.U. Transactions of the International Astronomical Union.
Veröff. Univ.Berlin-Babelsberg. Veröffentlichungen der Universitätssternwarte zu Berlin-Babelsberg.
Veröff. König. Astron. Rechen-Instit. Veröffentlichungen des Königlichen Astronomischen Rechen-Instituts zu Berlin.
Wash. Nat. Ac. Proc. Proceedings of the National Academy of Science of the United States of America.

FUNDAMENTAL, SPHERICAL AND DYNAMICAL ASTRONOMY

[The following standard textbooks are arranged in approximate order of less to more advanced.]

1. W. M. SMART, *Foundations of Astronomy* (Longmans, Green, 1944).
2. M. DAVIDSON, *Elements of Mathematical Astronomy* (Hutchinson's Scientific and Technical Publications, 1949).
3. C. W. C. BARLOW & G. H. BRYAN, *Elementary Mathematical Astronomy* (University Tutorial Press, 1944).
4. W. M. SMART, *Text-book on Spherical Astronomy* (Cambridge University Press, 1944).
5. J. J. NASSAU, *A Textbook of Practical Astronomy* (McGraw-Hill, N.Y., 1948).
6. E. LOOMIS, *An Introduction to Practical Astronomy* (Harper, 1899).
7. G. F. CHAMBERS, *A Handbook of Descriptive and Practical Astronomy* (Oxford, 1890, 3 vols).
8. W. W. CAMPBELL, *The Elements of Practical Astronomy* (Macmillan, 1899).
9. W. CHAUVENET, *Manual of Spherical and Practical Astronomy* (Lippincott, Philadelphia, 1906, 2 vols).
10. R. BALL, *A Treatise on Spherical Astronomy* (Cambridge, 1908).
11. S. NEWCOMB, *A Compendium of Spherical Astronomy* (Macmillan, 1906).
12. H. C. PLUMMER, *An Introductory Treatise on Dynamical Astronomy* (Cambridge, 1918).
13. F. R. MOULTON, *An Introduction to Celestial Mechanics* (Macmillan, 1935).
14. J. C. WATSON, *Theoretical Astronomy* (Lippincott, Philadelphia, 1900).
15. J. BAUSCHINGER, *Tafeln zur theoretischen Astronomie* (Leipzig, 1901). [Invaluable collection of astronomical data.]
16. J. BAUSCHINGER, *Die Bahnbestimmung der Himmelskörper* (Engelmann, Leipzig, 1928).
17. W. KLINKERFUES, *Theoretische Astronomie* (Vieweg, Braunschweig, 3rd edn, revised by H. BUCHHOLZ, 1912).

DESCRIPTIVE AND GENERAL ASTRONOMY

[Including some miscellaneous works which cannot suitably be placed in other divisions of the Bibliography.]

18. R. S. BALL (rev., T. E. R. PHILLIPS), *A Popular Guide to the Heavens* (Philip, 1925).

19. R. H. BAKER, *An Introduction to Astronomy* (Van Nostrand, N.Y., 1940).

20. W. BARTKY, *Highlights of Astronomy* (University of Chicago Press, 1935).

21. E. A. BEET, *A Text Book of Elementary Astronomy* (Cambridge, 1945).

22. M. DAVIDSON (ed.), *Astronomy for Everyman* (Dent, 1953).

23. J. C. DUNCAN, *Astronomy* (Harper, 1935).

24. P. EMANUELLI, *Il Cielo e le sue Meraviglie* (Milan, 1934). [150 fine astronomical photographs with descriptive text.]

25. D. S. EVANS, *Frontiers of Astronomy* (Sigma Books, Lond., 1946).

26. E. A. FATH, *The Elements of Astronomy* (McGraw-Hill, N.Y., 1936).

27. G. E. HALE, *Ten Years' Work of a Mountain Observatory* (*Publ. Carneg Instn.*, 235, 1916). [The story of the early days at Mount Wilson.]

28. H. SPENCER JONES, *General Astronomy* (Arnold, 3rd edn., 1951).

29. S. A. MITCHELL & C. G. ABBOTT, *The Fundamentals of Astronomy* (Chapman & Hall, 1929).

30. F. R. MOULTON, *Astronomy* (*Macmillan*, 1931).

31. T. E. R. PHILLIPS & W. H. STEAVENSON (eds.), *Splendour of the Heavens* (Hutchinson, 1923).

32. H. N. RUSSELL, R. S. DUGAN & J. Q. STEWART, *Astronomy* (Ginn, Boston, 2 vols, 1945). [Revision of C. A. YOUNG's *A Text-Book of General Astronomy*, 1st edn., 1888.]

33. J. B. SIDGWICK, *The Heavens Above: A Rationale of Astronomy* (Oxford University Press, Lond., 1948, and N.Y., 1950).

34. W. T. SKILLING & R. S. RICHARDSON, *Astronomy* (McGraw-Hill, 1946).

35. W. STOCKLEY, *Stars and Telescopes* (McGraw-Hill, 1936).

36. E. O. TANCOCK, *The Elements of Descriptive Astronomy* (Oxford, 2nd edn., 1919). [Elementary.]

37. R. L. WATERFIELD, *A Hundred Years of Astronomy* (Duckworth, 1938).

OPTICS, INCLUDING TELESCOPIC FUNCTION AND ABERRATIONS

38. M. A. AINSLIE, Presidential Address to the B.A.A. (*J.B.A.A.*, 40, No. 1, 5).

39. F. C. CHAMPION, *University Physics* (Blackie, 1941).

40. W. R. DAWES, Catalogue of Micrometrical Measurements of Double Stars (Introduction) (*Mem. R.A.S.*, 35, No. 4, 137). [The 'Introduction' is reprinted in *M.N.*, 27, No. 6, 217.]

41. A. GLEICHEN (trans. EMSLEY & SWAINE), *The Theory of Modern Optical Instruments* (H.M.S.O., 1918).

42. R. A. HOUSTOUN, *A Treatise on Light* (Longmans, Green, 1933).

43. M. R. JARRY-DESLOGES, Influence des divers éléments d'un objectif (ouverture, distance focale, grossissement) sur la qualité des images télescopiques (*C.R.*, 177, 248).

44. B. K. JOHNSON, *Practical Optics* (Hatton Press, Lond., 1947).

45. [T. LEWIS] On the Class of Double Stars which can be observed with Refractors of Various Apertures (*Obs.*, 37, No. 479, 378).

46. L. C. MARTIN, *An Introduction to Applied Optics* (Pitman, 1930, 2 vols).

47. G. S. MONK, *Light: Principles and Experiments* (McGraw-Hill, 1937).

48. W. H. PICKERING, Definition and Resolution (*Pop. Astr.*, **28**, No. 9, 510).
49. W. H. PICKERING, Artificial Disks (*Harv. Ann.*, **32**, 149).
50. T. PRESTON, *The Theory of Light* (Macmillan, 1890).
51. M. VON ROHR (trans. R. KANTHACK), *Geometrical Investigation of the Formation of Images in Optical Instruments* (H.M.S.O., 1920).
52. F. E. ROSS, Astrometry with Mirrors and Lenses (*Ap. J.*, **77**, 243).
53. A. SCHUSTER, *An Introduction to the Theory of Optics* (Arnold, 1904).
54. K. SCHWARZSCHILD, Untersuchungen zur geometrischen Optik, II (*Astr. Mitt. Göttingen*, **10**, 1905).
55. J. P. C. SOUTHALL, *Mirrors, Prisms and Lenses—A Text-Book of Geometrical Optics* (Macmillan, 1933).
56. W. H. STEAVENSON, Note on the Light-grasp of Refractors (*J.B.A.A.*, **25**, No. 4, 186).
57. W. H. STEAVENSON, Further Note on the Light-grasp of Refractors (*J.B.A.A.*, **25**, 229).
58. H. DENNIS TAYLOR, An Experiment with a 12½-in. Refractor, whereby the Light lost through the Secondary Spectrum is separated out and rendered approximately measurable (*Mem. R.A.S.*, **51**, 77). [Reprinted, B. 152, 3rd ed.]
59. H. DENNIS TAYLOR, The Secondary Colour Aberrations of the Refracting Telescope in Relation to Vision (*M.N.*, **54**, No. 2, 67). [Reprinted, B. 152, 3rd edn.]
60. H. DENNIS TAYLOR, Description of a Perfectly Achromatic Refractor (*M.N.*, **54**, No. 5, 328). [Reprinted, B. 152, 3rd edn.]
61. P. J. TREANOR, On the Telescopic Resolution of Unequal Binaries (*Obs.*, **66**, No. 831, 255).
62. E. T. WHITTAKER, *The Theory of Optical Instruments* (Cambridge, 1907).

OPTICAL MATERIALS

63. CARTWRIGHT, HAWLEY & TURNER, Reducing the Reflection from Glass by Evaporated Films (*Phys. Rev.*, **55**, 595).
64. CARTWRIGHT, HAWLEY & TURNER, Reducing the Reflection from Glass by Multilayer Films (*Phys. Rev.*, **55**, 675).
65. CARTWRIGHT, HAWLEY & TURNER, Multilayer Films of High Reflecting Power (*Phys. Rev.*, **55**, 1128).
66. G. E. HALE, On the Comparative Value of Refracting and Reflecting Telescopes for Astrophysical Investigations (*Ap. J.*, **5**, No. 2, 119).
67. J. STRONG et al., *Modern Physical Laboratory Practice* (Blackie, 1949).
68. A. F. TURNER, Anti-reflection Films on Glass Surfaces (*J. Appl. Phys.*, **12**, 351).
69. V. VAND, The Cleaning of Mirrors before Aluminising (*J.B.A.A.*, **53**, No. 7, 209).
70. H. C. VOGEL, The Absorption of Light as a Determining Factor in the Selection of the Size of the Objective for the Great Refractor of the Potsdam Observatory (*Ap. J.*, **5**, 75).
71. *German Vacuum Evaporation Methods for Producing First Surface Mirrors, Semi-transparent Mirrors, and Non-reflecting Films* (Report PB.4158, Office of Technical Publications, Dept. of Commerce, Washington, 1945).

OCULARS

72. L. BELL, Ghosts and Oculars (*Proc. Amer. Acad. Arts Sci.*, **56**, No. 2, 45).
73. F. J. HARGREAVES, A Modified Newtonian Telescope (*J.B.A.A.*, **52**, No. 7, 226).
74. E. W. TAYLOR, The Inverting Eyepiece and its Evolution (*J. Sci. Instr.*, **22**, No. 3, 43).
75. F. H. THORNTON, Mounting the Barlow Lens (*J.B.A.A.*, **60**, No. 3, 77).

TELESCOPES, AND GENERAL WORKS ON OPTICAL INSTRUMENTS

76. J. G. BAKER, A Method of Making Aspherical Surfaces of Revolution by means of Spherical Surfaces alone (*Pop. Astr.*, **48**, 78).
77. J. G. BAKER, The Solid-Glass Schmidt Camera and a New Type Nebular Spectrograph (*Proc. Amer. Phil. Soc.*, **82**, No. 3, 323).
78. J. G. BAKER, A Family of Flat-Field Cameras, Equivalent in Performance to the Schmidt Camera (*Proc. Amer. Phil. Soc.*, **82**, No. 3, 339).
79. L. BELL, *The Telescope* (McGraw-Hill, 1922).
80. E. B. BROWN, *Optical Instruments* (Chemical Publishing Co., N.Y., 1946).
81. C. R. BURCH, On the Optical See-Saw Diagram (*M.N.*, **102**, No. 3, 159). [Advanced treatment of the Schmidt-Cassegrain.]
82. A. C. CLARKE, The Astronomer's New Weapons—Electronic Aids to Astronomy (*J.B.A.A.*, **55**, No. 6, 143).
83. H. J. COOPER (ed.), *Scientific Instruments* (Hutchinson's Scientific and Technical Publications, 1946; vol. 2, 1948).
84. H. W. COX, The Construction of Schmidt Cameras (*J. Sci. Instr.*, **16**, No. 8, 257).
85. H. W. & L. A. COX, The Construction of a Schmidt Camera (*J.B.A.A.*, **48**, No. 8, 308).
86. H. W. &. L. A. COX, Further Notes on Schmidt Cameras (*J.B.A.A.*, **50**, No: 2, 61).
87. H. W. COX, Schmidt Camera Tests (*J.B.A.A.*, **51**, 152, 174).
88. H. W. COX, New Astronomical Reflectors (*J.B.A.A.*, **55**, No. 6, 137).
89. H. E. DALL, Diffraction Effects due to Axial Obstructions in Telescopes (*J.B.A.A.*, **48**, No. 4, 163).
90. A. DANJON & A. COUDER, *Lunettes et Télescopes* (Editions de la Revue d'Optique Théorique et Instrumentale, Paris, 1935).
91. C. DÉVÉ (trans. T. L. TIPPELL), *Optical Workshop Principles* (Hilger, Lond., 1945).
92. G. Z. DIMITROFF & J. G. BAKER, *Telescopes and Accessories* (Blakiston, Philadelphia, 1945).
93. H. DRAPER, On the Construction of a Silvered Glass Telescope, 15½ inches in Aperture, and its Use in Celestial Photography (*Smithson. Contr.*, 34).
94. T. DUNHAM, The Use of Schmidt Cameras in Plane Grating Spectrographs (*Phys. Rev.*, 2nd Series, **46**, 326).
95. T. DUNHAM, Schmidt Spectrograph Cameras (*Publ. A.A.S.*, **8**, 110).
96. W. F. A. ELLISON, *The Amateur's Telescope* (Belfast, 1932).
97. D. O. HENDRIX, An Extremely Fast Schmidt Camera (*Publ. A.S.P.*, **51**, No. 301, 158).

98. D. O. HENDRIX & W. H. CHRISTIE, Some Applications of the Schmidt Principle in Optical Design (*Sci. Amer.*, **161**, 118).

99. F. J. HARGREAVES, A New Plea for the Reflector (*J.B.A.A.*, **46**, No. 5, 193).

100. F. J. HARGREAVES, *The Optical Adjustment of a Newtonian reflector* (B.A.A. Sectional Notes, No. 2).

101. A. G. INGALLS (ed.), *Amateur Telescope Making, Vol. 1* (Scientific American, 4th edn., 23rd ptg., 1978).

102. A. G. INGALLS* (ed.), *Amateur Telescope Making, Vol. 2* (Scientific American, 17th ptg., 1978).

103. E. H. LINFOOT & HAWKINS, An Improved Type of Schmidt Camera (*Obs.*, **66**, No. 831, 239).

104. E. H. LINFOOT, On Some Optical Systems Employing Aspherical Surfaces (*M.N.*, **103**, No. 4, 210). [Advanced theory.]

105. E. H. LINFOOT, The Schmidt-Cassegrain Systems and Their Application to Astronomical Photography (*M.N.*, **104**, No. 1, 48). [Advanced theory.]

106. E. H. LINFOOT, Some Recent Applications of Optics to Astronomy (*M.N.*, **108**, No. 1, 81).

107. E. H. LINFOOT, Achromatised Plate-Mirror Systems (*Proc. Phys. Soc. Lond.*, **57**, No. 3, 199).

108. H. A. LOWER, Notes on the Construction of a $f/1$ Schmidt Camera (B. 102).

109. D. D. MAKSUTOV, New Catadioptric Meniscus Systems (*J. Opt. Soc. Amer.*, **34**, No. 5, 270).

110. G. MATTHEWSON, *Constructing an Astronomical Telescope* (Blackie, 1947).

111. G. MCHARDIE, *Preparation of Mirrors for Astronomical Telescopes* (Blackie, 1937).

112. R. E. PRESSMAN, An Experimental Compound Reflecting Telescope (*J.B.A.A.*, **57**, No. 6, 224).

113. G. W. RITCHEY, On the Modern Reflecting Telescope and the Making and Testing of Optical Mirrors (*Smithson. Contr.*, 34).

114. H. N. RUSSELL, The Schmidt Camera—Introductory (B. 102).

115. R. A. SAMPSON, On a Cassegrain Reflector with Corrected Field (*Phil. Trans.*, A **213**, 27).

116. N. J. SCHELL & T. G. BEEDE, [A 'criss-cross' off-axis telescope] (*Sci. Amer.*, **162**, No. 5, 314).

117. B. SCHMIDT, Ein lichtstarkes komafreies Spiegelsystem (*Central Zeitung fur, Optik und Mechanik*, **52**, 1931).

118. C. H. SMILEY, The Schmidt Camera (*Pop. Astr.*, **44**, No. 438, 415).

119. C. H. SMILEY, The Schmidt Camera—Chromatic Aberration and Other Factors Governing Design (*Pop. Astr.*, **48**, No. 474, 175).

120. C. H. SMILEY, The Construction of a Schmidt Camera (*Publ. A.A.S.*, **9**, Nos. 19–20, 1937).

121. C. H. SMILEY, A Note on the Schmidt Camera (*J.B.A.A.*, **49**, No. 1, 34).

122. O. STRUVE, A New Slit Spectrograph for Diffuse Galactic Nebulae (*Ap. J.*, **86**, No. 5, 613).

123. C. J. TENUKEST, R. SCHAEFER & H. PINNOCK, Notes on the Construction of a 6-inch Maksutov Telescope (*J.B.A.A.*, **56**, No. 7, 130).

* 102A. A. G. INGALLS (ed.), *Amateur Telescope Making, Vol. 3* (Scientific American, 9th. ptg., 1977).

124. J. TEXEREAU, *La Construction du Télescope d'Amateur* (Soc. Astr. de France, 1951). Eng. trans.: *How to Make a Telescope* (Doubleday, 1963).
125. A. J. THOMPSON, *Making Your Own Telescope* (Sky Publishing Corp., Cambridge, Mass., 1947).
126. W. H. THORNTHWAITE, *Hints on Reflecting and Refracting Telescopes* (6th edn., 1895).
127. F. TWYMAN, *Prism and Lens Making* (Hilger & Watts, London, 1952).
128. A. DE VANY, A Rapid Method of Making a Schmidt Correcting Lens (*Pop. Astr.*, **47**, 197).
129. R. L. WALAND, Note on Figuring Schmidt Correcting Lenses (*J. Sci. Instr.*, **15**, No. 10, 339).
130. D. O. WOODBURY, *The Glass Giant of Palomar* (Heinemann, 1940).
131. F. B. WRIGHT, An Aplanatic Reflector with a Flat Field Related to the Schmidt Telescope (*Publ. A.S.P.*, **47**, No. 280, 300).
132. F. B. WRIGHT, Theory and Design of Aplanatic Reflectors Employing a Correcting Lens (*B*. 102).
133. C. G. WYNNE, New Wide-aperture Catadioptric Systems (*Obs.*, **68**, No. 843, 52).

DEWING-UP AND TUBE CURRENTS

134. J. FRANKLIN-ADAMS, Prevention of Dew-deposit on Glass Surfaces (*M.N.*, **72**, No. 2, 165).
135. A. F. LINDEMANN, A Revolving Eyepiece, electrically warmed (*M.N.*, **58**, 362).

TESTING

136. J. A. ANDERSON & R. W. PORTER, Ronchi's Method of Optical Testing (*Ap. J.*, **70**, No. 3, 175).
137. M. G. BIGOURDAN, *Les Méthodes d'Examen des Lunettes et des Télescopes* (Gauthier-Villars, Paris, 1915).
138. C. R. BURCH, On the Phase-Contrast Test of F. Zernike (*M.N.*, **94**, No. 5, 384).
139. C. R. BURCH, On Reflection Compensators for Testing Paraboloids (*M.N.*, **96**, No. 5, 438).
140. C. R. BURCH, Note on Compensation of Spherical Aberration in Telescopes (*J.B.A.A.*, **48**, No. 3, 115).
141. H. W. COX, The Testing of Wide-Angle Mirrors (*J.B.A.A.*, **56**, No. 6, 110).
142. H. E. DALL, A Null Test for Paraboloids (*J.B.A.A.*, **57**, No. 5, 201).
143. P. FOX, An Investigation of the Forty-inch Objective of the Yerkes Observatory [by the Hartmann test] (*Ap. J.*, **27**, No. 4, 237).
144. F. J. HARGREAVES, A Spurious Aberration in Reflecting Telescopes (*J.B.A.A.*, **59**, No. 6, 180).
145. F. J. HARGREAVES, A Note on the Auto-Collimation Test for Concave Mirrors (*J.B.A.A.*, **45**, No. 2, 72).
146. F. J. HARGREAVES, A Further Note on the Auto-Collimation Test (*J.B.A.A.*, **45**, No. 8, 320).
147. F. J. HARGREAVES, A Note on the Hartmann Criterion (*J.B.A.A.*, **51**, No. 4, 121).

148. J. HARTMANN, An Improvement of the Foucault Knife-edge Test in the Investigation of Telescope Objectives (*Ap. J.*, **27**, No. 4, 254).

149. W. H. NEWMAN, A Method of Measuring the Focal Lengths of Zones of a Reflecting Telescope (*J.B.A.A.*, **49**, No. 5, 186).

150. J. S. PLASKETT, The Character of the Star Images in Spectrographic Work (*Ap. J.*, **25**, No. 3, 195). [Hartmann test.]

151. G. W. RITCHEY, On Methods of Testing Optical Mirrors during Construction (*Ap. J.*, **19**, 59).

152. H. DENNIS TAYLOR, *The Adjustment and Testing of Telescope Objectives* (Sir Howard Grubb, Parsons, 4th edn., 1946). [Last revised edn., 1921.]

153. J. V. THOMSON, A Compact Knife-Edge Apparatus for the Foucault Test (*J.B.A.A.*, **54**, No. 4/5, 87).

154. F. ZERNIKE, Diffraction Theory of the Knife-edge Test and its Improved Form, the Phase-Contrast Method (*M.N.*, **94**, No. 5, 377).

TELESCOPE MOUNTINGS, INCLUDING DRIVES

155. M. A. AINSLIE, A Compact 'Reflex' Mounting for a Refractor (*J.B.A.A.*, **46**, No. 9, 324).

156. A. COLEMAN, Equatorial Head and Clock Drive for Camera for Small Scale Photographs of the Eclipse (*J.B.A.A.*, **37**, No. 6, 216).

157. E. C. COLLINSON, [Report of B.A.A. Meeting] (*J.B.A.A.*, **37**, No. 5, 157).

158. M. A. CORNU, On the Law of Diurnal Rotation of the Optical Field of the Siderostat and Heliostat (*Ap. J.*, **11**, No. 2, 148).

159. W. H. COX, A Variable-Frequency Telescope Drive (*J.B.A.A.*, **48**, No. 6, 248).

160. DAUNT, A Home-made Wooden Equatorial Stand (*J.B.A.A.*, **17**, No. 4 169).

161. C. D. P. DAVIES, The Coelostat (*J.B.A.A.*, **12**, 359).

162. D. W. DEWHIRST, Automatic Guiding of Solar Telescopes (*J.B.A.A.*, **56**, No. 8, 144).

163. R. M. FRY, The Adjustment of a Polar Axis by Star Trails near the Pole (*J.B.A.A.*, **45**, No. 6, 234).

164. D. R. GLASSPOOL, An Altazimuth Stand for 10½-inch Reflector (*J.B.A.A.*, **55**, No. 1, 18).

165. H. E. HANSON, The New Telescope Drive at the Harvard Observatory (*Rev. Sci. Instr.*, **10**, 184).

166. F. J. HARGREAVES, A Modification of the 'Gerrish' Drive (*J.B.A.A.*, **41**, No. 3, 112).

167. F. L. HAUGHTON, Telescope Drives (*J.B.A.A.*, **50**, No. 4, 147).

168. [G. A. HOLE], Mr G. A. Hole's Observatory at Patcham, Sussex (*J.B.A.A.*, **51**, No. 9, 329).

169. G. F. KELLAWAY, A Note on the Gerrish Drive (*J.B.A.A.*, **50**, No. 6, 215).

170. F. B. LINDSAY, A New Mounting and Housing for a Telescope (*Pop. Astr.*, **39**, No. 8, 497).

171. MCMATH, HULBERT & MCMATH, Some New Methods in Astronomical Photography (*Publ. Michigan Obs.*, **4**, No. 4, 53).

172. MCMATH & GREIG, A New Method of Driving Equatorial Telescopes (*Publ. Michigan Obs.*, **5**, No. 10, 123).

173. P. R. McMATH, The Tower Telescope of the McMath-Hulbert Observatory (*Publ. Michigan Obs.*, **7**, No. 1, 1).

174. G. W. MOFFITT, Frequency Requirements and the Control of Frequency for Synchronous Motor Operation of Astronomical Telescopes (*Rev. Sci. Instr.*, **3**, No. 9, 499).

175. W. H. PICKERING, Reflectors versus Refractors (*Pop. Astr.*, **38**, No. 3, 134).

176. W. H. PICKERING, The Practical Use of Small Reflectors (*Pop. Astr.*, **34**, No. 9, 570).

177. F. J. SELLERS, A Simple, Inexpensive and Effective Clock Drive for Small Telescopes (*J.B.A.A.*, **40**, No. 6, 222).

178. F. H. THORNTON, A Simple, Controlled, Electric Drive (*J.B.A.A.*, **58**, No. 5, 175).

179. H. H. TURNER, The Coelostat (*J.B.A.A.*, **10**, 264).

180. H. H. TURNER, Some Notes on the Use and Adjustment of the Coelostat (*M.N.*, **56**, No. 8, 408).

181. C. L. TWEEDALE, A Cheap and Effective Clock-driven Portable Equatorial (*Engl. Mech.*, **75**, 310, 326, 448; **76**, 72, 206, 492; **77**, 28, 226; **78**, 326).

182. F. L. O. WADSWORTH, On a New Form of Mounting for Reflecting Telescopes devised by the late Arthur Cowper Ranyard (*Ap. J.*, **5**, No. 2, 132).

183. R. L. WATERFIELD, The Elimination of Atmospheric Refraction in the Adjustment of Equatorials (*J.B.A.A.*, **45**, No. 7, 282).

184. W. REES WRIGHT, An Altazimuth Mounting for Reflecting Telescopes (*J.B.A.A.*, **50**, No. 3, 113).

OBSERVING HUTS

185. H. E. BECKETT, The Painting of Observatory Domes as a Protection against Solar Heat (*Obs.*, **59**, No. 740, 14).

186. H. E. DALL, Thermal Effects of Observatory Paints (*J.B.A.A.*, **48**, No. 3, 116).

187. H. SPENCER JONES, [Letter commenting on B. 185] (*Obs.*, **59**, No. 740, 17).

MICROMETERS

188. C. R. DAVIDSON & L. S. T. SYMMS, A Comparison Image Micrometer. (*M.N.*, **98**, No. 3, 176).

189. R. M. FRY, The Determination of the Errors of a Cross-Wire Micrometer (*J.B.A.A.*, **46**, No. 3, 113).

190. F. J. HARGREAVES, A Comparison-Image Micrometer (*M.N.*, **92**, No. 1, 72).

191. F. J. HARGREAVES, A Modified Comparison-Image Micrometer (*M.N.*, **92**, No. 5, 453).

192. L. RICHARDSON, The Interferometer (*J.B.A.A.*, **35**, No. 3, 105).

193. L. RICHARDSON, A Home-made Interferometer (*J.B.A.A.*, **35**, No. 5, 155).

194. L. RICHARDSON, Results with a Home-made Interferometer (*J.B.A.A.*, **37**, No. 8, 311).

195. H. R. WRIGHT, Bright Wire Illumination (*J.B.A.A.*, **31**, No. 6, 229).

SPECTROSCOPY

196. J. EVERSHED, A New Arrangement of Prisms for a Solar Prominence Spectroscope (*J.B.A.A.*, **7**, No. 6, 331).

197. R. A. SAWYER, *Experimental Spectroscopy* (Chapman & Hall, 1945).

198. F. J. SELLERS, *A Solar Prominence Spectroscope, with instructions for making and using* (B.A.A. Sectional Notes, No. 1=*J.B.A.A.*, **45**, No. 9, 362).

199. W. REES WRIGHT, Two Spectroscopes for Prominence Observations (*J.B.A.A.*, **52**, No. 4, 130).

PHOTOGRAPHY

200. I. S. BOWEN & L. T. CLARK, Hypersensitization and Reciprocity Failure of Photographic Plates (*J. Opt. Soc. Amer.*, **30**, 508).

201. B. BURRELL, Photographs of the Great Aurora, 17th April 1947 (*J.B.A.A.*, **57**, No. 5, 205).

202. A. COLEMAN, The Photography of the Zodiacal Light (*J.B.A.A.*, **44**, No. 7, 262).

203. E. H. COLLINSON & J. P. M. PRENTICE, The Photography of Meteors (*J.B.A.A.*, **37**, No. 7, 266).

204. E. H. COLLINSON, An Automatic Meteor Camera (*J.B.A.A.*, **39**, No. 5, 150).

205. E. H. COLLINSON, An Improved Automatic Meteor Camera (*J.B.A.A.*, **44**, No. 4, 157).

206. E. H. COLLINSON, Meteor Photography, 1928–31 (*J.B.A.A.*, **46**, No. 3, 116).

207. L. J. COMRIE, Note on the Reduction of Photographic Plates (*J.B.A.A.*, **39**, No. 6, 203).

208. [L. J. COMRIE, Report of B.A.A. Meeting] (*J.B.A.A.*, **39**, No. 6, 173).

209. A. E. DOUGLASS, Zodiacal Light and Counterglow and the Photography of Large Areas and Faint Contrasts (*Photogr. J.*, New Series, **40**, No. 2, 44).

210. F. J. HARGREAVES, The Focussing and Squaring-on of Photographic Objectives (*J.B.A.A.*, **37**, No. 3, 85).

211. F. J. HARGREAVES, Camera Work in the Eclipse (*J.B.A.A.*, **37**, No. 6, 212).

212. J. J. HILL, Some Notes on Lunar Photography with a 4¼-inch Refractor (*J.B.A.A.*, **55**, No. 3, 77).

213. T. H. JAMES & G. C. HIGGINS, *Fundamentals of Photographic Theory* (Lond., 1948).

214. G. F. KELLAWAY, The Measurement of Photographic Plates by Projection (*J.B.A.A.*, **47**, 231).

215. E. S. KING, Forms of Images in Stellar Photography (*Harv. Ann.*, **41**, No. 6, 153).

216. E. SKINNER KING, *A Manual of Celestial Photography* (Eastern Science Supply Co., Boston, 1931). [Invaluable.]

217. F. A. LINDEMANN & G. M. B. DOBSON, Note on the Photography of Meteors (*M.N.*, **83**, No. 3, 163).

218. J. C. MABY, A Programme for the Observation and Photography of Sunspots (*J.B.A.A.*, **41**, No. 4, 190).

219. J. C. MABY, Sunspot Photography with a small Visual Refractor (*J.B.A.A.*, **47**, 321).

220. C. E. K. MEES, *The Theory of the Photographic Process* (Macmillan, N.Y., 1942).

221. G. MERTON, An Interpolation Method for Astronomical Photographs (*J.B.A.A.*, **39**, No. 6, 209).

222. G. MERTON, Photography and the Amateur Astronomer (*J.B.A.A.*, **63**, No. 1, 7). [B.A.A. Presidential Address, 1952; useful bibliography.]

223. P. M. MILLMAN, The Theoretical Frequency Distribution of Photographic Meteors (*Selected Papers from Wash. Nat. Ac. Proc.*, **19**, 34, 1933).

224. P. M. MILLMAN, Amateur Meteor Photography (*Pop. Astr.*, **41**, No. 6, 298).

225. P. M. MILLMAN, An Analysis of Meteor Spectra (*Harv. Ann.*, **82**, No. 6, 113).

226. P. M. MILLMAN & D. HOFFLEIT, A Study of Meteor Photographs Taken through a Rotating Shutter (*Harv. Ann.*, **105**, No. 31, 601).

227. C. B. NEBLETTE, *Photography—Its Principles and Practice* (Chapman & Hall, 1943).

228. D. NORMAN, The Development of Astronomical Photography (*Harv. Repr.*, 153).

229. J. PATON, Aurora Borealis: Photographic measurements of height (*Weather*, **1**, No. 6, 8, 1946).

230. J. PATON, Aurora Borealis (*Science News*, **11**, No. 9, 15, 1949).

231. H. PINNOCK, Photographing the Sun (*J.B.A.A.*, **55**, 124).

231A. T. RACKHAM, *Astronomical Photography at the Telescope* (Faber).

232. F. E. ROSS, *Physics of the Developed Photographic Image* (Eastman Kodak Monograph No. 5, N.Y., 1924).

233. F. SCHLESINGER, A Short Method for Deriving Positions of Asteroids, Comets etc., from Photographs (*A.J.*, **37**, 872).

234. F. SCHLESINGER, Some Aspects of Astronomical Photography of Precision (*M.N.*, **87**, No. 7, 506). [George Darwin Lecture, 1927.]

235. F. H. SEARES & P. J. VAN RHIJN, The Distribution of Stars according to Photographic Magnitude and Galactic Latitude (*Trans. I.A.U.*, **2**, 92). [See also *B.A.A.H.*, 1926, p. 34.]

236. E. W. H. SELWYN, *Photography in Astronomy* (Eastman Kodak, 1950).

237. J. SKYORA, La Photographie des Etoiles Filantes (*B.S.A.F.*, **38**, 64).

238. H. H. TURNER, On the Measurement of a Meteor Trail on a Photographic Plate (*M.N.*, **67**, No. 9, 562).

239. H. H. WATERS, The Photography of Meteors (*J.B.A.A.*, **46**, No. 4, 152).

240. H. H. WATERS, *Astronomical Photography for Amateurs* (Gall & Inglis, 1921). [Useful as introduction, but superseded by B. 216.]

241. F. L. WHIPPLE, The Harvard Photographic Meteor Programme (*Sky & Telesc.*, **8**, No. 4, 90=*Harv. Repr.*, 319).

242. F. L. WHIPPLE & P. J. RUBENSTEIN, The Limiting Magnitudes of Photographic Telescopes (*Pop. Astr.*, **50**, 24).

243. W. T. WHITNEY, The Determination of Meteor Velocities (*Pop. Astr.*, **45**, No. 3, 162).

244. H. E. WOOD, The Measurement of the Position of Objects on Photographic Plates (*J.B.A.A.*, **39**, No. 6, 196).

245. H. E. WOOD, Further Note on the Method of Dependences as Applied to the Reduction of Photographic Plates (*J.B.A.A.*, **43**, No. 1, 25).

246. W. H. WRIGHT, On Photographs of the Brighter Planets by Light of Different Colours (*M.N.*, **88**, No. 9,709). [George Darwin Lecture, 1928.]

247. [VARIOUS] *Photography as a Scientific Instrument* (Blackie, 1923).

248. —— *Elementary Photographic Chemistry* (Kodak, *N.D.*).

249. —— *Photographic Plates for Scientific and Technical Use* (Eastman Kodak, N.Y., 1949).

PHOTOMETRY

250. H. W. COX, Photoelectric Photometry (*J.B.A.A.*, **58**, No. 3, 101).

251. M. A. DANJON, Recherches de Photométrie Astronomique (*Ann. Obs. Strasbourg*, **2**, No. 1, 82).

252. G. M. B. DOBSON, I. O. GRIFFITH & D. N. HARRISON, *Photographic Photometry* (Clarendon Press, 1926).

253. D. S. EVANS, *Astronomical Photographic Photometry* (Chapman & Hall, 1952).

254. P. B. FELLGETT, An Exploration of Infra-red Stellar Magnitudes using the Photo-conductivity of Lead Sulphide (*M.N.*, **111**, No. 6, 537).

255. E. GRIFFITHS, *Methods of Measuring Temperature* (Griffin, Lond., 3rd edn., 1947).

256. J. B. IRWIN, The Observational Problem of Stellar Magnitudes and Colours (*Obs.*, **67**, No. 837, 62). [See also B. 270.]

257. G. E. KRON, The Construction and Use of a Photomultiplier Type of Photoelectric Photometer (*Bull. of the Panel on the Orbits of Eclipsing Binaries*, 4).

258. G. E. KRON et al., Photoelectric Photometry (*Bull. of the Panel on the Orbits of Eclipsing Binaries*, 6).

259. G. E. KRON, Application of the Multiplier Phototube to Astronomical Photoelectric Photometry (*Ap. J.*, **103**, No. 3, 326).

260. M. W. OVENDEN, A Visual Photometer for the Comparison of Stars of Different Colours (*J.B.A.A.*, **54**, No. 8/9, 172, 156).

261. E. PETTIT & S. B. NICHOLSON, Stellar Radiation Measurements (*Ap. J.*, **68**, No. 4, 279=*M.W.C.*, 369).

262. E. PETTIT & S. B. NICHOLSON, Lunar Radiation and Temperatures (*Ap. J.*, **71**, No. 2, 102=*M.W.C.*, 392).

263. E. PETTIT, Radiation Measurements on the Eclipsed Moon (*Ap. J.*, **91**, No. 4, 408=*M.W.C.*, 627).

264. E. C. PICKERING, Photometric Observations (*Harv. Ann.*, **11**, No. 1, 1).

265. E. C. PICKERING, A New Form of Stellar Photometer (*Ap. J.*, **2**, No. 2, 89).

266. E. C. PICKERING & O. C. WENDELL, Discussion of Observations made with the Meridian Photometer (*Harv. Ann.*, **23**, No. 1, 1).

267. J. STEBBINS, Measuring Starlight by Photocell (*Sci. Amer.*, **186**, No. 3, 56).

268. J. STEBBINS, The Photo-Electric Photometer of the Yerkes Observatory (*Ap. J.*, **74**, No. 5, 289).

269. J. STEBBINS, The Electrical Photometry of Stars and Nebulae (*M.N.*, **110**, No. 5, 416). [George Darwin Lecture, 1950.]

270. R. H. STOY & R. O. REDMAN, The Observational Problem of Stellar Magnitudes and Colours (*Obs.*, **66**, No. 834, 330). [See B. 256.]

271. V. VAND, Remarks on Making Thermocouples (*J.B.A.A.*, **53**, No. 2, 85).
272. H. F. WEAVER, The Development of Astronomical Photometry (*Pop. Astr.*, **54**, 211, 287, 339, 451, 504).
273. H. P. WILKINS, The Total Eclipse of the Moon, 1942 Aug 26 (*J.B.A.A.*, **52**, No. 9, 297).
274. H. P. WILKINS, Lunar Thermal Researches (*J.B.A.A.*, **53**, No. 2, 86).
275. H. P. WILKINS, A Thermal Eyepiece (*J.B.A.A.*, **54**, No. 2, 38).

ACCESSORY INSTRUMENTS

[Arranged according to instrument]

276. R. B. DUNN, *How to Build a Quartz Monochromator* (Sky Pub. Corp., Cambridge, Mass., 1952). [Reprint of series of articles in *Sky & Telesc.*, 1951.]
277. J. W. EVANS, The Quartz Polarising Monochromator (*Publ. A.S.P.*, **52**, No. 309, 305).
278. B. LYOT, Un monochromateur à grand champ utilisant les interférences en lumière polarisée (*C.R.*, **197**, 1593).
279. B. LYOT, Le Filtre Monochromatique Polarisant et ses Applications en Physique Solaire (*Ann. d'Astrophys.*, **7**, No. 1–2, 31).
280. Y. ÖHMAN, A New Monochromator (*Nat.*, **141**, No. 3560, 157).
281. E. PETTIT, The Interference Polarising Monochromator (*Publ. A.S.P.*, **53**, No. 313, 171).

282. M. A. ELLISON, The Formation of the Monochromatic Image in a Spectrohelioscope (*J.B.A.A.*, **50**, No. 2, 68).
283. M. A. ELLISON, The Construction of an Auto-collimating Spectrohelioscope (*J.B.A.A.*, **50**, No. 3, 107).
     G. E. HALE, The Spectrohelioscope and its Work:
284. I. (*Ap. J.*, **70**, No. 5, 265=*M.W.C.*, 388).
285. II. (*Ap. J.*, **71**, No. 2, 73=*M.W.C.*, 393).
286. III. (*Ap. J.*, **73**, No. 5, 379=*M.W.C.*, 425).
287. IV. (*Ap. J.*, **74**, No. 3, 214=*M.W.C.*, 434).
288. G. E. HALE, The Brightness of Prominences as shown by the Spectrohelioscope (*M.N.*, **95**, No. 5, 467).
289. F. J. SELLERS, A Note on the Spectrohelioscope and a Description of a Vibrating Slit Mechanism (*J.B.A.A.*, **48**, No. 6, 243).
290. F. J. SELLERS, Seventeenth Report of the Section for the Observation of the Sun, 1928–1950 (*Mem. B.A.A.*, **37**, No. 2, 1952).

291. A. F. COLLINS, A Mains Chronograph with provision for G.M.T. Signals (*J.B.A.A.*, **57**, No. 4, 176).
292. W. T. HAY, A Simple Chronograph (*J.B.A.A.*, **43**, No. 2, 80).
293. W. T. HAY, An Improved Simple Chronograph (*J.B.A.A.*, **44**, No. 8, 299).
294. R. L. WATERFIELD, A Cheap Chronograph (*J.B.A.A.*, **42**, No. 4, 147).
295. H. H. WATERS & D. A. CAMPBELL, A Simple Method of Recording Radio Time-Signals on a Chronograph (*J.B.A.A.*, **42**, No. 4, 147).

296. W. A. GRANGER, An Observing Ladder with Easily Adjustable Seats (*J.B.A.A.*, **60**, No. 4, 107).

297. F. J. SELLERS, An Efficient and Easily Constructed Observing Chair (*J.B.A.A.*, **43**, No. 7, 283; repr. with added Note *J.B.A.A.*, **63**, No. 1, 36).

298. B. LYOT, Quelques Observations de la Couronne Solaire et des Protubérances en 1935 (*B.S.A.F.*, **51**, 203).

299. B. LYOT, A Study of the Solar Corona and Prominences without Eclipses (*M.N.*, **99**, No. 8, 580). [George Darwin Lecture, 1939].

300. E. PETTIT & F. SLOCUM, Observations of Solar Prominences with a Lyot Telescope (*Publ. A.S.P.*, **45**, No. 266, 187).

301. A. M. SKELLETT, *The Coronaviser* (Washington, 1940).

302. A. M. SKELLETT, *Proposal of a Method for Observing the Solar Corona without an Eclipse* (Bell Telephone System Monograph B-807).

303. A. A. MICHELSON & F. G. PEASE, Measurement of the Diameter of α Orionis with the Interferometer (*Ap. J.*, **53**, No. 4, 249 = *M.W.C.*, 203).

VISION

304. W. DE. W. ABNEY, *Colour Vision* (Sampson Low, 1895).

305. E. M. ANTONIADI, On the Optical Character of Gemination (*J.B.A.A.*, **8**, No. 4, 176).

306. E. M. ANTONIADI, Further Considerations on Gemination (*J.B.A.A.*, **8**, No. 7, 308).

307. E. M. ANTONIADI, On Some Subjective Phenomena (*J.B.A.A.*, **9**, No. 6, 269).

308. E. M. ANTONIADI, Considerations on the Double Canals of Mars (*J.B.A.A.*, **10**, No. 7, 305; **11**, No. 1, 26).

309. M. A. DANJON, L'Acuité Visuelle et ses Variations (*Rev. d'Opt.*, **7**, 205).

310. J. E. EVANS & E. W. MAUNDER, Experiments as to the Actuality of the 'Canals' observed on Mars (*M.N.*, **63**, No. 8, 488).

311. H. SPENCER JONES, The Eye and Astronomical Observation (*Brit. J. Physiol. Optics*, **8**, No. 2, 93).

312. P. LOWELL, The Canals of Mars, Optically and Psychologically Considered: A Reply to Professor Newcomb (*Ap. J.*, **26**, No. 3, 131). [See B. 316, 317, 313.]

313. P. LOWELL, Reply to Professor Newcomb's Note (*Apl. J.*, **26**, No. 3, 142). [See B. 316.]

314. T. L. MACDONALD, The Observation of Planetary Detail—An Experiment (*J.B.A.A.*, **29**, No. 8, 225).

315. E. W. & A. S. D. MAUNDER, Some Experiments on the Limits of Vision for Lines and Spots as applicable to the Question of the Actuality of the Canals of Mars (*J.B.A.A.*, **13**, No. 9, 344).

316. S. NEWCOMB, The Optical and Psychological Principles involved in the Interpretation of the so-called Canals of Mars (*Ap. J.*, **26**, No. 1, 1). [See B. 312, 317, 313.]

317. S. NEWCOMB, Note on the Preceding Paper (*Ap. J.*, **26**, No. 3, 141). [Refers to B. 312.]

318. M. H. PIRENNE, *Vision and the Eye* (Pilot Press, Lond., 1948).

319. F. W. SHARPLEY, Night Vision (*J.B.A.A.*, **57**, No. 3, 148).

320. W. S. STILES & B. H. CRAWFORD, The Effect of a Glaring Light Source on Extrafoveal Vision (*Proc. Roy. Soc.*, Series B, **122**, No. 827, 255).

321. W. J. WRIGHT, *Photometry and the Eye* (Hatton Press, 1949).

PERSONAL EQUATION

322. R. L. DUNCOMBE, Personal Equation in Astronomy (*Pop. Astr.*, **53**, No. 2, 63, 110).

323. E. C. PHILLIPS, Second Note on Personal Equation in Observing Occultations (*Pop. Astr.*, **36**, 403).

324. W. H. PICKERING, Artificial Disks (*Harv. Ann.*, **32**, No. 2, 117).

325. T. H. SAFFORD, On the Various Forms of Personal Equation in Meridian Transits of Stars (*M.N.*, **57**, No. 7, 504).

326. T. H. SAFFORD, Additional Note on Personal Equation (*M.N.*, **58**, No. 2, 38).

ATMOSPHERE AND SEEING

327. J. A. ANDERSON, Astronomical Seeing (*J. Opt. Soc. Amer.*, **25**, No. 5, 152).

328. D. BRUNT, *Physical and Dynamical Meteorology* (Cambridge University Press, 2nd edn., 1939).

329. H. E. BUTLER, Observations of Stellar Scintillation (*Contr. Dunsink Obs.*, 4 =*Proc. R. Irish Acad.*, **54A**, 321, 1952).

330. H. E. BUTLER, Scintillation and Atmospheric Seeing (*Irish Astr. J.*, **1**, No. 8, 225).

331. A. E. DOUGLASS, Atmosphere, Telescope and Observer (*Pop. Astr.*, **5**, No. 2, 64).

332. A. E. DOUGLASS, Scales of Seeing (*Pop. Astr.*, **6**, No. 4, 193).

333. A. E. DOUGLASS, The Study of Atmospheric Currents by the Aid of Large Telescopes, and the effect of such currents on the Quality of the Seeing (*Meteorological Jnl.*, U.S.A., March 1895).

334. G. M. B. DOBSON & A. R. MEETHAM, Atmospheric Pollution (*Nat.*, **151**, No. 3829, 324).

335. M. A. ELLISON & H. SEDDON, Some Experiments on the Scintillation of Stars and Planets (*M.N.*, **112**, No. 1, 73).

336. D. W. HORNER, *Meteorology for All* (Witherby, Lond., 1919). [Elementary introduction.]

337. W. J. HUMPHREYS, *Physics of the Air* (McGraw-Hill, 1940).

338. A. E. LEVIN, The Correction of Declination and Hour Angle for Atmospheric Refraction (*Mem. B.A.A.*, **24**, No. 1, 18).

339. W. H. PICK, *A Short Course in Elementary Meteorology* (H.M.S.O., 1938).

TIME AND CLOCKS

340. J. E. HASWELL, *Horology* (Chapman & Hall, 1947).

341. F. HOPE-JONES, *Electric Clocks* (N.A.G. Press, Lond., *N.D.* [1931]).

342. F. HOPE-JONES, *Electrical Timekeeping* (N.A.G. Press, Lond., 1940).

343. W. NOBLE, The Transit Instrument (*J.B.A.A.*, **3**, No. 4, 177).

344. M. W. OVENDEN, The Fundamental Principles of the Quartz Oscillation Clock (*J.B.A.A.*, **60**, No. 1, 31).

345. H. T. STETSON, *A Manual of Laboratory Astronomy* (Eastern Science Supply Co., Boston, 1928).

346. G. A. TOMLINSON, Recent Developments in Precision Time-keeping (*Obs.*, **57**, 189).

347. —— *Information Sheet No. 2* (Time Dept., Royal Greenwich Observatory, Abinger, Surrey).

348. —— U.S. Dept. of Commerce *Circular Letter L.C.886* (National Bureau of Standards, Washington, D.C.).

MATHEMATICAL: COMBINATION OF OBSERVATIONS, COMPUTATION, TABLES

349. A. C. AITKEN, *Statistical Mathematics* (Oliver & Boyd, 1939). [Useful general introduction.]

350. D. BRUNT, *The Combination of Observations* (Cambridge, 1931).

351. J. L. COOLIDGE, *An Introduction to Mathematical Probability* (Oxford University Press, 1925).

352. H. R. HULME & L. S. T. SYMMS, The Law of Error and the Combination of Observations (*M.N.*, **99**, No. 8, 642).

353. C. G. LAMB, *Elements of Statistics* (Longmans, 1952). [Recommended modern treatment.]

354. H. C. LEVINSON, *The Science of Chance* (Faber, 1952). [Introductory.]

355. M. J. MORONEY, *Facts from Figures* (Pelican Books, 1951).

356. H. C. PLUMMER, *Probability and Frequency* (Macmillan, 1939). [Concise treatment of theory of error, with numerous references to astronomical observations.]

357. ALCOCK & JONES, *The Nomogram* (Pitman, 1950).

358. J. LIPKA, *Graphical and Mechanical Computation* (Chapman & Hall, 1918).

359. L. J. COMRIE, Memoranda for Observers: Computing Section (*J.B.A.A.*, **32**, No. 3, 94). [Valuable advice regarding many of the Tables listed below.]

360. L. J. COMRIE, Mathematical Tables (*B.A.A.H.*, 1929). [Likewise gives valuable advice on the use of Tables.]

361. A. FLETCHER, J. C. P. MILLER & L. ROSENHEAD, *An Index of Mathematical Tables* (Scientific and Computing Service, Lond., 1946).

362. H. ANDOYER, *Nouvelles Tables Trigonométriques Fondamentales—Valeurs Naturelles* (Paris, 1905).

363. H. ANDOYER, *Nouvelles Tables Trigonométriques Fondamentales— Logarithmes* (Paris, 1911).

364. V. BAGAY, *Nouvelles Tables Astronomiques et Hydrographiques* (Paris, 1829). [Seven-figure.]

365. J. BAUSCHINGER & J. T. PETERS, *Logarithmic-Trigonometrical Tables with Eight Decimal Places* (Leipzig, 2 vols., 1910, 1911).

366. H. BRANDENBURG, *Siebenstellige trigonometrische Tafel* (Leipzig, 1931).

367. C. BREMIKER, *Tables of the Common Logarithms of Numbers and Trigonometrical Functions to Six Places of Decimals* (Nutt, Lond., 1888) [One of the best 6-figure sets.]

368. C. C. BRUHNS, *A New Manual of Logarithms to Seven Places of Decimals* (Leipzig, 1920).

369. F. CASTLE, *Five-figure Logarithmic and Other Tables* (Macmillan, 1942).

370. B. COHN, *Tables of Addition and Subtraction Logarithms with Six Decimals* (Leipzig, 1909; Scientific and Computing Service, 1939).

371. L. J. COMRIE, *Chambers's Four-figure Mathematical Tables* (Chambers, 1947).

372. L. J. COMRIE, *Chambers's Six-figure Mathematical Tables* (Chambers, 1949). [Vol. 1, logarithmic values; vol. 2, natural values.]

373. L. J. COMRIE, *Chambers's Shorter Six-figure Mathematical Tables* (Chambers, 1950).

374. L. J. COMRIE, *Four-figure Tables of the Natural and Trigonometrical Functions with the Argument in Time* (Lond., 1931). [Obtainable from Scientific and Computing Service Ltd.]

375. L. J. COMRIE, *Barlow's Tables of Squares, Cubes, Square Roots, Cube Roots and Reciprocals of all Integers up to 12,500* (Spon, Lond., 1947).

376. A. C. D. CROMMELIN, Tables for Facilitating the Computation of the Perturbations of Periodic Comets by the Planets (*M.N.*, **64**, No. 5, 149).

377. H. R. DESVALLÉES, *Tables Logarithmiques et Trigonométriques à Quatre Décimales* (Gauthier-Villars, Paris, 1919). [One of the best 4-figure works. See B. 359 for errata.]

378. L. J. FOXWELL, Table giving $\log \frac{1}{r^3}$ with argument r^2 (*Mem. B.A.A.*, **30**, No. 1).

379. E. GIFFORD, *Natural Sines to Every Second of Arc and Eight Places of Decimals* (Manchester, 1914).

380. E. GIFFORD, *Natural Tangents to Every Second of Arc and Eight Places of Decimals* (Manchester, 1920). [From 0° to 45°.]

381. G. J. HOÜEL, *Tables de Logarithmes à Cinq Décimales* (Gauthier-Villars, Paris, 1921).

382. O. LOHSE (ed. P. V. NEUGEBAUER), *Tafeln für numerisches Rechnen mit Maschinen* (Engelmann, Leipzig, 2nd ed., 1935). [Conveniently set out 5-figure natural functions.]

383. J. C. P. MILLER, *Tables for Converting Rectangular to Polar Co-ordinates* (Scientific and Computing Service, 1939).

384. L. M. MILNE-THOMSON, *Standard Table of Square Roots . . . to eight significant figures of all four-figure numbers* (Bell, 1929).

385. L. M. MILNE-THOMSON & L. J. COMRIE, *Standard Four-figure Mathematical Tables* (Macmillan, 1931). [Good 4-figure work.]

386. J. PETERS, *Sechsstellige Tafel der Trigonometrischen Funktionen* (Dümmlers, Berlin, 1929). [Natural trigonometrical functions to 6 figures.]

387. J. PETERS, *Sechsstellige Werte der Kreis- und Evolventenfunktionen* (Dümmlers, Berlin, 1937).

388. J. PETERS, *Logarithmic Table to Seven Places of Decimals of the Trigonometrical Functions for every Second of the Quadrant* (Engelmann, Leipzig, 1911). [One of the best 7-figure Tables.]

389. J. PETERS, *Seven-place Values of Trigonometrical Functions for Every Thousandth of a Degree* (Van Nostrand, N.Y., 1942).

390. J. PETERS, *Fünfstellige Logarithmentafel der Trigonometrische Funktionen* (Berlin, 1912).

391. J. PRYDE (ed.), *Chambers's Seven-figure Mathematical Tables* (Chambers, 1948).

392. E. SANG, *Logarithmic Tables* (Lond., 1915). [Of numbers from 20,000 to 200,000 to 7 decimals.]

393. H. L. F. SCHRÖN, *Seven-figure Logarithms for Numbers from 1–108,000 and for Sines, Cosines, Tangents and Cotangents for Every Ten Seconds of the Quadrant* (Lond., 1865).

394. R. SHORTREDE, *Logarithmic Tables* (Edinburgh, 1844). [Of numbers from 1 to 120,000 to 7 decimals.]

395. B. STRÖMGREN, Tables giving tan $\frac{v}{2}$ and tan^2 $\frac{v}{2}$ in Parabolic Motion, with Argument $M=(t-T)q^{-\frac{3}{2}}$, to facilitate the Computation of Ephemerides from Parabolic Elements (*Mem. B.A.A.*, **27**, No. 2).

396. J. ZECH, *Tafeln der Additions- und Subtractions-Logarithmen für Sieben Stellen* (Berlin, 1910).

397. —— *British Association Tables* [various] (Cambridge University Press, and British Association, Lond.).

398. —— *Five-figure Tables of Natural Trigonometrical Functions* (H.M.S.O., 1947).

399. —— *Seven-figure Trigonometrical Tables for Every Second of Time* (H.M.S.O., 1939).

400. —— *Tavole Logarithmiche a cinque cifre decimali* (Genoa, 2nd edn., 1916; abridged edn., 1920). [Very comprehensive 5-figure Tables.]

CONVERSION OF COORDINATES

401. W. HEATH, On a Mechanical Method of Transforming Spherical Co-ordinates (*J.B.A.A.*, **26**, No. 4, 160).

402. W. HEATH, On a Mechanical Method of Transforming Spherical Co-ordinates (*J.B.A.A.*, **28**, No. 2, 64).

403. W. HEATH, [Report of B.A.A. Meeting] (*J.B.A.A.*, **34**, No. 4, 128).

404. E. PIO, *Tavole per la trasformazione delle coordinate equatoriali in Co-ordinate Galattiche referite al polo 12^h 44^m +26°.8* (Rome, 1929).

AMATEUR OBSERVATIONAL ASTRONOMY

[With the exception of B. 408 the following are rather elementary or dated, or both, and are chiefly of value for the lists of telescopically interesting objects which they contain. A number of Star Atlases (q.v.) contain similar lists. See also B. 31.]

405. C. E. BARNES, *1001 Celestial Wonders as observed with Home-built Instruments* (California, 1929).

406. W. F. DENNING, *Telescopic Work for Starlight Evenings* (Taylor & Francis, Lond., 1891).

407. F. M. GIBSON, *The Amateur Telescopist's Handbook* (Longmans, Green, 1894).

408. R. HENSELING (ed.), *Astronomisches Handbuch* (Stuttgart, 1924).

409. W. NOBLE, *Hours with a Three-inch Telescope* (Longmans, Green, 1886).

410. W. T. OLCOTT, *In Starland with a Three-Inch Telescope* (Putnam, 1909).
411. R. A. PROCTOR (rev. W. H. STEAVENSON), *Half-Hours with the Telescope* (Longmans, Green, 1926).
412. G. P. SERVISS, *Astronomy with an Opera Glass* (Lond., 1902).
413. G. P. SERVISS, *Pleasures of the Telescope* (Hirschfield, Lond., 1902).
414. J. B. SIDGWICK, *Introducing Astronomy* (Faber, 1951).
415. W. H. SMYTH, *A Cycle of Celestial Objects* (1st edn., 1844, 2 vols [vol. 2 being the Bedford Catalogue]; 2nd edn., edited G. F. CHAMBERS, Oxford, 1881).
416. T. W. WEBB (rev. ESPIN), *Celestial Objects for Common Telescopes* (Longmans, Green, 1917, 2 vols.; rev. and enl. repub. of 6th edn. (MAYALL & MAYALL), Dover, 1962).

OBSERVATIONAL AIDS

417. *Astronomical Ephemeris* (London, annually).
418. *Connaissance des Temps, ou des Mouvements Célestes . . . publiée par le Bureau des Longitudes* (Gauthier-Villars, Paris, annually).
419. *Astronomical Ephemeris* (Washington, annually).
420. *Berliner Astronomisches Jahrbuch* (Akademie Verlag, Berlin).
421. *Handbook of the British Astronomical Association* (B.A.A., Computing Section, annually).
422. *Journal of the British Astronomical Association* (B.A.A., approximately monthly).
423. *British Astronomical Association Circulars* (B.A.A., at irregular intervals).
424. *The British Astronomical Association: Its Nature, Aims and Methods* (London, 1948).
425. B. J. BOK, Report on Astronomy (*Pop. Astr.*, **47**, No. 7, 356).

STAR ATLASES AND CHARTS

[See also various entries under 'Star Catalogues'.]

426. F. W. A. ARGELANDER,*Atlas des Nördlichen Gestirnten Himmels 1855.0* (Bonn, 1863; ed. KÜSTNER, Bonn, 1899). [40 charts covering the sky N of Dec −2° to mag 9 on scale of 2 cms/degree.]
427. E. E. BARNARD (ed. E. B. FROST & M. R. CALVERT), *A Photographic Atlas of Selected Regions of the Milky Way* (Publ. Carneg. Instn. 247, 1927). [Part I, photographs and descriptions; Part II, charts and tables.]
428. A. BEČVÁŘ, *Atlas Coeli Skalnaté Pleso 1950.0* (Praha, 1948; Sky Publishing Corporation, Harvard, 1949). [16 maps covering the whole sky to mag 7·75 on scale of 7·5 mm/degree, including doubles, variables, nebulae, etc.]
429. A. BEČVÁŘ, *Atlas Coeli Skalnaté Pleso II* (Praha, 1951). [Catalogue of interesting objects, and Tables.]
430. M. BEYER & K. GRAFF, *Stern-Atlas* (Hamburg, 1925, repr. 1952). [27 charts covering the whole sky N of Dec −23° to mag 9 (plus brighter nebulae and clusters) on scale of 1 cm/degree, epoch 1855. *B.A.A.H.*, 1926 contains a Table of Precession Corrections, 1855 to 1926, for application to the positions of the Beyer-Graff Atlas.]

* Best known as *Bonner Durchmusterung*.

431. G. BISHOP, *Ecliptic Chart* (Lond., 1848, etc). [24 charts, to mag 10, extending 3° on each side of the ecliptic, epoch 1825.]

432. K. F. BOTTLINGER, *Galaktischer Atlas* (Julius Springer, Berlin, 1937). [8 charts showing stars and nebulae to about mag 5·5 in galactic coordinates based on Pole at $12^h 40^m$, $+28°$.]

433. E. DELPORTE, *Atlas Céleste* (I.A.U., Cambridge, 1930). [26 maps covering whole sky to mag 6; I.A.U. constellation boundaries; lists of mags, spectroscopic types, and positions at 1875 and 1925 of all stars to mag 4·5, and principal doubles, variables, nebulae, etc.]

434. E. DELPORTE, *Délimitation Scientifique des Constellations* (*Tables et Cartes*) (I.A.U., Cambridge, 1930). [Virtually a quotation from B. 433.]

435. J. FRANKLIN-ADAMS, *Photographic Chart of the Sky* (R.A.S., 1914). [206 sheets, each 16° square, covering the whole sky to mag 15·5 on scale of about 1 in/$1°36$. Charts 1–67, Dec −90° to −30°; charts 68–139, −15° to +15°; charts 140–206, +30° to +90°. See also *M.N.*, **64**, 608, 1904; *ibid.*, **97**, 89, 1936.]

436. E. HEIS, *Atlas Coelestis Eclipticus: Octo Continens Tabulas ad Delineandum Lumen Zodiacale* (1878). [8 charts of the zodiacal region, especially made for the Zodiacal Light observer.]

437. *Mappa Coelestis Nova* (Sky Publishing Corpn., Harvard, 1949). [Single chart, about 33 × 30 ins, covering sky N of Dec −42° to mag 5; stars of spectral types B–M printed in different colours; also novae, radiants, nebulae, clusters.]

438. *New Popular Star Atlas, Epoch 1950* (Gall & Inglis, *N.D.*). [Virtually a cheaper and simplified B. 439, and without the latter's 'Handbook' section; stars to mag 5·5.]

439. A. P. NORTON, *A Star Atlas and Reference Handbook, Epoch 1950* (Gall & Inglis, 17th edn., 1978). [The stand-by of every amateur; some 7000 stars to mag 6·5, plus nebulae, clusters, etc, and a mass of general information and reference material of use to the amateur.]

440. J. PALISA & M. WOLF, *Palisa-Wolf Charts* (Vienna, 1908–31). [210 photographic charts in 11 Series; epoch 1875; scale 36 mm/degree.]

441. *Palomar Sky Atlas.* [Two-colour photographic survey of the whole sky visible from Mt Palomar, made with the 48-in Schmidt; about 2000 plates anticipated, each 14 ins square; to mag 20; sponsored by the National Geographical Society of America.]

442. W. PECK, *The Observer's Atlas of the Heavens* (Gall & Inglis, 1898). [Same general type as B. 439, but superseded by the latter. 30 charts, plus positions and mags of over 1400 doubles, variables, nebulae, etc.]

443. F. E. ROSS & M. R. CALVERT, *Atlas of the Northern Milky Way* (University of Chicago Press, 1934). [39 plates, each about 13¼ ins square, exposed in a D=5, F=35-in camera at Mt Wilson and Flagstaff.]

444. E. SCHÖNFELD, *Atlas der Himmelszone zwischen 1° und 23° südlicher Declination, 1855* (*Bonner Sternkarten, Zweite Serie*) (Bonn, 1887). [24 charts continuing B. 426 to Dec −23°.]

445. P. STUKER, *Sternatlas für Freunde der Astronomie* (Stuttgart, 1925). [Photographic; to mag 7·5; epoch 1900.]

* Best known as *Bonner Durchmusterung.*

446. O. THOMAS, *Atlas der Sternbilder* (Salzburg, 1945). [32 main charts; also useful section on objects of interest.]

447. H. B. WEBB, *Atlas of the Stars* (privately printed, N.Y., 2nd edn., 1945). [110 charts covering the sky N of Dec −23° to about mag 9·5 on scale 1 cm/degree; epoch 1920.0, with coordinate intersections for 2000.0. Very useful supplement to B. 439.]

STAR CATALOGUES—POSITIONAL

448. A. KOPF, Star Catalogues, expecially those of Fundamental Character (*M.N.*, 96, No. 8, 714). [George Darwin Lecture, 1936; an extremely useful summary of 19th- and 20th-century work.]

[The following selection of the catalogues of the last 120 years, devoted primarily to star positions, is arranged chronologically.]

449. F. BAILY, *A Catalogue of Those Stars in the Histoire Céleste Française of J. Lalande* [etc] (British Association, Lond., 1837). [47,390 stars reduced to epoch 1800.0 from Lalande's observations.]

450. S. GROOMBRIDGE (ed. G. B. AIRY), *A Catalogue of Circumpolar Stars* (Murray, Lond., 1838). [4243 stars, epoch 1810.0; see also B. 467.]

451. F. BAILY, *British Association Catalogue* (Lond., 1845). [8377 stars, epoch 1850.]

452. M. WEISSE, *Positiones mediae stellarum fixarum* [etc] (Petropoli, 1846). [Weisse's reductions of 31,085 stars within 15° of the equator to epoch 1825.0, using Bessel's observations; see also B. 453. Abbrev: *WB*.]

453. M. WEISSE, *Positiones mediae stellarum fixarum* [etc] (Petropoli, 1863) [Continuation of B. 452; 31,445 stars between Dec +15° and +45°. Abbrev: *WB2*.]

454. W. OELTZEN, *Argelanders Zonen-Beobachtungen vom 45° bis 80° nördlicher Declination in mittleren Positionen für 1842.0* (Wien, 1851–52). [Oeltzen's reductions of 26,425 stars from Argelander's observations. Abbrev: *OA(N)*.]

455. W. OELTZEN, *Argelanders Zonen-Beobachtungen vom 15° bis 31° südlicher Declination in mittleren Positionen für 1853.0* (Wien, 1857–58). [Continuation of B. 454. Abbrev: *OA(S)*.]

456. F. W. A. ARGELANDER, *Bonner Durchmusterung des nördlichen Himmels* (Bonn, 1859–62; reprint 1903; 3 vols). [Approximate positions and visual mags to nearest 0·1 mag, to about mag 9·5. Arranged in successive 1°-wide Dec zones from +90° to −2°. Charts, scale 2 cms/degree, have been issued in photostat. Abbrev: *BD*. See B. 459, 462.]

457. HEIS, *Atlas* [and] *Catalogus Coelestis Novus* (Cologne, 1872). [Abbrev: H′.]

458. J. BIRMINGHAM, The Red Stars: Observations and Catalogue (*Trans. Roy. Irish Acad.*, 26, 1877; ed. T. E. ESPIN, Dublin, 1890). [Abbrevs: *B* and *E-B* respectively.]

459. E. SCHÖNFELD, *Durchmusterung* (1886). [Continuation of B. 456 to Dec. −23°. Abbrev: *BD*.]

460. B. A. GOULD, The Argentine General Catalogue (*Result. Obs. Nac. Argent.*, 14, Cordoba, 1886). [32,448 southern stars plus additional stars in clusters. Abbrevs: *CGA, AGC*.]

* Best known as *Bonner Durchmusterung*.

461. *Astronomische Gesellschaft Katalog* (Leipzig, 1890 etc). [The most complete catalogue of precision; epoch 1875. Abbrevs: *AG, AGC, CAG.* See B. 485.]

462. J. M. THOME, Cordoba Durchmusterung (*Result. Obs. Nac. Argent.*, **16**, 1892 etc). [Continuation of B. 459 to Dec −52° in 10° vols. To mag 10 approximately; with visual mags and charts. Abbrev: *CD.*]

463. *The Astrographic Catalogue.* [Initiated at the Paris Congress, 1887; work shared by observatories over the world, began 1892, still in progress. Positions taken from photographic charts, scale 6 cms/degree. Abbrev: *AC.*]

464. D. GILL & J. C. KAPTEYN, The Cape Photographic Durchmusterung, 1875.0 (*Ann. Cape Obs.*, **3**, 1896– 5, 1900). [Mags and approximate positions from Dec −19° to −90°, to mag 9. Abbrev: *CPD.*]

465. J. SCHEINER, Photographische Himmelskarte (*Publ. Astrophys. Obs. Potsdam*, 1899–1903). [Mags and positions, epoch 1900.0.]

466. S. NEWCOMB, Catalogue of Fundamental Stars for the Epochs 1875 and 1900 (*Astr. Pap., Wash.*, **8**, No. 2, 77, 1905).

467. F. W. DYSON & W. G. THACKERAY, *New Reduction of Groombridge's Circumpolar Catalogue, Epoch 1810.0* (H.M.S.O., 1905). [See B. 450.]

468. H. B. HEDRICK, Catalogue of Zodiacal Stars, for the Epochs 1900 and 1920 (*Astr. Pap., Wash.*, **8**, No. 3, 405, 1905). [1607 stars to mag. 7·5. Abbrev: *WZC*. See also B. 484.]

469. J. BOSSERT, *Catalogue d'Étoiles Brillantes, 1900.0* (Gauthier-Villars, Paris, 1906). [3799 stars in 1° NPD zones.]

470. J. & R. AMBRONN, *Sternverzeichnis enthaltend alle Sterne bis zur 6.5 Grösse* (Julius Springer, Berlin, 1907). [Positions of 7796 stars to mag 6·5 for epoch 1900, and proper motions of 2226 stars.]

471. A. AUWERS, Neue Fundamentalkatalog des Berliner Astronomischen Jahrbuchs (*Veröff. König. Astron. Rechen-Instit.*, **33**, 1910). [925 stars; one of the best fundamental catalogues; abbrev: *NFK*. See also B. 480.]

472. L. BOSS, *Preliminary General Catalogue of 6188 Stars for the Epoch 1900* (Publ. Carneg. Instn., 115, 1910). [Accurate positions and proper motions of all naked-eye stars; abbrev: *PGC.*]

473. T. W. BACKHOUSE, *Catalogue of 9842 Stars, or all Stars Very Conspicuous to the Naked Eye* (Sunderland, 1911). [Epoch 1900; useful for amateurs.]

474. *Geschichte des Fixstern-Himmels* (Karlsruhe, 1922–23). [Collection of pre-1900 observations reduced to 1875.0.]

475. E. C. PICKERING & J. C. KAPTEYN, Durchmusterung of Selected Areas between $\delta=0$ and $\delta=+90°$ (*Harv. Ann.*, **101**, 1918).

476. E. C. PICKERING, J. C. KAPTEYN & P. J. VAN RHIJN, Durchmusterung of Selected Areas between $\delta=-15°$ and $\delta=-30°$ (*Harv. Ann.*, **102**, 1923).

477. E. C. PICKERING, J. C. KAPTEYN & P. J. VAN RHIJN, Durchmusterung of Selected Areas between $\delta=-45°$ and $\delta=-90°$ (*Harv. Ann.*, **103**, 1924).

478. L. BOSS & B. BOSS, *San Luis Catalogue of 15,333 Stars for the Epoch 1920.0* (Publ. Carneg. Instn., 386, 1928). [To mag 7; mostly southern.]

479. R. SCHORR & W. KRUSE, *Index der Sternörter 1900–25* (Bergedorf, 1928). [Monumental analysis of over 400 catalogues. Vol. 1, northern stars; vol. 2, southern.]

480. *Vierter Fundamentalkatalog* (Karlsruhe, 1970). [Abbrev: *FK4*. See B.471. (Berlin, 1934). [Abbrev: *FK3*. See B. 471.]

481. F. SCHLESINGER, Catalogue of Bright Stars containing all important data known in June 1930 (*Publ. Yale Obs.*, 1930). [9110 stars to visual mag 6·5, and some fainter. Abbrev: *BS*.]

482. B. BOSS, *Albany Catalogue of 20,811 Stars for the Epoch 1910* (Carnegie Institution of Washington, 1931).

483. B. BOSS, *General Catalogue of 33,342 Stars for the Epoch 1950* (Carnegie Institution of Washington, 1937, 5 vols.).

484. J. ROBERTSON, Catalogue of 3539 Zodiacal Stars for the Equinox 1950.0 (*Astr. Pap., Wash.*, **10**, No. 2, 175, 1940). [Revision and enlargement of B. 468; abbrev: *NZC*.]

485. R. SCHORR & A. KOHLSCHÜTTER, *Zweiter Katalog der Astronomische Gesellschaft, Äquinoktium 1950* (Hamburg-Bergedorf, 1951). [Vols 1–8, Dec +90° to −2°. See B. 461. Abbrev: *AGK2*.]

486. *Apparent Places of Fundamental Stars* (H.M.S.O., annually since 1941). [Mean and Apparent places of the 1535 stars of the *FK3* (B. 480) and its Supplement.]

STAR CATALOGUES—MOTIONS AND PARALLAXES

487. W. S. ADAMS & A. H. JOY, The Radial Velocities of 1013 Stars (*Ap. J.*, **57**, No. 3, 149=*M.W.C.*, 258).

488. J. BOSSERT, *Catalogue des mouvements propres des 5671 étoiles* (Paris, 1919).

489. W. W. CAMPBELL & J. H. MOORE, Radial Velocities of Stars brighter than Visual Magnitude 5·51 (*Publ. Lick. Obs.*, **16**, 1928).

490. W. S. EICHELBERGER, Positions and Proper Motions of 1504 Standard Stars, 1925.0 (*Astr. Pap., Wash.*, **10**, Part 1, 1925).

491. H. KNOX-SHAW & H. G. SCOTT BARRETT, *The Radcliffe Catalogue of Proper Motions in Selected Areas 1 to 115* (Oxford University Press, 1934). [To mag 14 in the Selected Areas on and N of the equator.]

492. J. H. MOORE, General Catalogue of Radial Velocities of Stars, Nebulae and Clusters (*Publ. Lick Obs.*, **18**, 1932).

493. J. S. PLASKETT & J. A. PEARCE, A Catalogue of the Radial Velocities of O and B Type Stars (*Publ. Dom. Astrophys. Obs.*, **5**, No. 2, 99).

494. J. G. PORTER, E. I. YOWELL & E. S. SMITH, A Catalog of 1474 Stars with proper motion exceeding four-tenths of a second per year (*Publ. Cincinn. Obs.*, **20**).

495. F. SCHLESINGER, *General Catalogue of Stellar Parallaxes* (Yale, 1924). [Includes all determinations available in 1924 Jan.]

496. F. SCHLESINGER & L. F. JENKINS, General Catalogue of Stellar Parallaxes (*Publ. Yale Obs.*, 1935). [B. 495 revised to 1935.]

497. R. SCHORR, *Eigenbewegungs-Lexikon* (Hamburg Observatory, Bergedorf, 1 vol, 1923; 2 vols, 1936). [2nd edn. includes all proper motions available at the end of 1935; 94,741 stars, with mags and spectral types. Vol. 1, N stars; vol. 2, S stars. Abbrev: *EBL*.]

STAR CATALOGUES—PHOTOMETRIC

498. S. I. BAILEY, A Catalogue of 7922 Stars observed with the Meridian Photometer, 1889–91 (*Harv. Ann.*, **34**). [The 'Southern Meridian Photometry', abbrev: *SMP*. Continuation of B. 510 to the S Pole.]

499. A. BRUNN, *Atlas photométrique des Constellations de +90° à −30°* (privately printed, France, 1949). [55 sheets showing all *BD* stars to mag 7·5, scale 1 cm/degree, epoch 1900. Against each star is printed its visual mag, to 0·01 mag for stars of mag 6·50 and brighter, to 0·1 mag for those fainter than 6·50. Other data include photometric mags of all extragalactic nebulae brighter than mag 12·0 photographic.]

500. S. CHAPMAN & P. J. MELOTTE, Photographic Magnitudes of 262 Stars within 25′ of the North Pole (*M.N.*, 74, No. 1, 40).

501. B. G. FESSENKOFF, *Photometrical Catalogue of 1155 Stars* (Kharkow, 1926).

502. W. FLEMING, Spectra and Photographic Magnitudes of Stars in Standard Regions (*Harv. Ann.*, 71, No. 2, 27).

503. B. A. GOULD, *Uranometria Argentina* (Buenos Aires, 1879). [Visually determined mags, and positions, of stars to mag 7 South of Dec +10°. Abbrev: *UA, G*.]

504. [Harvard], Stars near the North Pole (*Harv. Ann.*, 48, No. 1, 1).

505. Harvard Standard Regions (*Harv. Ann.*, 71, No. 4, 233).

506. H. S. LEAVITT, The North Polar Sequence (*Harv. Ann.*, 71, No. 3, 47).

507. Magnitudes of Stars of the North Polar Sequence (*B.A.A.H.*, 1926, 32).

508. G. MÜLLER & P. KEMPF, Photometrische Durchmusterung des nördlichen Himmels (*Publ. Astrophys. Obs. Potsdam*, Nos. 31, 43, 44, 51, 52, 1894–1907).

509. E. C. PICKERING, Adopted Photographic Magnitudes of 96 Polar Stars (*Harv. Circ.*, 170).

510. E. C. PICKERING, Observations with the Meridian Photometer, 1879–82 (*Harv. Ann.*, 14, 1884). [The Harvard Photometry (*HP*): magnitudes of 4260 stars, including all brighter than mag 6 N of Dec −30°.]

511. E. C. PICKERING, Revised Harvard Photometry (*Harv. Ann.*, 50). [Positions, visual mags, and spectral types of 9110 stars, mostly mag 6·5 and brighter. Abbrevs: *HR, RHP*.]

512. E. C. PICKERING, A Catalogue of 36,682 Stars Fainter than Magnitude 6·50 . . . forming a Supplement to the Revised Harvard Photometry (*Harv. Ann.*, 54).

513. C. PRITCHARD, *Uranometria Nova Oxoniensis* (Oxford, 1885). [Wedge-photometer redeterminations of Argelander's *Uranometria Nova* magnitudes; 2784 entries. Abbrev: *UO*.]

514. F. H. SEARES, Magnitudes of the North Polar Sequence. (Report of the Commission de photométrie stellaire) (*Trans. I.A.U.*, 1, 69).

515. F. H. SEARES, J. C. KAPTEYN & P. J. VAN RHIJN, *Mt Wilson Catalogue of Photographic Magnitudes in Selected Areas 1–139* (Carnegie Institution, Washington, 1930).

516. F. H. SEARES, F. E. ROSS & M. C. JOYNER, *Magnitudes and Colours of Stars North of +80°* (Publ. Carneg. Instn., 532, 1941).

STAR CATALOGUES—SPECTROSCOPIC

[Many of the catalogues mentioned elsewhere in this Bibliography quote spectroscopic types. Specially demanding mention, however, are the following.]

517. E. C. PICKERING, The Draper Catalogue of Stellar Spectra (*Harv. Ann.*, 27, 1890).

518. A. C. MAURY, Spectra of Bright Stars (*Harv. Ann.*, 28, 1897). [Together with B. 517 constitutes the 'old' Draper Catalogue of over 10,000 stars.]

519. A. J. CANNON & E. C. PICKERING, The Henry Draper Catalogue of Stellar Spectra (*Harv. Ann.*, 91–99, 1918–24).

520. A. J. CANNON, The Henry Draper Extension (*Harv. Ann.*, 100). [Together with B. 519 constitutes the 'new' Draper Catalogue (*HD*): mags, spectral types and positions of 225,000 stars to mag 10 approximately.]

521. A. SCHWASSMANN & P. J. VAN RHIJN, *Bergedorfer Spektral-Durchmusterung* (Bergedorf, 1935, 1938). [To mag 13 (photographic) in the northern Kapteyn areas.]

STAR CATALOGUES—VARIABLE STARS

522. A. J. CANNON,* Second Catalogue of Variable Stars (*Harv. Ann.*, 55, 1907). ['Second' with reference to B. 533; 1957 variables.]

523. S. C. CHANDLER,* Catalogue of Variable Stars (*A.J.*, 8, No. 11/12, 82, 1888). [225 variables.]

524. S. C. CHANDLER,* Second Catalogue of Variable Stars (*A.J.*, 13, No. 12, 89, 1893). [260 variables.]

525. S. C. CHANDLER,* Third Catalogue of Variable Stars (*A.J.*, 16, No. 9, 145, 1896). [393 variables.]

526. S. C. CHANDLER,* Revision of Elements of the Third Catalogue of Variable Stars (*A.J.*, 24, No. 1, 1, 1904).

527. J. G. HAGEN, *Atlas [et Catalogus] Stellarum Variabilium* (9 Series, 1899–1941). [Charts and lists of about 24,000 comparison stars for 488 variables. The 5th Series (Berlin, 1906) is particularly useful, including all variables wholly observable with the naked eye or binoculars (minima brighter than mag 7). See also B. 534.]

528. *Katalog und Ephemeriden Veränderlicher Sterne* (Vierteljahrsschrift der Astronomischen Gesellschaft, annually 1870–1926).

529. *Katalog und Ephemeriden Veränderlicher Sterne* (Berlin-Babelsberg, annually to 1941).

530. B. V. KUKARKIN & P. P. PARENAGO, *General Catalogue of Variable Stars* (Academy of Sciences of the U.S.S.R., 1948; annual Supplements 1949–52). [In Russian, but with English translations of the Introductions. The I.A.U.-recognised standard work.]

* Primarily of historical value, B. 528–9 and B. 530 now being the standard works.

531. R. PRAGER, *Geschichte und Literatur des Lichtwechsels der Veränderlichen Sterne* (Ferd. Dümmlers Verlagsbuchhandlung, Berlin; vol. 1, 1934; vol. 2, 1936). [See also B. 532.]

532. R. PRAGER, History and Bibliography of the Light Variations of Variable Stars (*Harv. Ann.*, **111**, 1941). [Vol. 3 of B. 531.]

533. Provisional Catalogue of Variable Stars* (*Harv. Ann.*, **48**, No. 3, 91, 1903). [1227 variables.]

534. J. STEIN & J. JUNKERS, [Index to the 9 Series of B. 527.] (*Ricerche Astronomiche*, **4**; Specola Vaticana, 1941).

STAR CATALOGUES—BINARY STARS

535. R. G. AITKEN, *New General Catalogue of Double Stars within 120° of the North Pole* (Publ. Carneg. Instn., 417, 2 vols, 1932). [Includes all measures of 17,181 doubles prior to 1927; epochs 1900·0 and 1950·0. Abbrev: *ADS*.]

536. S. W. BURNHAM, *A New General Catalogue of Double Stars within 121° of the North Pole* (Publ. Carneg. Instn., 5, 2 vols, 1906). [Measures etc of 13,665 doubles; epochs 1880.0 and 1900.0. Part I, The Catalogue; Part II, Notes to the Catalogue. Abbrev: *BGC*.]

537. W. W. CAMPBELL & H. D. CURTIS, First Catalogue of Spectroscopic Binaries (*L.O.B.*, **3**, No. 79, 136). [Complete to 1905; 140 stars.]

538. W. W. CAMPBELL, Second Catalogue of Spectroscopic Binary Stars (*L.O.B.*, **6**, No. 181, 17). [To 1910; 306 stars. See also B. 537, 553–555.]

539. F. W. DYSON, *Catalogue of Double Stars from observations made at The Royal Observatory Greenwich with the 28-inch Refractor, 1893–1919* (H.M.S.O., 1921).

540. J. F. W. HERSCHEL, Descriptions and approximate Places of 321 new Double and Triple Stars (*Mem. R.A.S.*, **2**, No. 29, 459).

541. J. F. W. HERSCHEL, Approximate Places and Descriptions of 295 new Double and Triple Stars (*Mem. R.A.S.*, **3**, No. 3, 47).

542. J. F. W. HERSCHEL, Third Series of Observations . . . Catalogue of 384 new Double and Multiple Stars; completing a first 1000 of Those Objects (*Mem. R.A.S.*, **3**, No. 13, 177).

543. J. F. W. HERSCHEL, Fourth Series of Observations . . . containing the Mean Places . . . of 1236 Double Stars [etc] (*Mem. R.A.S.*, **4**, No. 17, 331).

544. J. F. W. HERSCHEL, Fifth Catalogue of Double Stars . . . Places, Descriptions, and measured Angles of Position of 2007 of those objects [etc] (*Mem. R.A.S.*, **6**, No. 1, 1).

545. J. F. W. HERSCHEL, Sixth Catalogue of Double Stars . . . 286 of those objects (*Mem. R.A.S.*, **9**, No. 7, 193).

546. J. F. W. HERSCHEL, Seventh Catalogue of Double Stars (*Mem. R.A.S.*, **38**, 1870).

547. J. F. W. HERSCHEL (ed. R. MAIN & C. PRITCHARD), Catalogue of 10,300 Multiple and Double Stars (*Mem. R.A.S.*, **40**, 1874).

548. W. HERSCHEL, Catalogue of Double Stars (*Phil. Trans.*, **72**, 112, 1782).

* Primarily of historical value, B. 528–9 and B. 530 now being the standard works.

549. W. HERSCHEL, Catalogue of Double Stars (*Phil. Trans.*, **75**, 40, 1785).

550. W. HERSCHEL, On the places of 145 new Double Stars (*Mem. R.A.S.*, **1**, 166, 1821).

551. R. JONCKHEERE, Catalogue and Measures of Double Stars discovered visually from 1905–1916 within 105° of the North Pole and under 5″ Separation (*Mem. R.A.S.*, **17**, 1917). [Virtually a Supplement to B. 536.]

552. T. LEWIS, Measures of the Double Stars contained in the Mensurae Micrometricae of F. G. W. Struve (*Mem. R.A.S.*, **56**). [See B. 558.]

553. J. H. MOORE, Third Catalogue of Spectroscopic Binary Stars (*L.O.B.*, **11**, No. 355, 141). [To 1924; 1054 stars. See also B. 537–8, 554–5.]

554. J. H. MOORE, Fourth Catalogue of Spectroscopic Binary Stars (*L.O.B.*, **18**, No. 483, 1). [375 stars.]

555. J. H. MOORE & F. J. NEUBAUER, Fifth Catalogue of the Orbital Elements of Spectroscopic Binary Stars (*L.O.B.*, No. 521).

556. J. SOUTH & J. F. W. HERSCHEL, *Observations of the Apparent Distances and Positions of 380 Double and Triple Stars, made in the Years 1821, 1822 and 1823* (Lond., 1825).

557. F. G. W. STRUVE, *Catalogus Novus Stellarum Duplicium et Multiplicium* (Dorpat, 1827). [The Dorpat Catalogue (Σ).]

558. F. G. W. STRUVE, *Stellarum Duplicium et Multiplicium Mensurae Micrometricae* (Petropoli, 1837). [See B. 552.]

559. F. G. W. STRUVE, *Stellarum Fixarum imprimis Duplicium et Multiplicium Positiones Mediae pro Epocha 1830.0* (Petropoli, 1852).

560. O. STRUVE, *Revised Pulkova Catalogue* (Pulkova, 1850). [Abbrev: OΣ; Part II denoted by OΣΣ.]

NEBULAE AND CLUSTERS—CATALOGUES

[See also under 'Star Atlases and Charts'.]

561. S. I. BAILEY, Globular Clusters—A Provisional Catalogue (*Harv. Ann.*, **76**, No. 4, 43). [113 clusters.]

562. E. E. BARNARD, On the Dark Markings of the Sky, with a Catalogue of 182 Such Objects (*Ap. J.*, **49**, No. 1, 1). [See *Ap. J.* **49**, No. 5, 360 for list of errata.]

563. H. D. CURTIS, Descriptions of 762 Nebulae and Clusters Photographed with the Crossley Reflector (*Publ. Lick Obs.*, **13**, No. 1, 9).

564. J. L. E. DREYER, New General Catalogue of Nebulae and Clusters of Stars (*Mem. R.A.S.*, **49**, 1888). [NGC. Based on B. 565.]
—— Index Catalogue (*Mem. R.A.S.*, **51**, 1895). [IC. Extension of NGC.]
—— Second Index Catalogue (*Mem. R.A.S.*, **59**, 1908). [Extension of NGC.]*

565. J. F. W. HERSCHEL, *General Catalogue of Nebulae and Clusters of Stars of the Epoch 1860* (Lond., 1864).

* A one-volume edition of *New General Catalogue of Nebulae and Clusters of Stars, the Index Catalogue and the Second Index Catalogue* was issued by the R.A.S. in 1953.

566. W. HERSCHEL, Catalogue of 1000 new Nebulae and Clusters of Stars (*Phil. Trans.*, **76**, 457, 1786).

567. W. HERSCHEL, Catalogue of a second 1000 of New Nebulae and Clusters of Stars (*Phil. Trans.*, **79**, 212, 1789).

568. W. HERSCHEL, Catalogue of 500 new Nebulae, Nebulous Stars, Planetary Nebulae, and Clusters of Stars (*Phil. Trans.*, 1802, 477).

569. J. HOLETSCHEK, Catalogue of Nebular Magnitudes (*Ann. K.K. Univ.-Stern.*, **20**, 114, Vienna, 1907).

570. P. J. MELOTTE, A Catalogue of Star Clusters shown on Franklin-Adams Chart Plates (*Mem. R.A.S.*, **60**, No. 5, 175). [245 objects.]

571. C. MESSIER, *Catalogue of 103 Nebulae and Clusters* (1771–84). [Abbrev: M. See also B. 572.]

572. H. SHAPLEY & H. DAVIES, Messier's 'Catalogue of Nebulae and Clusters' (*Obs.* **41**, No. 529, 318). [Reprint of B. 571, with corresponding NGC (B. 564) numbers.]

573. C. WIRTZ, Flächenhelligkeiten von 566 Nebelflecken und Sternhaufen (*Lund. Medd.*, **2**, No. 29, 1923).

ADDITIONAL REFERENCES FROM TEXT

574. F. J. HARGREAVES [Report of Presidential Address to the B.A.A., 1944] (*J.B.A.A.*, **55**, No. 1, 1).

575. E. PETTIT & S. B. NICHOLSON, Measurements of the Radiation from the Planet Mars (*Pop. Astr.*, **32**, No. 10, 601).

576. J. G. PORTER, *Comets and Meteor Streams* (Chapman & Hall, 1952). [Excellent modern treatment, with valuable bibliographies.]

SUPPLEMENTARY BIBLIOGRAPHY

[Arranged alphabetically unless otherwise stated.]

DESCRIPTIVE AND GENERAL ASTRONOMY

S1. S. MITTON (ed.), *The Cambridge Encyclopedia of Astronomy* (Crown, 1977). [One of the best recent descriptive astronomies.]

S2. R. H. STOY (ed.), *Everyman's Astronomy* (Dent, 1974).

OPTICS, INCLUDING TELESCOPIC FUNCTION AND ABERRATIONS

S3. U.S. NAVY (Bureau of Naval Personnel), *Opticalman 3 & 2* (1966) (reprinted by Dover, 1969, as *Basic Optics and Optical Instruments*).

TELESCOPES, AND GENERAL WORKS ON OPTICAL INSTRUMENTS

S4. N. E. HOWARD, *Standard Handbook for Telescope Making* (Crowell, 1959).

S5. H. C. KING, *The History of the Telescope* (Dover, 1979).

S6. H. E. PAUL, *Telescopes for Skygazing* (Amphoto, 3rd edn., 2nd ptg., 1977).

PHOTOGRAPHY

S7. R. N. MAYALL & M. W. MAYALL, *Skyshooting—Photography for Amateur Astronomers* (Dover, 1968).

S8. H. E. PAUL, *Outer Space Photography for the Amateur* (Amphoto, 4th edn., 3rd ptg., 1979).

PHOTOMETRY

S9. F. B. WOOD, *Photoelectric Astronomy for Amateurs* (Macmillan, 1963).

AMATEUR OBSERVATIONAL ASTRONOMY

S10. R. BURNHAM, JR., *Burnham's Celestial Handbook—An Observer's Guide to the Universe Beyond the Solar System* (Dover, 1978, 3 vols.). [The best of such handbooks.]

S11. G. D. ROTH, *Astronomy—A Handbook for Amateur Astronomers* (Springer 1967).

557

BIBLIOGRAPHY

Star Atlases and Charts

S12. s. mitton (ed.), *Star Atlas* (Crown, 1979). [8 maps, over 4000 stars to 6ᵐ0 with some introduction, similar to B. 438.

S13. c. papadopoulos, *True Visual Magnitude Photographic Star Atlas* (Pergamon, 1978, 1980).

S14. *Smithsonian Astrophysical Observatory Star Atlas* (M.I.T. Press, 1966). [152 maps, 258,997 stars of the entire sky to 9ᵐ0.]

S14a. *AAVSO Star Atlas* (Sky Publishing, 1980). [Similar to the *Smithsonian*.]

Star Catalogues—Positional
[Arranged chronologically.]

S15. c. d. perinne, *Cordoba Durchmusterung* (Cordoba Observatory, 1910). [Continuation of B. 462 from Dec −52 to the South Pole, to mag 10.]

S16. *Dritter Katalog der Astronomischen Gesellschaft* (Hamburg-Bergedorf, 1970, 8 vols.). [See B. 461; abbrev *AGK3*.]

S17. d. hoffleit, *Bright Star Catalogue of the Yale University Observatory* (Yale University Observatory, 3rd rev. edn., 1964). [Contains all known data about all stars visible to naked eye.]

S18. *Smithsonian Astrophysical Observatory Star Catalog* (U.S. Government Printing Office, 1966, 4 vols.). [The catalogue to above S14.]

S19. h. vehrenberg, *Photographic Star Atlas* (Treugesell Verlag, 1972). [464 maps covering the whole sky to 13ᵐ.]

S20. h. vehrenberg, *Atlas Stellarum* (Treugesell Verlag, 1974). [486 maps of the whole sky down to 14ᵐ.]

S21. h. vehrenberg, *Atlas of Selected Areas* (Treugesell Verlag, 1976). [206 maps of the famous Kapteyn's areas down to 15ᵐ9.]

Star Catalogues—Spectroscopic

S22. w. w. morgan, p. c. keenan & e. kellman, *An Atlas of Stellar Spectra* (University of Chicago Press, 1943).

S23. w. c. seitter, *Atlas of Objective Prism Spectra* (*Bonner Spectral Atlas*) (Dümmler, 1970, 1975, 2 vols.).

S24. h. arp, *Atlas of Peculiar Galaxies* (Caltech, 1965).

Star Catalogues—Variable Stars

S25. b. v. kukarkin *et al.*, *General Catalogue of Variable Stars* (USSR Academy of Sciences, Moscow, 1969, 1975, 3 vols., 2 supplements).

S26. a. r. sandage, *Hubble Atlas of Galaxies* (Carnegie Institution, 1961).

Nebulae and Clusters—Catalogues

S27. j. w. sulentic & w. g. tifft, *The Revised New General Catalogue of Nonstellar Astronomical Objects* (University of Arizona Press, 1976). [See also B. 564.]

RADIO TIME SIGNALS*

The list of radio signals below is thought to be of use to observers. The stations transmit Co-ordinated Universal Time (U.T.C.). The transmission also carries a coded correction so that it is possible to convert U.T.C. to U.T.$_1$: however, observers should always report their observations in U.T.C.

Station (Country)	Call Sign	Transmission Frequencies (kHz)	Times of Transmission	Details of Signal
Rugby (England)	MSF (i)	60	00 m 00 s–59 m 55 s in each hour (except 1000–1400 on first Tuesday of each month)	second marker 100 ms ⎤ interruption of minute marker 500 ms ⎦ carrier wave
	(ii)	2500 5000 10000	alternate 5 minutes starting at 00h 00m	second marker 5 ms pulse minute marker 100 ms pulse The 5 seconds prior to transmission period—call sign in morse code
Nauen (DDR)	DIZ	4525	continuous except from 0815–0945	second marker 100 ms pulse minute marker 500 ms pulse (except last 2 seconds of each hour when second markers are 500 ms)
Liblice (Czechoslovakia)	OMA (i)	2500	continuous except 0600–1200 on last Wednesday of each month	00m, 15m, 30m, 45m—call sign for 1 minute and 1 kHz tone for 4 minutes. For rest of period second marker 5 ms pulse minute marker 100 ms pulse
	(ii)	50	continuous	second marker 100 ms ⎤ interruption of minute marker 500 ms ⎦ carrier wave
Olifansfontein (South Africa)	ZUO	2500 5000 ⎤ 100 MHz ⎦	1800–1400 continuous	00m–04m second marker 5 ms pulse 04m–05m call sign and time at next minute; cycle repeated every 5 minutes
Tokyo (Japan)	JJY	2500 5000 10000 15000	continuous except 25m–35m	second marker 5 ms pulse minute marker 660 ms pulse 34m–35m ⎤ call sign and time (in morse) 59m–60m ⎦ followed by voice announcement
Lyndhurst (Australia)	VNG	(i) 4500 (ii) 7500 (iii) 12000	(i) 0945–2130 (ii) 0000–2230 2245–0000 (iii) 2145–0930	second marker 50 ms pulse minute marker 500 ms pulse (except 55s–58s of each minute and 50s–58s of every fifth minute—second marker 5 ms)
Fort Collins (U.S.A.)	WWV	2500 5000 10000 15000	continuous	second marker 5 ms pulse (29s and 59s omitted) minute marker 800 ms pulse male voice announcement 52s–60s
Kauai (Hawaii)	WWVH	2500 5000 10000 15000	continuous	second marker 6 ms pulse (29s and 59s omitted) minute marker 800 ms pulse female voice announcement 52s–60s
Ottawa (Canada)	CHU	3330 7335 14670	continuous	second marker—01s–28s ⎤ 300ms pulse 30s–50s ⎦ minute marker 500 ms pulse 51s–59s—voice announcement, station and time
Caracas (Venezuela)	YVTO	6100	0030–0130 1200–2200	01s–52s ⎤ second marker 100 ms pulse 57s–59s ⎦ 52s–57s voice announcement of time minute marker 500 ms pulse
Radio Relogio Federal (Brazil)	ZYZ	590 4905	continuous	01s–57s second marker 58s–60s prolonged marker

* Reprinted with the permission of the British Astronomical Association.

INDEX

A CATALOGUE OF
SELECTED DOVER BOOKS
IN ALL FIELDS OF INTEREST

A CATALOGUE OF SELECTED DOVER
BOOKS IN ALL FIELDS OF INTEREST

CELESTIAL OBJECTS FOR COMMON TELESCOPES, T. W. Webb. The most used book in amateur astronomy: inestimable aid for locating and identifying nearly 4,000 celestial objects. Edited, updated by Margaret W. Mayall. 77 illustrations. Total of 645pp. 5⅜ x 8½.
20917-2, 20918-0 Pa., Two-vol. set $9.00

HISTORICAL STUDIES IN THE LANGUAGE OF CHEMISTRY, M. P. Crosland. The important part language has played in the development of chemistry from the symbolism of alchemy to the adoption of systematic nomenclature in 1892. ". . . wholeheartedly recommended,"—Science. 15 illustrations. 416pp. of text. 5⅝ x 8¼.
63702-6 Pa. $6.00

BURNHAM'S CELESTIAL HANDBOOK, Robert Burnham, Jr. Thorough, readable guide to the stars beyond our solar system. Exhaustive treatment, fully illustrated. Breakdown is alphabetical by constellation: Andromeda to Cetus in Vol. 1; Chamaeleon to Orion in Vol. 2; and Pavo to Vulpecula in Vol. 3. Hundreds of illustrations. Total of about 2000pp. 6⅛ x 9¼.
23567-X, 23568-8, 23673-0 Pa., Three-vol. set $26.85

THEORY OF WING SECTIONS: INCLUDING A SUMMARY OF AIR-FOIL DATA, Ira H. Abbott and A. E. von Doenhoff. Concise compilation of subatomic aerodynamic characteristics of modern NASA wing sections, plus description of theory. 350pp. of tables. 693pp. 5⅜ x 8½.
60586-8 Pa. $7.00

DE RE METALLICA, Georgius Agricola. Translated by Herbert C. Hoover and Lou H. Hoover. The famous Hoover translation of greatest treatise on technological chemistry, engineering, geology, mining of early modern times (1556). All 289 original woodcuts. 638pp. 6¾ x 11.
60006-8 Clothbd. $17.50

THE ORIGIN OF CONTINENTS AND OCEANS, Alfred Wegener. One of the most influential, most controversial books in science, the classic statement for continental drift. Full 1966 translation of Wegener's final (1929) version. 64 illustrations. 246pp. 5⅜ x 8½.
61708-4 Pa. $3.00

THE PRINCIPLES OF PSYCHOLOGY, William James. Famous long course complete, unabridged. Stream of thought, time perception, memory, experimental methods; great work decades ahead of its time. Still valid, useful; read in many classes. 94 figures. Total of 1391pp. 5⅜ x 8½.
20381-6, 20382-4 Pa., Two-vol. set $13.00

YUCATAN BEFORE AND AFTER THE CONQUEST, Diego de Landa. First English translation of basic book in Maya studies, the only significant account of Yucatan written in the early post-Conquest era. Translated by distinguished Maya scholar William Gates. Appendices, introduction, 4 maps and over 120 illustrations added by translator. 162pp. 5⅜ x 8½.
23622-6 Pa. $3.00

THE MALAY ARCHIPELAGO, Alfred R. Wallace. Spirited travel account by one of founders of modern biology. Touches on zoology, botany, ethnography, geography, and geology. 62 illustrations, maps. 515pp. 5⅜ x 8½.
20187-2 Pa. $6.95

THE DISCOVERY OF THE TOMB OF TUTANKHAMEN, Howard Carter, A. C. Mace. Accompany Carter in the thrill of discovery, as ruined passage suddenly reveals unique, untouched, fabulously rich tomb. Fascinating account, with 106 illustrations. New introduction by J. M. White. Total of 382pp. 5⅜ x 8½. (Available in U.S. only)   23500-9 Pa. $4.00

THE WORLD'S GREATEST SPEECHES, edited by Lewis Copeland and Lawrence W. Lamm. Vast collection of 278 speeches from Greeks up to present. Powerful and effective models; unique look at history. Revised to 1970. Indices. 842pp. 5⅜ x 8½.
20468-5 Pa. $8.95

THE 100 GREATEST ADVERTISEMENTS, Julian Watkins. The priceless ingredient; His master's voice; 99 44/100% pure; over 100 others. How they were written, their impact, etc. Remarkable record. 130 illustrations. 233pp. 7⅞ x 10 3/5.
20540-1 Pa. $5.00

CRUICKSHANK PRINTS FOR HAND COLORING, George Cruickshank. 18 illustrations, one side of a page, on fine-quality paper suitable for watercolors. Caricatures of people in society (c. 1820) full of trenchant wit. Very large format. 32pp. 11 x 16.
23684-6 Pa. $5.00

THIRTY-TWO COLOR POSTCARDS OF TWENTIETH-CENTURY AMERICAN ART, Whitney Museum of American Art. Reproduced in full color in postcard form are 31 art works and one shot of the museum. Calder, Hopper, Rauschenberg, others. Detachable. 16pp. 8¼ x 11.
23629-3 Pa. $2.50

MUSIC OF THE SPHERES: THE MATERIAL UNIVERSE FROM ATOM TO QUASAR SIMPLY EXPLAINED, Guy Murchie. Planets, stars, geology, atoms, radiation, relativity, quantum theory, light, antimatter, similar topics. 319 figures. 664pp. 5⅜ x 8½.
21809-0, 21810-4 Pa., Two-vol. set $10.00

EINSTEIN'S THEORY OF RELATIVITY, Max Born. Finest semi-technical account; covers Einstein, Lorentz, Minkowski, and others, with much detail, much explanation of ideas and math not readily available elsewhere on this level. For student, non-specialist. 376pp. 5⅜ x 8½.
60769-0 Pa. $4.00

THE COMPLETE BOOK OF DOLL MAKING AND COLLECTING, Catherine Christopher. Instructions, patterns for dozens of dolls, from rag doll on up to elaborate, historically accurate figures. Mould faces, sew clothing, make doll houses, etc. Also collecting information. Many illustrations. 288pp. 6 x 9. 22066-4 Pa. $4.00

THE DAGUERREOTYPE IN AMERICA, Beaumont Newhall. Wonderful portraits, 1850's townscapes, landscapes; full text plus 104 photographs. The basic book. Enlarged 1976 edition. 272pp. 8¼ x 11¼. 23322-7 Pa. $6.00

CRAFTSMAN HOMES, Gustav Stickley. 296 architectural drawings, floor plans, and photographs illustrate 40 different kinds of "Mission-style" homes from *The Craftsman* (1901-16), voice of American style of simplicity and organic harmony. Thorough coverage of Craftsman idea in text and picture, now collector's item. 224pp. 8⅛ x 11. 23791-5 Pa. $6.00

PEWTER-WORKING: INSTRUCTIONS AND PROJECTS, Burl N. Osborn. & Gordon O. Wilber. Introduction to pewter-working for amateur craftsman. History and characteristics of pewter; tools, materials, step-by-step instructions. Photos, line drawings, diagrams. Total of 160pp. 7⅞ x 10¾. 23786-9 Pa. $3.50

THE GREAT CHICAGO FIRE, edited by David Lowe. 10 dramatic, eye-witness accounts of the 1871 disaster, including one of the aftermath and rebuilding, plus 70 contemporary photographs and illustrations of the ruins—courthouse, Palmer House, Great Central Depot, etc. Introduction by David Lowe. 87pp. 8¼ x 11. 23771-0 Pa. $4.00

SILHOUETTES: A PICTORIAL ARCHIVE OF VARIED ILLUSTRATIONS, edited by Carol Belanger Grafton. Over 600 silhouettes from the 18th to 20th centuries include profiles and full figures of men and women, children, birds and animals, groups and scenes, nature, ships, an alphabet. Dozens of uses for commercial artists and craftspeople. 144pp. 8⅜ x 11¼. 23781-8 Pa. $4.00

ANIMALS: 1,419 COPYRIGHT-FREE ILLUSTRATIONS OF MAMMALS, BIRDS, FISH, INSECTS, ETC., edited by Jim Harter. Clear wood engravings present, in extremely lifelike poses, over 1,000 species of animals. One of the most extensive copyright-free pictorial sourcebooks of its kind. Captions. Index. 284pp. 9 x 12. 23766-4 Pa. $7.50

INDIAN DESIGNS FROM ANCIENT ECUADOR, Frederick W. Shaffer. 282 original designs by pre-Columbian Indians of Ecuador (500-1500 A.D.). Designs include people, mammals, birds, reptiles, fish, plants, heads, geometric designs. Use as is or alter for advertising, textiles, leathercraft, etc. Introduction. 95pp. 8¾ x 11¼. 23764-8 Pa. $3.50

SZIGETI ON THE VIOLIN, Joseph Szigeti. Genial, loosely structured tour by premier violinist, featuring a pleasant mixture of reminiscenes, insights into great music and musicians, innumerable tips for practicing violinists. 385 musical passages. 256pp. 5⅝ x 8¼. 23763-X Pa. $3.50

TONE POEMS, SERIES II: TILL EULENSPIEGELS LUSTIGE STREICHE, ALSO SPRACH ZARATHUSTRA, AND EIN HELDEN-LEBEN, Richard Strauss. Three important orchestral works, including very popular *Till Eulenspiegel's Marry Pranks*, reproduced in full score from original editions. Study score. 315pp. 9⅜ x 12¼. (Available in U.S. only)
23755-9 Pa. $7.50

TONE POEMS, SERIES I: DON JUAN, TOD UND VERKLARUNG AND DON QUIXOTE, Richard Strauss. Three of the most often performed and recorded works in entire orchestral repertoire, reproduced in full score from original editions. Study score. 286pp. 9⅜ x 12¼. (Available in U.S. only)
23754-0 Pa. $7.50

11 LATE STRING QUARTETS, Franz Joseph Haydn. The form which Haydn defined and "brought to perfection." (*Grove's*). 11 string quartets in complete score, his last and his best. The first in a projected series of the complete Haydn string quartets. Reliable modern Eulenberg edition, otherwise difficult to obtain. 320pp. 8⅜ x 11¼. (Available in U.S. only)
23753-2 Pa. $6.95

FOURTH, FIFTH AND SIXTH SYMPHONIES IN FULL SCORE, Peter Ilyitch Tchaikovsky. Complete orchestral scores of Symphony No. 4 in F Minor, Op. 36; Symphony No. 5 in E Minor, Op. 64; Symphony No. 6 in B Minor, "Pathetique," Op. 74. Bretikopf & Hartel eds. Study score. 480pp. 9⅜ x 12¼.
23861-X Pa. $10.95

THE MARRIAGE OF FIGARO: COMPLETE SCORE, Wolfgang A. Mozart. Finest comic opera ever written. Full score, not to be confused with piano renderings. Peters edition. Study score. 448pp. 9⅜ x 12¼. (Available in U.S. only)
23751-6 Pa. $11.95

"IMAGE" ON THE ART AND EVOLUTION OF THE FILM, edited by Marshall Deutelbaum. Pioneering book brings together for first time 38 groundbreaking articles on early silent films from *Image* and 263 illustrations newly shot from rare prints in the collection of the International Museum of Photography. A landmark work. Index. 256pp. 8¼ x 11.
23777-X Pa. $8.95

AROUND-THE-WORLD COOKY BOOK, Lois Lintner Sumption and Marguerite Lintner Ashbrook. 373 cooky and frosting recipes from 28 countries (America, Austria, China, Russia, Italy, etc.) include Viennese kisses, rice wafers, London strips, lady fingers, hony, sugar spice, maple cookies, etc. Clear instructions. All tested. 38 drawings. 182pp. 5⅜ x 8.
23802-4 Pa. $2.50

THE ART NOUVEAU STYLE, edited by Roberta Waddell. 579 rare photographs, not available elsewhere, of works in jewelry, metalwork, glass, ceramics, textiles, architecture and furniture by 175 artists—Mucha, Seguy, Lalique, Tiffany, Gaudin, Hohlwein, Saarinen, and many others. 288pp. 8⅜ x 11¼.
23515-7 Pa. $6.95

THE AMERICAN SENATOR, Anthony Trollope. Little known, long unavailable Trollope novel on a grand scale. Here are humorous comment on American vs. English culture, and stunning portrayal of a heroine/villainess. Superb evocation of Victorian village life. 561pp. 5⅜ x 8½.
23801-6 Pa. $6.00

WAS IT MURDER? James Hilton. The author of *Lost Horizon* and *Goodbye, Mr. Chips* wrote one detective novel (under a pen-name) which was quickly forgotten and virtually lost, even at the height of Hilton's fame. This edition brings it back—a finely crafted public school puzzle resplendent with Hilton's stylish atmosphere. A thoroughly English thriller by the creator of Shangri-la. 252pp. 5⅜ x 8. (Available in U.S. only)
23774-5 Pa. $3.00

CENTRAL PARK: A PHOTOGRAPHIC GUIDE, Victor Laredo and Henry Hope Reed. 121 superb photographs show dramatic views of Central Park: Bethesda Fountain, Cleopatra's Needle, Sheep Meadow, the Blockhouse, plus people engaged in many park activities: ice skating, bike riding, etc. Captions by former Curator of Central Park, Henry Hope Reed, provide historical view, changes, etc. Also photos of N.Y. landmarks on park's periphery. 96pp. 8½ x 11. 23750-8 Pa. $4.50

NANTUCKET IN THE NINETEENTH CENTURY, Clay Lancaster. 180 rare photographs, stereographs, maps, drawings and floor plans recreate unique American island society. Authentic scenes of shipwreck, lighthouses, streets, homes are arranged in geographic sequence to provide walking-tour guide to old Nantucket existing today. Introduction, captions. 160pp. 8⅞ x 11¾. 23747-8 Pa. $6.95

STONE AND MAN: A PHOTOGRAPHIC EXPLORATION, Andreas Feininger. 106 photographs by *Life* photographer Feininger portray man's deep passion for stone through the ages. Stonehenge-like megaliths, fortified towns, sculpted marble and crumbling tenements show textures, beauties, fascination. 128pp. 9¼ x 10¾. 23756-7 Pa. $5.95

CIRCLES, A MATHEMATICAL VIEW, D. Pedoe. Fundamental aspects of college geometry, non-Euclidean geometry, and other branches of mathematics: representing circle by point. Poincare model, isoperimetric property, etc. Stimulating recreational reading. 66 figures. 96pp. 5⅝ x 8¼.
63698-4 Pa. $2.75

THE DISCOVERY OF NEPTUNE, Morton Grosser. Dramatic scientific history of the investigations leading up to the actual discovery of the eighth planet of our solar system. Lucid, well-researched book by well-known historian of science. 172pp. 5⅜ x 8½. 23726-5 Pa. $3.00

THE DEVIL'S DICTIONARY. Ambrose Bierce. Barbed, bitter, brilliant witticisms in the form of a dictionary. Best, most ferocious satire America has produced. 145pp. 5⅜ x 8½. 20487-1 Pa. $1.75

HISTORY OF BACTERIOLOGY, William Bulloch. The only comprehensive history of bacteriology from the beginnings through the 19th century. Special emphasis is given to biography-Leeuwenhoek, etc. Brief accounts of 350 bacteriologists form a separate section. No clearer, fuller study, suitable to scientists and general readers, has yet been written. 52 illustrations. 448pp. 5⅝ x 8¼. 23761-3 Pa. $6.50

THE COMPLETE NONSENSE OF EDWARD LEAR, Edward Lear. All nonsense limericks, zany alphabets, Owl and Pussycat, songs, nonsense botany, etc., illustrated by Lear. Total of 321pp. 5⅜ x 8½. (Available in U.S. only) 20167-8 Pa. $3.00

INGENIOUS MATHEMATICAL PROBLEMS AND METHODS, Louis A. Graham. Sophisticated material from Graham *Dial*, applied and pure; stresses solution methods. Logic, number theory, networks, inversions, etc. 237pp. 5⅜ x 8½. 20545-2 Pa. $3.50

BEST MATHEMATICAL PUZZLES OF SAM LOYD, edited by Martin Gardner. Bizarre, original, whimsical puzzles by America's greatest puzzler. From fabulously rare *Cyclopedia,* including famous 14-15 puzzles, the Horse of a Different Color, 115 more. Elementary math. 150 illustrations. 167pp. 5⅜ x 8½. 20498-7 Pa. $2.50

THE BASIS OF COMBINATION IN CHESS, J. du Mont. Easy-to-follow, instructive book on elements of combination play, with chapters on each piece and every powerful combination team—two knights, bishop and knight, rook and bishop, etc. 250 diagrams. 218pp. 5⅜ x 8½. (Available in U.S. only) 23644-7 Pa. $3.50

MODERN CHESS STRATEGY, Ludek Pachman. The use of the queen, the active king, exchanges, pawn play, the center, weak squares, etc. Section on rook alone worth price of the book. Stress on the moderns. Often considered the most important book on strategy. 314pp. 5⅜ x 8½. 20290-9 Pa. $3.50

LASKER'S MANUAL OF CHESS, Dr. Emanuel Lasker. Great world champion offers very thorough coverage of all aspects of chess. Combinations, position play, openings, end game, aesthetics of chess, philosophy of struggle, much more. Filled with analyzed games. 390pp. 5⅜ x 8½. 20640-8 Pa. $4.00

500 MASTER GAMES OF CHESS, S. Tartakower, J. du Mont. Vast collection of great chess games from 1798-1938, with much material nowhere else readily available. Fully annotated, arranged by opening for easier study. 664pp. 5⅜ x 8½. 23208-5 Pa. $6.00

A GUIDE TO CHESS ENDINGS, Dr. Max Euwe, David Hooper. One of the finest modern works on chess endings. Thorough analysis of the most frequently encountered endings by former world champion. 331 examples, each with diagram. 248pp. 5⅜ x 8½. 23332-4 Pa. $3.50

SECOND PIATIGORSKY CUP, edited by Isaac Kashdan. One of the greatest tournament books ever produced in the English language. All 90 games of the 1966 tournament, annotated by players, most annotated by both players. Features Petrosian, Spassky, Fischer, Larsen, six others. 228pp. 5⅜ x 8½.   23572-6 Pa. $3.50

ENCYCLOPEDIA OF CARD TRICKS, revised and edited by Jean Hugard. How to perform over 600 card tricks, devised by the world's greatest magicians: impromptus, spelling tricks, key cards, using special packs, much, much more. Additional chapter on card technique. 66 illustrations. 402pp. 5⅜ x 8½. (Available in U.S. only)   21252-1 Pa. $3.95

MAGIC: STAGE ILLUSIONS, SPECIAL EFFECTS AND TRICK PHOTOGRAPHY, Albert A. Hopkins, Henry R. Evans. One of the great classics; fullest, most authorative explanation of vanishing lady, levitations, scores of other great stage effects. Also small magic, automata, stunts. 446 illustrations. 556pp. 5⅜ x 8½.   23344-8 Pa. $5.00

THE SECRETS OF HOUDINI, J. C. Cannell. Classic study of Houdini's incredible magic, exposing closely-kept professional secrets and revealing, in general terms, the whole art of stage magic. 67 illustrations. 279pp. 5⅜ x 8½.   22913-0 Pa. $3.00

HOFFMANN'S MODERN MAGIC, Professor Hoffmann. One of the best, and best-known, magicians' manuals of the past century. Hundreds of tricks from card tricks and simple sleight of hand to elaborate illusions involving construction of complicated machinery. 332 illustrations. 563pp. 5⅜ x 8½.   23623-4 Pa. $6.00

MADAME PRUNIER'S FISH COOKERY BOOK, Mme. S. B. Prunier. More than 1000 recipes from world famous Prunier's of Paris and London, specially adapted here for American kitchen. Grilled tournedos with anchovy butter, Lobster a la Bordelaise, Prunier's prized desserts, more. Glossary. 340pp. 5⅜ x 8½. (Available in U.S. only)   22679-4 Pa. $3.00

FRENCH COUNTRY COOKING FOR AMERICANS, Louis Diat. 500 easy-to-make, authentic provincial recipes compiled by former head chef at New York's Fitz-Carlton Hotel: onion soup, lamb stew, potato pie, more. 309pp. 5⅜ x 8½.   23665-X Pa. $3.95

SAUCES, FRENCH AND FAMOUS, Louis Diat. Complete book gives over 200 specific recipes: bechamel, Bordelaise, hollandaise, Cumberland, apricot, etc. Author was one of this century's finest chefs, originator of vichyssoise and many other dishes. Index. 156pp. 5⅜ x 8.

23663-3 Pa. $2.50

TOLL HOUSE TRIED AND TRUE RECIPES, Ruth Graves Wakefield. Authentic recipes from the famous Mass. restaurant: popovers, veal and ham loaf, Toll House baked beans, chocolate cake crumb pudding, much more. Many helpful hints. Nearly 700 recipes. Index. 376pp. 5⅜ x 8½.

23560-2 Pa. $4.00

"OSCAR" OF THE WALDORF'S COOKBOOK, Oscar Tschirky. Famous American chef reveals 3455 recipes that made Waldorf great; cream of French, German, American cooking, in all categories. Full instructions, easy home use. 1896 edition. 907pp. 6⅝ x 9⅜.　20790-0 Clothbd. $15.00

COOKING WITH BEER, Carole Fahy. Beer has as superb an effect on food as wine, and at fraction of cost. Over 250 recipes for appetizers, soups, main dishes, desserts, breads, etc. Index. 144pp. 5⅜ x 8½. (Available in U.S. only)　23661-7 Pa. $2.50

STEWS AND RAGOUTS, Kay Shaw Nelson. This international cookbook offers wide range of 108 recipes perfect for everyday, special occasions, meals-in-themselves, main dishes. Economical, nutritious, easy-to-prepare: goulash, Irish stew, boeuf bourguignon, etc. Index. 134pp. 5⅜ x 8½.
23662-5 Pa. $2.50

DELICIOUS MAIN COURSE DISHES, Marian Tracy. Main courses are the most important part of any meal. These 200 nutritious, economical recipes from around the world make every meal a delight. "I . . . have found it so useful in my own household,"—*N.Y. Times.* Index. 219pp. 5⅜ x 8½.　23664-1 Pa. $3.00

FIVE ACRES AND INDEPENDENCE, Maurice G. Kains. Great back-to-the-land classic explains basics of self-sufficient farming: economics, plants, crops, animals, orchards, soils, land selection, host of other necessary things. Do not confuse with skimpy faddist literature; Kains was one of America's greatest agriculturalists. 95 illustrations. 397pp. 5⅜ x 8½.
20974-1 Pa. $3.95

A PRACTICAL GUIDE FOR THE BEGINNING FARMER, Herbert Jacobs. Basic, extremely useful first book for anyone thinking about moving to the country and starting a farm. Simpler than Kains, with greater emphasis on country living in general. 246pp. 5⅜ x 8½.
23675-7 Pa. $3.50

A GARDEN OF PLEASANT FLOWERS (PARADISI IN SOLE: PARADISUS TERRESTRIS), John Parkinson. Complete, unabridged reprint of first (1629) edition of earliest great English book on gardens and gardening. More than 1000 plants & flowers of Elizabethan, Jacobean garden fully described, most with woodcut illustrations. Botanically very reliable, a "speaking garden" of exceeding charm. 812 illustrations. 628pp. 8½ x 12¼.　23392-8 Clothbd. $25.00

ACKERMANN'S COSTUME PLATES, Rudolph Ackermann. Selection of 96 plates from the *Repository of Arts,* best published source of costume for English fashion during the early 19th century. 12 plates also in color. Captions, glossary and introduction by editor Stella Blum. Total of 120pp. 8⅜ x 11¼.　23690-0 Pa. $4.50

MUSHROOMS, EDIBLE AND OTHERWISE, Miron E. Hard. Profusely illustrated, very useful guide to over 500 species of mushrooms growing in the Midwest and East. Nomenclature updated to 1976. 505 illustrations. 628pp. 6½ x 9¼.   23309-X Pa. $7.95

AN ILLUSTRATED FLORA OF THE NORTHERN UNITED STATES AND CANADA, Nathaniel L. Britton, Addison Brown. Encyclopedic work covers 4666 species, ferns on up. Everything. Full botanical information, illustration for each. This earlier edition is preferred by many to more recent revisions. 1913 edition. Over 4000 illustrations, total of 2087pp. 6⅛ x 9¼.   22642-5, 22643-3, 22644-1 Pa., Three-vol. set $24.00

MANUAL OF THE GRASSES OF THE UNITED STATES, A. S. Hitchcock, U.S. Dept. of Agriculture. The basic study of American grasses, both indigenous and escapes, cultivated and wild. Over 1400 species. Full descriptions, information. Over 1100 maps, illustrations. Total of 1051pp. 5⅜ x 8½.   22717-0, 22718-9 Pa., Two-vol. set $12.00

THE CACTACEAE,, Nathaniel L. Britton, John N. Rose. Exhaustive, definitive. Every cactus in the world. Full botanical descriptions. Thorough statement of nomenclatures, habitat, detailed finding keys. The one book needed by every cactus enthusiast. Over 1275 illustrations. Total of 1080pp. 8 x 10¼.   21191-6, 21192-4 Clothbd., Two-vol. set $35.00

AMERICAN MEDICINAL PLANTS, Charles F. Millspaugh. Full descriptions, 180 plants covered: history; physical description; methods of preparation with all chemical constituents extracted; all claimed curative or adverse effects. 180 full-page plates. Classification table. 804pp. 6½ x 9¼.
23034-1 Pa. $10.00

A MODERN HERBAL, Margaret Grieve. Much the fullest, most exact, most useful compilation of herbal material. Gigantic alphabetical encyclopedia, from aconite to zedoary, gives botanical information, medical properties, folklore, economic uses, and much else. Indispensable to serious reader. 161 illustrations. 888pp. 6½ x 9¼. (Available in U.S. only)
22798-7, 22799-5 Pa., Two-vol. set $11.00

THE HERBAL or GENERAL HISTORY OF PLANTS, John Gerard. The 1633 edition revised and enlarged by Thomas Johnson. Containing almost 2850 plant descriptions and 2705 superb illustrations, Gerard's *Herbal* is a monumental work, the book all modern English herbals are derived from, the one herbal every serious enthusiast should have in its entirety. Original editions are worth perhaps $750. 1678pp. 8½ x 12¼.
23147-X Clothbd. $50.00

MANUAL OF THE TREES OF NORTH AMERICA, Charles S. Sargent. The basic survey of every native tree and tree-like shrub, 717 species in all. Extremely full descriptions, information on habitat, growth, locales, economics, etc. Necessary to every serious tree lover. Over 100 finding keys. 783 illustrations. Total of 986pp. 5⅜ x 8½.
20277-1, 20278-X Pa., Two-vol. set $10.00

AMERICAN BIRD ENGRAVINGS, Alexander Wilson et al. All 76 plates. from Wilson's *American Ornithology* (1808-14), most important ornithological work before Audubon, plus 27 plates from the supplement (1825-33) by Charles Bonaparte. Over 250 birds portrayed. 8 plates also reproduced in full color. 111pp. 9⅜ x 12½. 23195-X Pa. $6.00

CRUICKSHANK'S PHOTOGRAPHS OF BIRDS OF AMERICA, Allan D. Cruickshank. Great ornithologist, photographer presents 177 closeups, groupings, panoramas, flightings, etc., of about 150 different birds. Expanded *Wings in the Wilderness*. Introduction by Helen G. Cruickshank. 191pp. 8¼ x 11. 23497-5 Pa. $6.00

AMERICAN WILDLIFE AND PLANTS, A. C. Martin, et al. Describes food habits of more than 1000 species of mammals, birds, fish. Special treatment of important food plants. Over 300 illustrations. 500pp. 5⅜ x 8½. 20793-5 Pa. $4.95

THE PEOPLE CALLED SHAKERS, Edward D. Andrews. Lifetime of research, definitive study of Shakers: origins, beliefs, practices, dances, social organization, furniture and crafts, impact on 19th-century USA, present heritage. Indispensable to student of American history, collector. 33 illustrations. 351pp. 5⅜ x 8½. 21081-2 Pa. $4.00

OLD NEW YORK IN EARLY PHOTOGRAPHS, Mary Black. New York City as it was in 1853-1901, through 196 wonderful photographs from N.-Y. Historical Society. Great Blizzard, Lincoln's funeral procession, great buildings. 228pp. 9 x 12. 22907-6 Pa. $7.95

MR. LINCOLN'S CAMERA MAN: MATHEW BRADY, Roy Meredith. Over 300 Brady photos reproduced directly from original negatives, photos. Jackson, Webster, Grant, Lee, Carnegie, Barnum; Lincoln; Battle Smoke, Death of Rebel Sniper, Atlanta Just After Capture. Lively commentary. 368pp. 8⅜ x 11¼. 23021-X Pa. $8.95

TRAVELS OF WILLIAM BARTRAM, William Bartram. From 1773-8, Bartram explored Northern Florida, Georgia, Carolinas, and reported on wild life, plants, Indians, early settlers. Basic account for period, entertaining reading. Edited by Mark Van Doren. 13 illustrations. 141pp. 5⅜ x 8½. 20013-2 Pa. $4.50

THE GENTLEMAN AND CABINET MAKER'S DIRECTOR, Thomas Chippendale. Full reprint, 1762 style book, most influential of all time; chairs, tables, sofas, mirrors, cabinets, etc. 200 plates, plus 24 photographs of surviving pieces. 249pp. 9⅞ x 12¾. 21601-2 Pa. $6.50

AMERICAN CARRIAGES, SLEIGHS, SULKIES AND CARTS, edited by Don H. Berkebile. 168 Victorian illustrations from catalogues, trade journals, fully captioned. Useful for artists. Author is Assoc. Curator, Div. of Transportation of Smithsonian Institution. 168pp. 8½ x 9½.

23328-6 Pa. $5.00

THE SENSE OF BEAUTY, George Santayana. Masterfully written discussion of nature of beauty, materials of beauty, form, expression; art, literature, social sciences all involved. 168pp. 5⅜ x 8½. 20238-0 Pa. $2.50

ON THE IMPROVEMENT OF THE UNDERSTANDING, Benedict Spinoza. Also contains Ethics, Correspondence, all in excellent R. Elwes translation. Basic works on entry to philosophy, pantheism, exchange of ideas with great contemporaries. 402pp. 5⅜ x 8½. 20250-X Pa. $4.50

THE TRAGIC SENSE OF LIFE, Miguel de Unamuno. Acknowledged masterpiece of existential literature, one of most important books of 20th century. Introduction by Madariaga. 367pp. 5⅜ x 8½.
20257-7 Pa. $3.50

THE GUIDE FOR THE PERPLEXED, Moses Maimonides. Great classic of medieval Judaism attempts to reconcile revealed religion (Pentateuch, commentaries) with Aristotelian philosophy. Important historically, still relevant in problems. Unabridged Friedlander translation. Total of 473pp. 5⅜ x 8½. 20351-4 Pa. $5.00

THE I CHING (THE BOOK OF CHANGES), translated by James Legge. Complete translation of basic text plus appendices by Confucius, and Chinese commentary of most penetrating divination manual ever prepared. Indispensable to study of early Oriental civilizations, to modern inquiring reader. 448pp. 5⅜ x 8½. 21062-6 Pa. $4.00

THE EGYPTIAN BOOK OF THE DEAD, E. A. Wallis Budge. Complete reproduction of Ani's papyrus, finest ever found. Full hieroglyphic text, interlinear transliteration, word for word translation, smooth translation. Basic work, for Egyptology, for modern study of psychic matters. Total of 533pp. 6½ x 9¼. (Available in U.S. only) 21866-X Pa. $4.95

THE GODS OF THE EGYPTIANS, E. A. Wallis Budge. Never excelled for richness, fullness: all gods, goddesses, demons, mythical figures of Ancient Egypt; their legends, rites, incarnations, variations, powers, etc. Many hieroglyphic texts cited. Over 225 illustrations, plus 6 color plates. Total of 988pp. 6⅛ x 9¼. (Available in U.S. only)
22055-9, 22056-7 Pa., Two-vol. set $12.00

THE ENGLISH AND SCOTTISH POPULAR BALLADS, Francis J. Child. Monumental, still unsuperseded; all known variants of Child ballads, commentary on origins, literary references, Continental parallels, other features. Added: papers by G. L. Kittredge, W. M. Hart. Total of 2761pp. 6½ x 9¼.
21409-5, 21410-9, 21411-7, 21412-5, 21413-3 Pa., Five-vol. set $37.50

CORAL GARDENS AND THEIR MAGIC, Bronsilaw Malinowski. Classic study of the methods of tilling the soil and of agricultural rites in the Trobriand Islands of Melanesia. Author is one of the most important figures in the field of modern social anthropology. 143 illustrations. Indexes. Total of 911pp. of text. 5⅝ x 8¼. (Available in U.S. only)
23597-1 Pa. $12.95

THE PHILOSOPHY OF HISTORY, Georg W. Hegel. Great classic of Western thought develops concept that history is not chance but a rational process, the evolution of freedom. 457pp. 5⅜ x 8½.   20112-0 Pa. $4.50

LANGUAGE, TRUTH AND LOGIC, Alfred J. Ayer. Famous, clear introduction to Vienna, Cambridge schools of Logical Positivism. Role of philosophy, elimination of metaphysics, nature of analysis, etc. 160pp. 5⅜ x 8½. (Available in U.S. only)   20010-8 Pa. $1.75

A PREFACE TO LOGIC, Morris R. Cohen. Great City College teacher in renowned, easily followed exposition of formal logic, probability, values, logic and world order and similar topics; no previous background needed. 209pp. 5⅜ x 8½.   23517-3 Pa. $3.50

REASON AND NATURE, Morris R. Cohen. Brilliant analysis of reason and its multitudinous ramifications by charismatic teacher. Interdisciplinary, synthesizing work widely praised when it first appeared in 1931. Second (1953) edition. Indexes. 496pp. 5⅜ x 8½.   23633-1 Pa. $6.00

AN ESSAY CONCERNING HUMAN UNDERSTANDING, John Locke. The only complete edition of enormously important classic, with authoritative editorial material by A. C. Fraser. Total of 1176pp. 5⅜ x 8½.
20530-4, 20531-2 Pa., Two-vol. set $14.00

HANDBOOK OF MATHEMATICAL FUNCTIONS WITH FORMULAS, GRAPHS, AND MATHEMATICAL TABLES, edited by Milton Abramowitz and Irene A. Stegun. Vast compendium: 29 sets of tables, some to as high as 20 places. 1,046pp. 8 x 10½.   61272-4 Pa. $14.95

MATHEMATICS FOR THE PHYSICAL SCIENCES, Herbert S. Wilf. Highly acclaimed work offers clear presentations of vector spaces and matrices, orthogonal functions, roots of polynomial equations, conformal mapping, calculus of variations, etc. Knowledge of theory of functions of real and complex variables is assumed. Exercises and solutions. Index. 284pp. 5⅝ x 8¼.   63635-6 Pa. $4.50

THE PRINCIPLE OF RELATIVITY, Albert Einstein et al. Eleven most important original papers on special and general theories. Seven by Einstein, two by Lorentz, one each by Minkowski and Weyl. All translated, unabridged. 216pp. 5⅜ x 8½.   60081-5 Pa. $3.00

THERMODYNAMICS, Enrico Fermi. A classic of modern science. Clear, organized treatment of systems, first and second laws, entropy, thermodynamic potentials, gaseous reactions, dilute solutions, entropy constant. No math beyond calculus required. Problems. 160pp. 5⅜ x 8½.
60361-X Pa. $2.75

ELEMENTARY MECHANICS OF FLUIDS, Hunter Rouse. Classic undergraduate text widely considered to be far better than many later books. Ranges from fluid velocity and acceleration to role of compressibility in fluid motion. Numerous examples, questions, problems. 224 illustrations. 376pp. 5⅝ x 8¼.   63699-2 Pa. $5.00

AN AUTOBIOGRAPHY, Margaret Sanger. Exciting personal account of hard-fought battle for woman's right to birth control, against prejudice, church, law. Foremost feminist document. 504pp. 5⅜ x 8½.

20470-7 Pa. $5.50

MY BONDAGE AND MY FREEDOM, Frederick Douglass. Born as a slave, Douglass became outspoken force in antislavery movement. The best of Douglass's autobiographies. Graphic description of slave life. Introduction by P. Foner. 464pp. 5⅜ x 8½. 22457-0 Pa. $5.00

LIVING MY LIFE, Emma Goldman. Candid, no holds barred account by foremost American anarchist: her own life, anarchist movement, famous contemporaries, ideas and their impact. Struggles and confrontations in America, plus deportation to U.S.S.R. Shocking inside account of persecution of anarchists under Lenin. 13 plates. Total of 944pp. 5⅜ x 8½.

22543-7, 22544-5 Pa., Two-vol. set $9.00

LETTERS AND NOTES ON THE MANNERS, CUSTOMS AND CONDITIONS OF THE NORTH AMERICAN INDIANS, George Catlin. Classic account of life among Plains Indians: ceremonies, hunt, warfare, etc. Dover edition reproduces for first time all original paintings. 312 plates. 572pp. of text. 6⅛ x 9¼. 22118-0, 22119-9 Pa.. Two-vol. set $10.00

THE MAYA AND THEIR NEIGHBORS, edited by Clarence L. Hay, others. Synoptic view of Maya civilization in broadest sense, together with Northern, Southern neighbors. Integrates much background, valuable detail not elsewhere. Prepared by greatest scholars: Kroeber, Morley, Thompson, Spinden, Vaillant, many others. Sometimes called Tozzer Memorial Volume. 60 illustrations, linguistic map. 634pp. 5⅜ x 8½.

23510-6 Pa. $7.50

HANDBOOK OF THE INDIANS OF CALIFORNIA, A. L. Kroeber. Foremost American anthropologist offers complete ethnographic study of each group. Monumental classic. 459 illustrations, maps. 995pp. 5⅜ x 8½.

23368-5 Pa. $10.00

SHAKTI AND SHAKTA, Arthur Avalon. First book to give clear, cohesive analysis of Shakta doctrine, Shakta ritual and Kundalini Shakti (yoga). Important work by one of world's foremost students of Shaktic and Tantric thought. 732pp. 5⅜ x 8½. (Available in U.S. only)

23645-5 Pa. $7.95

AN INTRODUCTION TO THE STUDY OF THE MAYA HIEROGLYPHS, Syvanus Griswold Morley. Classic study by one of the truly great figures in hieroglyph research. Still the best introduction for the student for reading Maya hieroglyphs. New introduction by J. Eric S. Thompson. 117 illustrations. 284pp. 5⅜ x 8½. 23108-9 Pa. $4.00

A STUDY OF MAYA ART, Herbert J. Spinden. Landmark classic interprets Maya symbolism, estimates styles, covers ceramics, architecture, murals, stone carvings as artforms. Still a basic book in area. New introduction by J. Eric Thompson. Over 750 illustrations. 341pp. 8⅜ x 11¼.

21235-1 Pa. $6.95

THE STANDARD BOOK OF QUILT MAKING AND COLLECTING, Marguerite Ickis. Full information, full-sized patterns for making 46 traditional quilts, also 150 other patterns. Quilted cloths, lame, satin quilts, etc. 483 illustrations. 273pp. 6⅞ x 9⅝. 20582-7 Pa. $4.50

ENCYCLOPEDIA OF VICTORIAN NEEDLEWORK, S. Caulfield, Blanche Saward. Simply inexhaustible gigantic alphabetical coverage of every traditional needlecraft—stitches, materials, methods, tools, types of work; definitions, many projects to be made. 1200 illustrations; double-columned text. 697pp. 8⅛ x 11. 22800-2, 22801-0 Pa., Two-vol. set $12.00

MECHANICK EXERCISES ON THE WHOLE ART OF PRINTING, Joseph Moxon. First complete book (1683-4) ever written about typography, a compendium of everything known about printing at the latter part of 17th century. Reprint of 2nd (1962) Oxford Univ. Press edition. 74 illustrations. Total of 550pp. 6⅛ x 9¼. 23617-X Pa. $7.95

PAPERMAKING, Dard Hunter. Definitive book on the subject by the foremost authority in the field. Chapters dealing with every aspect of history of craft in every part of the world. Over 320 illustrations. 2nd, revised and enlarged (1947) edition. 672pp. 5⅜ x 8½. 23619-6 Pa. $7.95

THE ART DECO STYLE, edited by Theodore Menten. Furniture, jewelry, metalwork, ceramics, fabrics, lighting fixtures, interior decors, exteriors, graphics from pure French sources. Best sampling around. Over 400 photographs. 183pp. 8⅜ x 11¼. 22824-X Pa. $5.00

Prices subject to change without notice.

Available at your book dealer or write for free catalogue to Dept. GI, Dover Publications, Inc., 180 Varick St., N.Y., N.Y. 10014. Dover publishes more than 175 books each year on science, elementary and advanced mathematics, biology, music, art, literary history, social sciences and other areas.